STUDENT'S SOLUTIONS MANUAL

BEVERLY FUSFIELD

CALCULUS & ITS APPLICATIONS
FOURTEENTH EDITION

CALCULUS & ITS APPLICATIONS, BRIEF VERSION
FOURTEENTH EDITION

Larry J. Goldstein
Goldstein Educational Technologies

David C. Lay
University of Maryland

David I. Schneider
University of Maryland

Nakhlé H. Asmar
University of Missouri

ISBN-13: 978-0-13-446323-0
ISBN-10: 0-13-446323-4

CONTENTS

Chapter 0 Functions

0.1 Functions and Their Graphs

1.

3.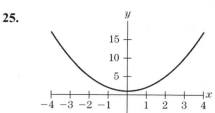

5.

7. $[2, 3)$

9. $[-1, 0)$

11. $(-\infty, 3)$

13. $f(x) = x^2 - 3x$

$f(0) = 0^2 - 3(0) = 0$

$f(5) = 5^2 - 3(5) = 25 - 15 = 10$

$f(3) = 3^2 - 3(3) = 9 - 9 = 0$

$f(-7) = (-7)^2 - 3(-7) = 49 + 21 = 70$

15. $f(x) = x^2 - 2x$

$f(a+1) = (a+1)^2 - 2(a+1)$
$= (a^2 + 2a + 1) - 2a - 2 = a^2 - 1$

$f(a+2) = (a+2)^2 - 2(a+2)$
$= (a^2 + 4a + 4) - 2a - 4 = a^2 + 2a$

17. $f(x) = 3x + 2,\ h \neq 0$

$f(3+h) = 3(3+h) + 2 = 9 + 3h + 2 = 3h + 11$
$f(3) = 3(3) + 2 = 11$

$\dfrac{f(3+h) - f(3)}{h} = \dfrac{(3h+11) - 11}{h} = \dfrac{3h}{h} = 3$

19. **a.** $k(x) = x + 273$
$5933 = x + 273 \Rightarrow x = 5660$
The boiling point of tungsten is 5660°C.

b. $f(x) = \dfrac{9}{5}x + 32$
$f(x) = \dfrac{9}{5}(5660) + 32 = 10220$
The boiling point of tungsten is 10220°F.

21. $f(x) = \dfrac{8x}{(x-1)(x-2)}$
all real numbers such that $x \neq 1, 2$ or
$(-\infty, -1) \cup (-1, 2) \cup (2, \infty)$

23. $g(x) = \dfrac{1}{\sqrt{3-x}}$
all real numbers such that $x < 3$ or $(-\infty, -3)$

25.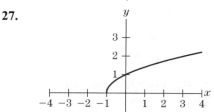

27.

29. function

31. not a function

33. not a function

35. $f(0) = 1;\ f(7) = -1$

37. positive

39. $[-1, 3]$

41. $(-\infty, -1] \cup [5, 9]$

43. $f(1) \approx .03;\ f(5) \approx .037$

45. $[0, .05]$

47. $f(x) = \left(x - \dfrac{1}{2}\right)(x + 2)$

$f(3) = \left(3 - \dfrac{1}{2}\right)(3 + 2) = \dfrac{25}{2}$

No, $(3, 12)$ is not on the graph.

49. $g(x) = \dfrac{3x - 1}{x^2 + 1}$

$g(1) = \dfrac{3(1) - 1}{(1)^2 + 1} = \dfrac{2}{2} = 1$

Yes, $(1, 1)$ is on the graph.

51. $f(x) = x^3$

$f(a+1) = (a+1)^3$

53. $f(x) = \begin{cases} \sqrt{x} & \text{for } 0 \le x < 2 \\ 1+x & \text{for } 2 \le x \le 5 \end{cases}$

$f(1) = \sqrt{1} = 1$
$f(2) = 1 + 2 = 3$
$f(3) = 1 + 3 = 4$

55. $f(x) = \begin{cases} \pi x^2 & \text{for } x < 2 \\ 1+x & \text{for } 2 \le x \le 2.5 \\ 4x & \text{for } 2.5 < x \end{cases}$

$f(1) = \pi(1)^2 = \pi$
$f(2) = 1 + 2 = 3$
$f(3) = 4(3) = 12$

57. a. $f(x) = \begin{cases} 0.06x & \text{for } 50 \le x \le 3000 \\ 0.02x + 15 & \text{for } 3000 < x \end{cases}$

b. $f(3000) = 0.06(3000) = 180$
$f(4500) = 0.02(4500) + 15 = 105$

59.

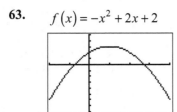

61. Entering **Y₁ = 1/X + 1** will graph the function

$f(x) = \dfrac{1}{x} + 1$. In order to graph the function

$f(x) = \dfrac{1}{x+1}$, you need to include parentheses

in the denominator: **Y₁ = 1/(X + 1)**.

63. $f(x) = -x^2 + 2x + 2$

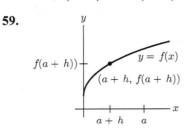

0.2   Some Important Functions

1. $y = 2x - 1$

x	y
1	1
0	−1
−1	−3

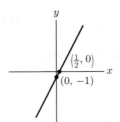

3. $y = 3x + 1$

x	y
1	4
0	1
−1	−2

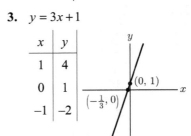

5. $y = -2x + 3$

x	y
−1	5
0	3
1	1

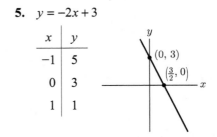

7. $x - y = 0$

x	y
1	1
0	0
−1	−1

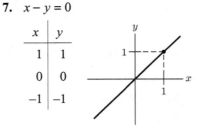

9. $x = 2y - 1$

x	y
3	2
1	1
−3	−1

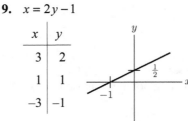

11. $f(x) = 9x + 3$
$f(0) = 9(9) + 3 = 3$
The y-intercept is $(0, 3)$.

$9x + 3 = 0 \Rightarrow 9x = -3 \Rightarrow x = -\dfrac{1}{3}$

The x-intercept is $\left(-\dfrac{1}{3}, 0\right)$.

13. $f(x) = 5$
The y-intercept is $(0, 5)$.
There is no x-intercept.

15. $x - 5y = 0$
$0 - 5y = 0 \Rightarrow y = 0$
The x- and y-intercept is $(0, 0)$.

17. a. Cost is $\$(24 + 200(.45)) = \$114$.

   b. $f(x) = .45x + 24$

19. Let x be the number of days of hospital confinement.
$f(x) = 700x + 1900$

21. $f(x) = \dfrac{50x}{105 - x}, \ 0 \le x \le 100$
From example 6, we know that $f(70) = 100$.
The cost to remove 75% of the pollutant is
$f(75) = \dfrac{50 \cdot 75}{105 - 75} = 125.$
The cost of removing an extra 5% is
$\$125 - \$100 = \$25$ million. To remove the final 5% the cost is
$f(100) - f(95) = 1000 - 475 = \$525$ million.
This costs 21 times as much as the cost to remove the next 5% after the first 70% is removed.

23. $f(x) = \left(\dfrac{K}{V}\right)x + \dfrac{1}{V}$

   a. $f(x) = .2x + 50$

   We have $\dfrac{K}{V} = .2$ and $\dfrac{1}{V} = 50$. If $\dfrac{1}{V} = 50$,

   then $V = \dfrac{1}{50}$. Now, $\dfrac{K}{V} = .2$ implies

   $\dfrac{K}{\frac{1}{50}} = .2$, so $K = \dfrac{1}{5} \cdot \dfrac{1}{50} = \dfrac{1}{250}.$

   b. $y = \left(\dfrac{K}{V}\right)x + \dfrac{1}{V}, \ \left(\dfrac{K}{V}\right) \cdot 0 + \dfrac{1}{V} = \dfrac{1}{V}$, so the

   y-intercept is $\left(0, \dfrac{1}{V}\right)$.

   Solving $\left(\dfrac{K}{V}\right)x + \dfrac{1}{V} = 0$, we get

   $\dfrac{K}{V}x = -\dfrac{1}{V} \Rightarrow x = -\dfrac{1}{K}$, so the x-intercept

   is $\left(-\dfrac{1}{K}, 0\right)$.

25. $y = 3x^2 - 4x$
$a = 3, b = -4, c = 0$

27. $y = 3x - 2x^2 + 1$
$a = -2, b = 3, c = 1$

29. $y = 1 - x^2$
$a = -1, b = 0, c = 1$

31. $f(x) = 2x^2 - 4x$
$a = 2, b = -4, c = 0$
vertex:

$\left(\dfrac{-(-4)}{2(2)}, f\left(\dfrac{-(-4)}{2(2)}\right)\right) = \left(1, f(1)\right) = \left(1, -2\right)$

x	y
0	0
2	0

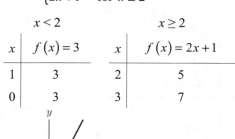

33. $f(x) = \begin{cases} 3 & \text{for } x < 2 \\ 2x + 1 & \text{for } x \ge 2 \end{cases}$

	$x < 2$			$x \ge 2$	
x	$f(x) = 3$		x	$f(x) = 2x + 1$	
1	3		2	5	
0	3		3	7	

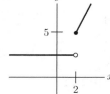

35. $f(x) = \begin{cases} 4 - x & \text{for } 0 \le x < 2 \\ 2x - 2 & \text{for } 2 \le x < 3 \\ x + 1 & \text{for } x \ge 3 \end{cases}$

$0 \le x < 2$

x	$f(x) = 4 - x$
0	4
1	3

$2 \le x < 3$

x	$f(x) = 2x - 2$
2	2
$\frac{5}{2}$	3

$x \ge 3$

x	$f(x) = x + 1$
3	4
4	5

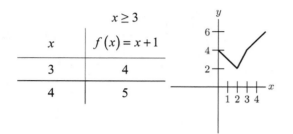

37. $f(x) = x^{100}, \ x = -1$

$f(-1) = (-1)^{100} = 1$

39. $f(x) = |x|, \ x = 10^{-2}$

$f(10^{-2}) = |10^{-2}| = 10^{-2}$

41. $f(x) = |x|, \ x = -2.5$

$f(-2.5) = |-2.5| = 2.5$

43.

```
Plot1 Plot2 Plot3
\Y1 = 3X^3+8
\Y2=
\Y3=
\Y4=
\Y5=
\Y6=
\Y7=
```

```
Y1(-11)
          -3985
Y1(10)
           3008
```

45.

```
Plot1 Plot2 Plot3
\Y1 = X^2/2+√(3)X-π
\Y2=■
\Y3=
\Y4=
\Y5=
\Y6=
```

```
Y1(-2)
      -4.605694269
Y1(20)
       231.4994235
```

0.3   The Algebra of Functions

1. $f(x) + g(x) = (x^2 + 1) + 9x = x^2 + 9x + 1$

3. $f(x)g(x) = (x^2 + 1)(9x) = 9x^3 + 9x$

5. $\dfrac{f(t)}{g(t)} = \dfrac{t^2 + 1}{9t} = \dfrac{t^2}{9t} + \dfrac{1}{9t} = \dfrac{t}{9} + \dfrac{1}{9t} = \dfrac{t^2 + 1}{9t}$

7. $\dfrac{2}{x-3} + \dfrac{1}{x+2} = \dfrac{2(x+2) + (x-3)}{(x-3)(x+2)}$

$= \dfrac{3x+1}{x^2 - x - 6}$

9. $\dfrac{x}{x-8} + \dfrac{-x}{x-4} = \dfrac{x(x-4) + (-x)(x-8)}{(x-8)(x-4)}$

$= \dfrac{4x}{x^2 - 12x + 32}$

11. $\dfrac{x+5}{x-10} + \dfrac{x}{x+10} = \dfrac{(x+5)(x+10) + x(x-10)}{(x-10)(x+10)}$

$= \dfrac{2x^2 + 5x + 50}{x^2 - 100}$

13. $\dfrac{x}{x-2} - \dfrac{5-x}{5+x} = \dfrac{x(5+x) - (5-x)(x-2)}{(x-2)(5+x)}$

$= \dfrac{2x^2 - 2x + 10}{x^2 + 3x - 10}$

15. $\dfrac{x}{x-2} \cdot \dfrac{5-x}{5+x} = \dfrac{-x^2 + 5x}{x^2 + 3x - 10}$

17. $\dfrac{\dfrac{x}{x-2}}{\dfrac{5-x}{5+x}} = \dfrac{x}{x-2} \cdot \dfrac{5+x}{5-x} = \dfrac{x^2 + 5x}{-x^2 + 7x - 10}$

19. $\dfrac{x+1}{(x+1)-2} \cdot \dfrac{5-(x+1)}{5+(x+1)} = \dfrac{x+1}{x-1} \cdot \dfrac{-x+4}{6+x}$

$= \dfrac{-x^2 + 3x + 4}{x^2 + 5x - 6}$

21. $\dfrac{\dfrac{5-(x+5)}{5+(x+5)}}{\dfrac{x+5}{(x+5)-2}} = \dfrac{5-(x+5)}{5+(x+5)} \cdot \dfrac{(x+5)-2}{x+5}$

$= \dfrac{-x}{10+x} \cdot \dfrac{x+3}{x+5}$

$= \dfrac{-x^2 - 3x}{x^2 + 15x + 50}$

23. $\dfrac{5 - \dfrac{1}{u}}{5 + \dfrac{1}{u}} = \dfrac{5u-1}{u} \cdot \dfrac{u}{5u+1} = \dfrac{5u-1}{5u+1}, \ u \ne 0$

25. $f\left(\dfrac{x}{1-x}\right) = \left(\dfrac{x}{1-x}\right)^6$

27. $h\left(\dfrac{x}{1-x}\right) = \left(\dfrac{x}{1-x}\right)^3 - 5\left(\dfrac{x}{1-x}\right)^2 + 1$

29. $g(t^3 - 5t^2 + 1) = \dfrac{t^3 - 5t^2 + 1}{1 - (t^3 - 5t^2 + 1)}$

$= \dfrac{t^3 - 5t^2 + 1}{-t^3 + 5t^2}$

31. $(x+h)^2 - x^2 = x^2 + 2xh + h^2 - x^2$

$= 2xh + h^2$

33. $\dfrac{\left[4(t+h) - (t+h)^2\right] - \left(4t - t^2\right)}{h}$

$= \dfrac{4t + 4h - (t^2 + 2th + h^2) - 4t + t^2}{h}$

$= \dfrac{4h - 2th - h^2}{h} = \dfrac{h(4 - 2t - h)}{h}$

$= 4 - 2t - h$

35. a. $C(A(t)) = 3000 + 80\left(20t - \dfrac{1}{2}t^2\right)$

$= 3000 + 1600t - 40t^2$

    b. $C(2) = 3000 + 1600(2) - 40(2)^2$

$= 3000 + 3200 - 160 = \$6040$

37. $h(x) = f(8x + 1) = \left(\dfrac{1}{8}\right)(8x + 1) = x + \dfrac{1}{8}$

$h(x)$ converts from British to U.S. sizes.

39. $f(x) + 1$:

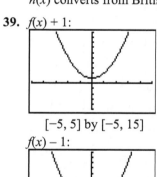

$[-5, 5]$ by $[-5, 15]$

$f(x) - 1$:

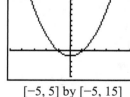

$[-5, 5]$ by $[-5, 15]$

$f(x) + 2$:

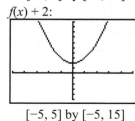

$[-5, 5]$ by $[-5, 15]$

$f(x) - 2$:

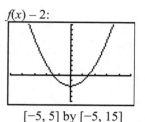

$[-5, 5]$ by $[-5, 15]$

The graph of $f(x) + c$ is the graph of $f(x)$ shifted up (if $c > 0$) or down (if $c < 0$) by $|c|$ units.

41. This is the graph of $f(x) = x^2$ shifted 2 units to the left and 1 unit down.

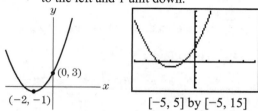

$[-5, 5]$ by $[-5, 15]$

43.

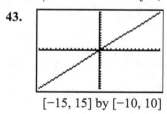

$[-15, 15]$ by $[-10, 10]$

$f(f(x)) = f\left(\dfrac{x}{x-1}\right) = \dfrac{\dfrac{x}{x-1}}{\dfrac{x}{x-1} - 1}$

$= \dfrac{x}{x - (x-1)} = x,\ x \neq 1$

0.4   Zeros of Functions—The Quadratic Formula and Factoring

1. $f(x) = 2x^2 - 7x + 6$

$2x^2 - 7x + 6 = 0$

$a = 2,\ b = -7,\ c = 6$

$\sqrt{b^2 - 4ac} = \sqrt{49 - 4(2)(6)} = \sqrt{1} = 1$

$x = \dfrac{-b \pm \sqrt{b^2 - 4ac}}{2a} = \dfrac{7 \pm 1}{4} = 2,\ \dfrac{3}{2}$

3. $f(t) = 4t^2 - 12t + 9$

$4t^2 - 12t + 9 = 0$

$t = \dfrac{-b \pm \sqrt{b^2 - 4ac}}{2a} = \dfrac{12 \pm \sqrt{(-12)^2 - 4(4)(9)}}{2(4)}$

$= \dfrac{12 \pm \sqrt{0}}{8} = \dfrac{3}{2}$

5. $f(x) = -2x^2 + 3x - 4$

$-2x^2 + 3x - 4 = 0$

$x = \dfrac{-b \pm \sqrt{b^2 - 4ac}}{2a} = \dfrac{-3 \pm \sqrt{3^2 - 4(-2)(-4)}}{2(-2)}$

$= \dfrac{-3 \pm \sqrt{-23}}{-4}$

$\sqrt{-23}$ is undefined, so $f(x)$ has no real zeros.

7. $5x^2 - 4x - 1 = 0$

$x = \dfrac{-b \pm \sqrt{b^2 - 4ac}}{2a} = \dfrac{4 \pm \sqrt{(-4)^2 - 4(5)(-1)}}{2(5)}$

$= \dfrac{4 \pm \sqrt{36}}{10} = \dfrac{4 \pm 6}{10} = 1, -\dfrac{1}{5}$

9. $15x^2 - 135x + 300 = 0$

$x = \dfrac{-b \pm \sqrt{b^2 - 4ac}}{2a}$

$= \dfrac{135 \pm \sqrt{(-135)^2 - 4(15)(300)}}{2(15)}$

$= \dfrac{135 \pm \sqrt{225}}{30} = \dfrac{135 \pm 15}{30} = 5, 4$

11. $\dfrac{3}{2}x^2 - 6x + 5 = 0$

$x = \dfrac{-b \pm \sqrt{b^2 - 4ac}}{2a} = \dfrac{6 \pm \sqrt{(-6)^2 - 4\left(\frac{3}{2}\right)(5)}}{2\left(\frac{3}{2}\right)}$

$= \dfrac{6 \pm \sqrt{6}}{3} = 2 + \dfrac{\sqrt{6}}{3}, 2 - \dfrac{\sqrt{6}}{3}$

13. $x^2 + 8x + 15 = (x + 5)(x + 3)$

15. $x^2 - 16 = (x - 4)(x + 4)$

17. $3x^2 + 12x + 12 = 3\left(x^2 + 4x + 4\right)$

$= 3(x + 2)(x + 2) = 3(x + 2)^2$

19. $30 - 4x - 2x^2 = -2\left(-15 + 2x + x^2\right)$

$= -2(x - 3)(x + 5)$

21. $3x - x^2 = x(3 - x)$

23. $6x - 2x^3 = -2x(x^2 - 3)$

$= -2x\left(x - \sqrt{3}\right)\left(x + \sqrt{3}\right)$

25. $x^3 - 1 = (x - 1)\left(x^2 + x + 1\right)$

27. $8x^3 + 27 = (2x + 3)\left(4x^2 - 6x + 9\right)$

29. $x^2 - 14x + 49 = (x - 7)^2$

31. $2x^2 - 5x - 6 = 3x + 4$

$2x^2 - 8x - 10 = 0$

$x = \dfrac{-b \pm \sqrt{b^2 - 4ac}}{2a} = \dfrac{8 \pm \sqrt{(-8)^2 - 4(2)(-10)}}{2(2)}$

$= \dfrac{8 \pm \sqrt{144}}{4} = \dfrac{8 \pm 12}{4} = 5, -1$

$y = 3x + 4 = 15 + 4 = 19$

$y = -3 + 4 = 1$

Points of intersection: (5, 19), (−1, 1)

33. $y = x^2 - 4x + 4$

$y = 12 + 2x - x^2$

$x^2 - 4x + 4 = 12 + 2x - x^2$

$2x^2 - 6x - 8 = 0$

$2(x^2 - 3x - 4) = 0$

$2(x - 4)(x + 1) = 0$

$x = 4, -1$

$y = x^2 - 4x + 4 = 4^2 - 4(4) + 4 = 4$

$y = (-1)^2 - 4(-1) + 4 = 9$

Points of intersection: (4, 4), (−1, 9)

35. $y = x^3 - 3x^2 + x$

$y = x^2 - 3x$

$x^3 - 3x^2 + x = x^2 - 3x$

$x^3 - 4x^2 + 4x = 0$

$x(x^2 - 4x + 4) = 0$

$x(x - 2)(x - 2) = 0 \Rightarrow x = 0, 2$

$y = x^2 - 3x = 0^2 - 3(0) = 0$

$y = 2^2 - 3(2) = 4 - 6 = -2$

Points of intersection: (0, 0), (2, −2)

37. $y = \frac{1}{2}x^3 + x^2 + 5$

$y = 3x^2 - \frac{1}{2}x + 5$

$\frac{1}{2}x^3 + x^2 + 5 = 3x^2 - \frac{1}{2}x + 5$

$\frac{1}{2}x^3 - 2x^2 + \frac{1}{2}x = 0$

$x\left(\frac{1}{2}x^2 - 2x + \frac{1}{2}\right) = 0$

$x = 0$ or $\frac{1}{2}x^2 - 2x + \frac{1}{2} = 0$

$x = \frac{-b \pm \sqrt{b^2 - 4ac}}{2a} = \frac{2 \pm \sqrt{(-2)^2 - 4\left(\frac{1}{2}\right)\left(\frac{1}{2}\right)}}{2\left(\frac{1}{2}\right)}$

$= 2 \pm \sqrt{3}$

$y = 3x^2 - \frac{1}{2}x + 5 = 3(0)^2 - \frac{1}{2}(0) + 5 = 5$

$y = 3\left(2 + \sqrt{3}\right)^2 - \frac{1}{2}\left(2 + \sqrt{3}\right) + 5 = 25 + \frac{23\sqrt{3}}{2}$

$y = 3\left(2 - \sqrt{3}\right)^2 - \frac{1}{2}\left(2 - \sqrt{3}\right) + 5 = 25 - \frac{23\sqrt{3}}{2}$

Points of intersection: $(0, 5)$,

$\left(2 - \sqrt{3}, \ 25 - \frac{23\sqrt{3}}{2}\right), \left(2 + \sqrt{3}, \ 25 + \frac{23\sqrt{3}}{2}\right)$

39. $\frac{21}{x} - x = 4$

$21 - x^2 = 4x$

$x^2 + 4x - 21 = 0$

$(x + 7)(x - 3) = 0 \Rightarrow x = -7, \ 3$

41. $x + \frac{14}{x + 4} = 5$

$x^2 + 4x + 14 = 5x + 20$

$x^2 - x - 6 = 0$

$(x - 3)(x + 2) = 0 \Rightarrow x = 3, \ -2$

43. $\frac{x^2 + 14x + 49}{x^2 + 1} = 0$

$x^2 + 14x + 49 = 0$

$(x + 7)^2 = 0 \Rightarrow x = -7$

45. $C(x) = 275 + 12x$

$R(x) = 32x - .21x^2$

$C(x) = R(x)$

$275 + 12x = 32x - .21x^2$

$.21x^2 - 20x + 275 = 0$

Thus

$x = \frac{20 \pm \sqrt{(-20)^2 - 4(.21)275}}{.42}$

$= 16,667$ or $78,571$ subscribers

47.

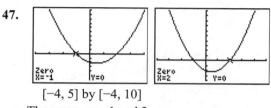

[−4, 5] by [−4, 10]

The zeros are −1 and 2.

49.

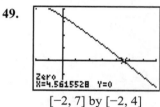

[−2, 7] by [−2, 4]

The zero is approximately 4.56.

51.

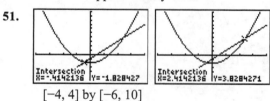

[−4, 4] by [−6, 10]

Approximate points of intersection:
(−0.41, −1.83) and (2.41, 3.83)

53.

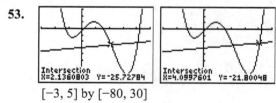

[−3, 5] by [−80, 30]

Approximate points of intersection:
(2.14, −25.73) and (4.10, −21.80)

Answers may vary for exercises 55−57.

55.

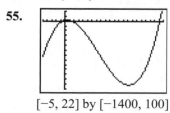

[−5, 22] by [−1400, 100]

57.

[−20, 4] by [−500, 2500]

0.5   Exponents and Power Functions

1. $3^3 = 27$

3. $1^{100} = 1$

5. $(.1)^4 = (.1)(.1)(.1)(.1) = .0001$

7. $-4^2 = -16$

9. $(16)^{1/2} = \sqrt{16} = 4$

11. $(.000001)^{1/3} = \sqrt[3]{.000001} = .01$

13. $6^{-1} = \dfrac{1}{6}$

15. $(.01)^{-1} = \dfrac{1}{.01} = 100$

17. $8^{4/3} = \left(\sqrt[3]{8}\right)^4 = 16$

19. $(25)^{3/2} = \left(\sqrt{25}\right)^3 = 125$

21. $(1.8)^0 = 1$

23. $16^{0.5} = 16^{1/2} = 4$

25. $4^{-1/2} = \dfrac{1}{\sqrt{4}} = \dfrac{1}{2}$

27. $(.01)^{-1.5} = \dfrac{1}{(.01)^{3/2}} = \dfrac{1}{.001} = 1000$

29. $5^{1/3} \cdot 200^{1/3} = 1000^{1/3} = 10$

31. $6^{1/3} \cdot 6^{2/3} = 6^1 = 6$

33. $\dfrac{10^4}{5^4} = 2^4 = 16$

35. $(2^{1/3} \cdot 3^{2/3})^3 = \left(\sqrt[3]{2}\sqrt[3]{9}\right)^3 = \left(\sqrt[3]{18}\right)^3 = 18$

37. $\left(\dfrac{8}{27}\right)^{2/3} = \dfrac{8^{2/3}}{27^{2/3}} = \dfrac{4}{9}$

39. $\dfrac{7^{4/3}}{7^{1/3}} = 7^{(4/3)-(1/3)} = 7^{3/3} = 7$

41. $(xy)^6 = x^6 y^6$

43. $\dfrac{x^4 \cdot y^5}{xy^2} = x^4 \cdot y^5 \cdot x^{-1} \cdot y^{-2} = x^3 y^3$

45. $x^{-1/2} = \dfrac{1}{\sqrt{x}}$

47. $\left(\dfrac{x^4}{y^2}\right)^3 = \dfrac{x^{4(3)}}{y^{2(3)}} = \dfrac{x^{12}}{y^6}$

49. $(x^3 y^5)^4 = x^{3(4)} \cdot y^{5(4)} = x^{12} y^{20}$

51. $x^5 \cdot \left(\dfrac{y^2}{x}\right)^3 = \dfrac{x^5 \cdot y^{2(3)}}{x^3} = x^5 \cdot y^6 \cdot x^{-3} = x^2 y^6$

53. $(2x)^4 = 2^4 \cdot x^4 = 16x^4$

55. $\dfrac{-x^3 y}{-xy} = \dfrac{x^3}{x} \cdot \dfrac{y}{y} = x^2$

57. $\dfrac{x^{-4}}{x^3} = \dfrac{1}{x^4} \cdot \dfrac{1}{x^3} = (-3)^3 \cdot x^3 = \dfrac{1}{x^7}$

59. $\sqrt[3]{x} \cdot \sqrt[3]{x^2} = x^{1/3} \cdot x^{2/3} = x$

61. $\left(\dfrac{3x^2}{2y}\right)^3 = \dfrac{3^3 \cdot x^6}{2^3 \cdot y^3} = \dfrac{27x^6}{8y^3}$

63. $\dfrac{2x}{\sqrt{x}} = 2x \cdot x^{-1/2} = 2\sqrt{x}$

65. $(16x^8)^{-3/4} = 16^{-3/4} \cdot x^{-6} = \dfrac{1}{8x^6}$

67. $\sqrt{x}\left(\dfrac{1}{4x}\right)^{5/2} = \dfrac{x^{1/2}}{4^{5/2} x^{5/2}} = \dfrac{x^{1/2} \cdot x^{-5/2}}{32}$
$= \dfrac{1}{32x^2}$

69. $\dfrac{(-27x^5)^{2/3}}{\sqrt[3]{x}} = \dfrac{(-27)^{2/3} x^{5(2/3)}}{x^{1/3}} = 9x^3$

For exercises 71–81, $f(x) = \sqrt[3]{x}$ and $g(x) = \dfrac{1}{x^2}$.

71. $f(x)g(x) = \sqrt[3]{x} \cdot \dfrac{1}{x^2} = x^{1/3} \cdot x^{-2} = x^{-5/3} = \dfrac{1}{x^{5/3}}$

73. $\dfrac{g(x)}{f(x)} = \dfrac{\frac{1}{x^2}}{\sqrt[3]{x}} = x^{-2} \cdot x^{-1/3} = x^{-7/3} = \dfrac{1}{x^{7/3}}$

75. $\left[f(x)g(x)\right]^3 = \left(\sqrt[3]{x} \cdot \dfrac{1}{x^2}\right)^3 = \left(x^{1/3} \cdot x^{-2}\right)^3$

$\qquad = \left(x^{-5/3}\right)^3 = x^{-5} = \dfrac{1}{x^5}$

77. $\sqrt{f(x)g(x)} = \left(\sqrt[3]{x} \cdot \dfrac{1}{x^2}\right)^{1/2} = \left(x^{1/3} \cdot x^{-2}\right)^{1/2}$

$\qquad = \left(x^{-5/3}\right)^{1/2} = x^{-5/6} = \dfrac{1}{x^{5/6}}$

79. $f(g(x)) = f\left(\dfrac{1}{x^2}\right) = f\left(x^{-2}\right) = \sqrt[3]{x^{-2}}$

$\qquad = \left(x^{-2}\right)^{1/3} = x^{-2/3} = \dfrac{1}{x^{2/3}}$

81. $f(g(x)) = f\left(\sqrt[3]{x}\right) = f\left(x^{1/3}\right) = \sqrt[3]{x^{1/3}}$

$\qquad = \left(x^{1/3}\right)^{1/3} = x^{1/9}$

83. $\sqrt{x} - \dfrac{1}{\sqrt{x}} = \dfrac{1}{\sqrt{x}}(x-1)$

85. $x^{-1/4} + 6x^{1/4} = x^{-1/4}\left(1 + 6\sqrt{x}\right)$

87. $\sqrt{a} \cdot \sqrt{b} = \sqrt{ab}$
$a^{1/2} \cdot b^{1/2} = (ab)^{1/2}$ (Law 5)

89. $f(x) = x^2 \Rightarrow f(4) = (4)^2 = 16$

91. $f(x) = x^{-1} \Rightarrow f(4) = (4)^{-1} = \dfrac{1}{4}$

93. $f(x) = x^{3/2} \Rightarrow f(4) = (4)^{3/2} = 8$

95. $f(x) = x^{-5/2} \Rightarrow f(4) = (4)^{-5/2} = \dfrac{1}{32}$

In exercises 97–103, use the compound interest formula $A = P\left(1 + \dfrac{r}{m}\right)^{mt}$, where P is the principal, r is the annual interest rate, m is the number of interest periods per year, and t is the number of years.

97. $A = 500\left(1 + \dfrac{.06}{1}\right)^{1(6)} \approx \$709.26$

99. $A = 50{,}000\left(1 + \dfrac{.095}{4}\right)^{4(10)} \approx \$127{,}857.61$

101. $A = 100\left(1 + \dfrac{.05}{12}\right)^{12(10)} \approx \$164.70$

103. $A = 1500\left(1 + \dfrac{.06}{365}\right)^{365(1)} \approx \$1592.75$

105. $A = 1000\left(1 + \dfrac{.068}{1}\right)^{1(18)} \approx \$3268.00$

107. $A = 500 + 500r + \dfrac{375}{2}r^2 + \dfrac{125}{4}r^3 + \dfrac{125}{64}r^4$

$\qquad = \dfrac{500}{256}\left(256 + 256r + 96r^2 + 16r^3 + r^4\right)$

109. If the speed is $2x$, then

$\qquad \dfrac{1}{20}(2x)^2 = \dfrac{1}{20}\left(4x^2\right) = 4\left(\dfrac{1}{20}x^2\right).$

111. $8.103\text{E}{-}4 = 8.103 \cdot 10^{-4} = .0008103$

113. $8.23\text{E}{-}6 = 8.23 \cdot 10^{-6} = .00000823$

0.6 Functions and Graphs in Applications

1.

$3x$

x

3.

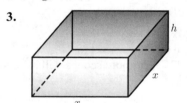

h

x

x

5.

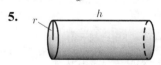

r h

7. $P = 2(x + 3x) = 8x$

$3x^2 = 25$

9. $A = \pi r^2$

$2\pi r = 15$

11. $V = x^2 h$
The surface area of the box is represented by
$S = x^2 + 4xh.$
$x^2 + 4xh = 65$

13. $\pi r^2 h = 100$

$$\text{Cost} = 5\pi r^2 + 6\pi r^2 + 7(2\pi rh)$$
$$= 11\pi r^2 + 14\pi rh$$

15. $2x + 3h = 5000$

$A = xh$

17. $C = 10(2\ell + 2h) + 8(2\ell) = 36\ell + 20h$

19. $8x = 40 \Rightarrow x = 5$

$$A = 3x^2 = 3(25) = 75 \ \text{cm}^2$$

21. a. $73 + 4x = 225 \Rightarrow x = 38$
When 38 T-shirts are sold, the cost will be $225.

    b. $C(50) - C(40)$
$$= \big(73 + 4(50)\big) - \big(73 + 4(40)\big)$$
$$= 273 - 233 = \$40$$
The cost will rise $40.

23. a. $.4x - 80 = 0 \Rightarrow x = \dfrac{80}{.4} = 200$

Sales will break-even when 200 scoops are sold.

    b. $30 = .4x - 80 \Rightarrow x = 275$
Sales of 275 scoops will generate a daily profit of $30.

    c. $40 = .4x - 80 \Rightarrow x = 300$
To raise the daily profit to $40,
$300 - 275 = 25$ more scoops will have to be sold.

25. a. $P(x) = R(x) - C(x) = 21x - 9x - 800$
$$= 12x - 800$$

    b. $P(120) = 1440 - 800 = \$640$

    c. $1000 = 12x - 800 \Rightarrow x = 150$
$$R(150) = 21(150) = \$3150$$

27. $f(6) = 270$ cents

29. A 100-inch^3 cylinder with radius 3 inches costs $1.62 to construct.

31. $f(3) = \$1.62$; $f(6) = \$2.70$, so the additional cost $= 2.70 - 1.62 = \$1.08$

33. From the graph, we see that revenue = $1800 and cost = $1200.

35. The cost is $1400 when production is 40 units.

37. $C(1000) = \$4000$

39. Find the y-coordinate of the point on the graph whose x-coordinate is 400.

41. The greatest profit, $52,500, occurs when 2500 units of goods are produced.

43. Find the x-coordinate of the point on the graph whose y-coordinate is 30,000.

45. Find $h(3)$. Find the y-coordinate of the point on the graph whose t-coordinate is 3.

47. Find the maximum value of $h(t)$. Find the y-coordinate of the highest point of the graph.

49. Solve $h(t) = 100$. Find the t-coordinates of the points whose y-coordinate is 100.

51. a.

$[0, 6]$ by $[-30, 120]$

    b. Using the Trace command or the Value command, the height is 96 feet.

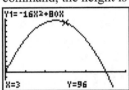

    c. Graphing $Y_2 = 64$ and using the Intersect command, the height is 64 feet when $x = 1$ and $x = 4$ seconds.

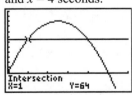

    d. Using the Trace command or the Zero command, the ball hits the ground when $x = 5$ seconds.

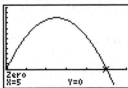

e. Using the Trace command or the Maximum command, the maximum height is reached when $x = 2.5$ seconds. The maximum height is 100 feet.

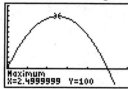

53. a.

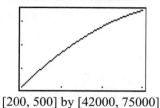

[200, 500] by [42000, 75000]

b. Graphing $Y_2 = 63,000$ and using the Intersect command, the revenue is $63,000 when sales are 350 bicycles per year.

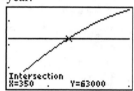

c. Using the Trace command or the Value command, the revenue is $68,000 when 400 bicycles are sold per year.

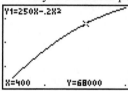

d.

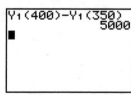

$R(400) - R(350) = 5000$
Revenue would decrease by $5000.

e.

Y₁(450)−Y₁(400)
4000

$R(450) - R(400) = 4000$
No, the store should not spend $5000 on advertising, since the revenues would only increase by $4000.

Chapter 0 Review Exercises

1. $f(x) = x^3 + \dfrac{1}{x}$

$f(1) = 1^3 + \dfrac{1}{1} = 2$

$f(3) = 3^3 + \dfrac{1}{3} = \dfrac{82}{3} = 27\dfrac{1}{3}$

$f(-1) = (-1)^3 + \dfrac{1}{(-1)} = -2$

$f\left(-\dfrac{1}{2}\right) = \left(-\dfrac{1}{2}\right)^3 - 2 = -\dfrac{17}{8} = -2\dfrac{1}{8}$

$f\left(\sqrt{2}\right) = \left(\sqrt{2}\right)^3 + \dfrac{1}{\sqrt{2}} = 2\sqrt{2} + \dfrac{1}{\sqrt{2}} = \dfrac{5\sqrt{2}}{2}$

2. $f(x) = 2x + 3x^2$

$f(0) = 2(0) + 3(0)^2 = 0$

$f\left(-\dfrac{1}{4}\right) = 2\left(-\dfrac{1}{4}\right) + 3\left(-\dfrac{1}{4}\right)^2 = -\dfrac{5}{16}$

$f\left(\dfrac{1}{\sqrt{2}}\right) = 2\left(\dfrac{1}{\sqrt{2}}\right) + 3\left(\dfrac{1}{\sqrt{2}}\right)^2 = \dfrac{3 + 2\sqrt{2}}{2}$

3. $f(x) = x^2 - 2$

$f(a - 2) = (a - 2)^2 - 2 = a^2 - 4a + 2$

4. $f(x) = \dfrac{1}{x + 1} - x^2$

$f(a + 1) = \dfrac{1}{(a + 1) + 1} - (a + 1)^2$

$= \dfrac{1}{a + 2} - (a^2 + 2a + 1)$

$= -\dfrac{a^3 + 4a^2 + 5a + 1}{a + 2}$

5. $f(x) = \dfrac{1}{x(x + 3)} \Rightarrow x \neq 0, -3$

6. $f(x) = \sqrt{x - 1} \Rightarrow x \geq 1$

7. $f(x) = \sqrt{x^2 + 1}$, all values of x

8. $f(x) = \dfrac{1}{\sqrt{3x}}$, $x > 0$

9. $h(x) = \dfrac{x^2 - 1}{x^2 + 1}$ p

$h\left(\dfrac{1}{2}\right) = \dfrac{\left(\frac{1}{2}\right)^2 - 1}{\left(\frac{1}{2}\right)^2 + 1} = -\dfrac{3}{5}$

Yes, the point $\left(\dfrac{1}{2}, -\dfrac{3}{5}\right)$ is on the graph.

10. $k(x) = x^2 + \dfrac{2}{x}$

$k(1) = 1^2 + \dfrac{2}{1} = 3$

No, the point $(1, -2)$ is not on the graph.

11. $5x^3 + 15x^2 - 20x = 5x(x^2 + 3x - 4)$
$= 5x(x - 1)(x + 4)$

12. $3x^2 - 3x - 60 = 3(x^2 - x - 20)$
$= 3(x - 5)(x + 4)$

13. $18 + 3x - x^2 = (-x - 3)(x - 6)$
$= (-1)(x - 6)(x + 3)$

14. $x^5 - x^4 - 2x^3 = x^3(x^2 - x - 2)$
$= x^3(x - 2)(x + 1)$

15. $y = 5x^2 - 3x - 2 \Rightarrow 5x^2 - 3x - 2 = 0.$

$x = \dfrac{-b \pm \sqrt{b^2 - 4ac}}{2a} = \dfrac{3 \pm \sqrt{(-3)^2 - 4(5)(-2)}}{2(5)}$

$= \dfrac{3 \pm 7}{10} \Rightarrow x = 1 \text{ or } x = -\dfrac{2}{5}$

16. $y = -2x^2 - x + 2 \Rightarrow -2x^2 - x + 2 = 0.$

$x = \dfrac{-b \pm \sqrt{b^2 - 4ac}}{2a} = \dfrac{1 \pm \sqrt{(-1)^2 - 4(-2)(2)}}{2(-2)}$

$= \dfrac{1 \pm \sqrt{17}}{-4} \Rightarrow x = \dfrac{-1 + \sqrt{17}}{4} \text{ or } x = \dfrac{-1 - \sqrt{17}}{4}$

17. Substitute $2x - 1$ for y in the quadratic equation, then find the zeros:

$5x^2 - 3x - 2 = 2x - 1 \Rightarrow 5x^2 - 5x - 1 = 0.$

$x = \dfrac{-b \pm \sqrt{b^2 - 4ac}}{2a} = \dfrac{5 \pm \sqrt{(-5)^2 - 4(5)(-1)}}{2(5)}$

$= \dfrac{5 \pm 3\sqrt{5}}{10}$

Now find the y-values for each x value:

$y = 2x - 1 = 2\left(\dfrac{5 + 3\sqrt{5}}{10}\right) - 1 = \dfrac{3\sqrt{5}}{5}$

$y = 2x - 1 = 2\left(\dfrac{5 - 3\sqrt{5}}{10}\right) - 1 = \dfrac{-3\sqrt{5}}{5}$

Points of intersection:

$\left(\dfrac{5 + 3\sqrt{5}}{10}, \dfrac{3\sqrt{5}}{5}\right), \left(\dfrac{5 - 3\sqrt{5}}{10}, -\dfrac{3\sqrt{5}}{5}\right)$

18. Substitute $x - 5$ for y in the quadratic equation, then find the zeros:

$-x^2 + x + 1 = x - 5 \Rightarrow x^2 - 6 = 0 \Rightarrow x = \pm\sqrt{6}$

Now find the y-values for each x value:

$y = x - 5 = \sqrt{6} - 5$

$y = -\sqrt{6} - 5$

Points of intersection:

$\left(\sqrt{6}, \sqrt{6} - 5\right), \left(-\sqrt{6}, -\sqrt{6} - 5\right)$

19. $f(x) + g(x) = \left(x^2 - 2x\right) + (3x - 1) = x^2 + x - 1$

20. $f(x) - g(x) = \left(x^2 - 2x\right) - (3x - 1)$
$= x^2 - 5x + 1$

21. $f(x)h(x) = \left(x^2 - 2x\right)\left(\sqrt{x}\right)$
$= x^2 \cdot x^{1/2} - 2x \cdot x^{1/2}$
$= x^{5/2} - 2x^{3/2}$

22. $f(x)g(x) = (x^2 - 2x)(3x - 1)$
$= 3x^3 - x^2 - 6x^2 + 2x$
$= 3x^3 - 7x^2 + 2x$

23. $\dfrac{f(x)}{h(x)} = \dfrac{x^2 - 2x}{\sqrt{x}} = x^{3/2} - 2x^{1/2}$

24. $g(x)h(x) = (3x - 1)\sqrt{x} = 3x \cdot x^{1/2} - x^{1/2}$
$= 3x^{3/2} - x^{1/2}$

25. $f(x) - g(x) = \dfrac{x}{x^2 - 1} - \dfrac{1 - x}{1 + x}$
$= \dfrac{x - (x - 1)(1 - x)}{x^2 - 1}$
$= \dfrac{x^2 - x + 1}{x^2 - 1} = \dfrac{x^2 - x + 1}{(x - 1)(x + 1)}$

26. $f(x) - g(x + 1) = \dfrac{x}{x^2 - 1} - \dfrac{1 - (x + 1)}{1 + (x + 1)}$
$= \dfrac{x(x + 2) - (-x)\left(x^2 - 1\right)}{\left(x^2 - 1\right)(x + 2)}$
$= \dfrac{x^3 + x^2 + x}{\left(x^2 - 1\right)(x + 2)}$

27. $g(x) - h(x) = \dfrac{1-x}{1+x} - \dfrac{2}{3x+1}$

$= \dfrac{(1-x)(3x+1) - 2(1+x)}{(1+x)(3x+1)}$

$= -\dfrac{3x^2 + 1}{(1+x)(3x+1)}$

$= -\dfrac{3x^2 + 1}{3x^2 + 4x + 1}$

28. $f(x) + h(x) = \dfrac{x}{x^2 - 1} + \dfrac{2}{3x+1}$

$= \dfrac{x(3x+1) + 2(x^2 - 1)}{(x^2 - 1)(3x+1)}$

$= \dfrac{5x^2 + x - 2}{(x^2 - 1)(3x+1)}$

29. $g(x) - h(x-3) = \dfrac{1-x}{1+x} - \dfrac{2}{3(x-3)+1}$

$= \dfrac{(1-x)(3x-8) - 2(1+x)}{(1+x)(3x-8)}$

$= \dfrac{-3x^2 + 9x - 10}{(1+x)(3x-8)}$

$= \dfrac{-3x^2 + 9x - 10}{3x^2 - 5x - 8}$

30. $f(x) + g(x) = \dfrac{x}{x^2 - 1} + \dfrac{1-x}{1+x} = \dfrac{x + (1-x)(x-1)}{x^2 - 1}$

$= \dfrac{-x^2 + 3x - 1}{x^2 - 1}$

For exercises 31−36, $f(x) = x^2 - 2x + 4$,

$g(x) = \dfrac{1}{x^2}$ and $h(x) = \dfrac{1}{\sqrt{x} - 1}$.

31. $f\big(g(x)\big) = f\left(\dfrac{1}{x^2}\right) = \left(\dfrac{1}{x^2}\right)^2 - 2\left(\dfrac{1}{x^2}\right) + 4$

$= \dfrac{1}{x^4} - \dfrac{2}{x^2} + 4$

32. $g\big(f(x)\big) = g(x^2 - 2x + 4) = \dfrac{1}{\left(x^2 - 2x + 4\right)^2}$

$= \dfrac{1}{x^4 - 4x^3 + 12x^2 - 16x + 16}$

33. $g\big(h(x)\big) = g\left(\dfrac{1}{\sqrt{x} - 1}\right) = \dfrac{1}{\left(\dfrac{1}{\sqrt{x}-1}\right)^2} = \dfrac{1}{\dfrac{1}{x-2\sqrt{x}+1}}$

$= x - 2\sqrt{x} + 1 = \left(\sqrt{x} - 1\right)^2$

34. $h\big(g(x)\big) = h\left(\dfrac{1}{x^2}\right) = \dfrac{1}{\sqrt{\dfrac{1}{x^2}} - 1} = \dfrac{1}{\dfrac{1}{|x|} - 1} = \dfrac{|x|}{1 - |x|}$

35. $f\big(h(x)\big) = f\left(\dfrac{1}{\sqrt{x} - 1}\right)$

$= \left(\dfrac{1}{\sqrt{x}-1}\right)^2 - 2\left(\dfrac{1}{\sqrt{x}-1}\right) + 4$

$= \dfrac{1}{\left(\sqrt{x}-1\right)^2} - \dfrac{2}{\sqrt{x}-1} + 4$

36. $h\big(f(x)\big) = h\left(x^2 - 2x + 4\right)$

$= \dfrac{1}{\sqrt{x^2 - 2x + 4} - 1}$

$= \left(\sqrt{x^2 - 2x + 4} - 1\right)^{-1}$

37. $(81)^{3/4} = \left(\sqrt[4]{81}\right)^3 = 27$

$8^{5/3} = \left(\sqrt[3]{8}\right)^5 = 2^5 = 32$

$(0.25)^{-1} = \left(\dfrac{1}{4}\right)^{-1} = 4$

38. $(100)^{3/2} = \left(\sqrt{100}\right)^3 = 1000$

$(.001)^{1/3} = \left(\sqrt[3]{.001}\right) = .1$

39. $C(x)$ = carbon monoxide level corresponding to population x

$P(t)$ = population of the city in t years

$C(x) = 1 + .4x$

$P(t) = 750 + 25t + .1t^2$

$C\big(P(t)\big) = 1 + .4\left(750 + 25t + .1t^2\right)$

$= 1 + 300 + 10t + .04t^2$

$= .04t^2 + 10t + 301$

40. $R(x) = 5x - x^2$

$f(d) = 6\left(1 - \dfrac{200}{d + 200}\right)$

$R\big(f(d)\big) = 5 \cdot 6\left(1 - \dfrac{200}{d + 200}\right)$

$\qquad\qquad - \left[6\left(1 - \dfrac{200}{d + 200}\right)\right]^2$

$= 30\left(1 - \dfrac{200}{d + 200}\right) - 36\left(1 - \dfrac{200}{d + 200}\right)^2$

41. $\left(\sqrt{x+1}\right)^4 = (x+1)^{4/2} = (x+1)^2 = x^2 + 2x + 1$

42. $\dfrac{xy^3}{x^{-5}y^6} = x \cdot x^5 \cdot y^3 \cdot y^{-6} = \dfrac{x^6}{y^3}$

43. $\dfrac{x^{3/2}}{\sqrt{x}} = x^{3/2} \cdot x^{-1/2} = x$

44. $\sqrt[3]{x}(8x^{2/3}) = x^{1/3} \cdot 8x^{2/3} = 8x$

45. a. $P = 15000,\ r = .04,\ m = 12$

$$A(t) = 15000\left(1 + \dfrac{.04}{12}\right)^{12t}$$
$$= 15000(1.00333)^{12t}$$

b. $A(2) = 15000\left(1 + \dfrac{.04}{12}\right)^{12\cdot 2} \approx 16247.14$

$A(5) = 15000\left(1 + \dfrac{.04}{12}\right)^{12\cdot 5} \approx 18314.94$

At the end of 2 years, the account balance is about $16,247. At the end of 5 years, the account balance is about $18,315.

46. a. $P = 7000,\ r = .09,\ m = 2$

$$A(t) = 7000\left(1 + \dfrac{.09}{2}\right)^{2t} = 7000(1.045)^{2t}$$

b. $A(10) = 7000(1.045)^{2\cdot 10} = 16882.00$
$A(20) = 7000(1.045)^{2\cdot 20} = 40714.55$

At the end of 10 years, the account balance is about $16,882. At the end of 20 years, the account balance is about $40,715.

47. a. $P = 15000,\ m = 1,\ t = 10$
$A(r) = 15000(1 + r)^{10}$

b. $A(.04) = 15000(1 + .04)^{10} = 22203.66$
$A(.06) = 15000(1 + .06)^{10} = 26862.72$

48. a. $P = 7000,\ m = 1,\ t = 20$
$A(r) = 7000(1 + r)^{20}$

b. $A(.07) = 7000(1 + .07)^{20} = 27087.79$
$A(.12) = 7000(1 + .12)^{20} = 67524.05$

Chapter 1 The Derivative

1.1 The Slope of a Straight Line

1. $y = 3 - 7x$; y-intercept: $(0, 3)$, slope: -7

3. $x = 2y - 3 \Rightarrow y = \dfrac{x+3}{2} \Rightarrow y = \dfrac{1}{2}x + \dfrac{3}{2}$;

y-intercept: $\left(0, \dfrac{3}{2}\right)$, slope: $\dfrac{1}{2}$

5. $y = \dfrac{x}{7} - 5 \Rightarrow y = \dfrac{1}{7}x - 5$; y-intercept: $(0, -5)$,

slope: $\dfrac{1}{7}$

7. slope $= -1$, $(7, 1)$ on line.
Let $(x, y) = (7, 1)$, $m = -1$.
$y - y_1 = m(x - x_1) \Rightarrow y - 1 = -(x - 7) \Rightarrow$
$y = -x + 8$

9. slope $= \dfrac{1}{2}$; $(2, 1)$ on line.

Let $(x_1, y_1) = (2, 1)$; $m = \dfrac{1}{2}$.

$y - y_1 = m(x - x_1) \Rightarrow y - 1 = \dfrac{1}{2}(x - 2) \Rightarrow$

$y = \dfrac{1}{2}x$

11. $\left(\dfrac{5}{7}, 5\right)$ and $\left(-\dfrac{5}{7}, -4\right)$ on line.

slope $= \dfrac{y_2 - y_1}{x_2 - x_1} = \dfrac{-4 - 5}{-\frac{5}{7} - \frac{5}{7}} = \dfrac{-9}{-\frac{10}{7}} = \dfrac{63}{10}$

Let $(x_1, y_1) = \left(\dfrac{5}{7}, 5\right)$, $m = \dfrac{63}{10}$.

$y - y_1 = m(x - x_1) \Rightarrow y - 5 = \dfrac{63}{10}\left(x - \dfrac{5}{7}\right)$

13. $(0, 0)$ and $(1, 0)$ on line.
slope $= \dfrac{y_2 - y_1}{x_2 - x_1} = \dfrac{0 - 0}{1 - 0} = 0$
$y - 0 = 0(x - 0) \Rightarrow y = 0$

15. Horizontal through $(2, 9)$.
Let $(x_1, y_1) = (2, 9)$, $m = 0$ (horizontal line).
$y - y_1 = m(x - x_1) \Rightarrow y - 9 = 0(x - 2) \Rightarrow$
$y = 9$

17. x-intercept is $-\pi$; y-intercept is 1.
The intercepts $(-\pi, 0)$ and $(0, 1)$ are on the line.
slope $= \dfrac{y_2 - y_1}{x_2 - x_1} = \dfrac{1 - 0}{0 - (-\pi)} = \dfrac{1}{\pi}$
y-intercept $(0, b) = (0, 1)$
$y = mx + b \Rightarrow y = \dfrac{x}{\pi} + 1$

19. Slope $= -2$; x-intercept is -2.
The x-intercept $(-2, 0)$ is on the line.
Let $(x_1, y_1) = (-2, 0)$, $m = -2$.
$y - y_1 = m(x - x_1) \Rightarrow y - 0 = -2(x + 2) \Rightarrow$
$y = -2x - 4$

21. Parallel to $y = x$; $(2, 0)$ on line.
Let $(x_1, y_1) = (2, 0)$; slope $= m = 1$.
$y - y_1 = m(x - x_1) \Rightarrow y - 0 = 1(x - 2) \Rightarrow$
$y = x - 2$

23. Parallel to $y = 3x + 7$; x-intercept is 2.
slope $= m = 3$. Let $(x_1, y_1) = (2, 0)$.
$y - y_1 = m(x - x_1) \Rightarrow y - 0 = 3(x - 2) \Rightarrow$
$y = 3x - 6$

25. Perpendicular to $y + x = 0$; $(2, 0)$ on line.
$y + x = 0 \Rightarrow y = -x \Rightarrow$ slope $= m_1 = -1$
$m_1 \cdot m_2 = -1 \Rightarrow -1 \cdot m_2 = -1 \Rightarrow m_2 = 1$
Let $(x_2, y_2) = (2, 0)$.
$y - y_2 = m_2(x - x_2) \Rightarrow y - 0 = x - 2 \Rightarrow$
$y = x - 2$

27. Start at $(1, 0)$, then move one unit right and one unit up to $(2, 1)$.

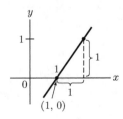

29. Start at $(1, -1)$, then move one unit up and three units to the left. Alternatively, move one unit down and three units to the right.

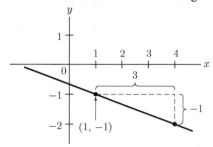

31. (a)–(C) x- and y-intercepts are 1.
(b)–(B) x-intercept is 1, y-intercept is -1.
(c)–(D) x- and y-intercepts are -1.
(d)–(A) x-intercept is -1, y-intercept is 1.

33. $m = \dfrac{1}{3}$, $h = 3$

If you move 3 units in the x-direction, then you must move 1 unit in the y-direction to return to the line.

35. $m = -3$, $h = .25$
If you move .25 unit in the x-direction, then you must move $-3 \cdot .25 = -.75$ unit in the y-direction.

37. Slope = 2, $(1, 3)$ on line.
$x_1 = 1$, $y_1 = 3$
If $x = 2$, then $y - 3 = 2(2 - 1) \Rightarrow y = 5$.
If $x = 3$, then $y - 3 = 2(3 - 1) \Rightarrow y = 7$.
If $x = 0$, then $y - 3 = 2(0 - 1) \Rightarrow y = 1$.
The points are $(2, 5)$, $(3, 7)$, and $(0, 1)$.

39. $f(1) = 0 \Rightarrow (1, 0)$ lies on the line.
$f(2) = 1 \Rightarrow (2, 1)$ lies on the line. Thus, the

slope of the line is $\dfrac{1 - 0}{2 - 1} = 1$. If $x = 3$ and

$y = f(3)$, then $1 = \dfrac{y - 1}{3 - 2} \Rightarrow 1 = y - 1 \Rightarrow y = 2$.
Thus $f(3) = 2$.

41. l_1

43. Slope = $m = -2$
y-intercept: $(0, -1)$
$y = mx + b$
$y = -2x - 1$

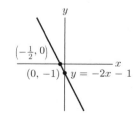

45. a is the x-coordinate of the point of intersection of $y = -x + 4$ and $y = 2$. Use substitution to find the x-coordinate.
$2 = -x + 4 \Rightarrow x = 2$
So $a = 2$. $f(a)$ is the y-coordinate of the intersection point. So $f(a) = 2$.

47. $C(x) = 12x + 1100$

   a. $C(10) = 12(10) + 1100 = \$1220$

   b. The marginal cost is the slope of line.
     Marginal cost = $m = \$12$/unit

   c. It would cost an additional \$12 to raise the daily production level from 10 units to 11 units.

49. Let x be the number of months since January 1, 2015. Then $(0, 2.19)$ is one point on the line. The slope is $-.04$ since the price fell \$.04 per month. Therefore, $P(x) = -.04x + 2.19$ gives the price of gasoline x months after January 1, 2015. On April 1, 2015, 3 months later, the cost of one gallon of gasoline is:
$P(3) = -.04(3) + 2.19 = \$2.07 /$ gallon. So, 15 gallons cost $15 \cdot 2.07 = \$31.05$.
On September 1, 2015, 8 months after January 1, the cost of one gallon of gasoline is:
$P(3) = -.04(8) + 2.19 = \$1.87 /$ gallon. So, 15 gallons cost $15 \cdot 1.87 = \$28.05$.

51. Let $x =$ the cost of order. Then
$C(x) = .03x + 5$.

53. The points $(2.10, 1500)$ and $(2.25, 1250)$ are on the line. The slope of the line is
$$m = \frac{y_2 - y_1}{x_2 - x_1} = \frac{1500 - 1250}{2.10 - 2.25} = -\frac{5000}{3}.$$
Let $(x_1, y_1) = (2.10, 1500)$. The equation of the line is
$$y - y_1 = m(x - x_1)$$
$$y - 1500 = -\frac{5000}{3}(x - 2.10)$$
$$y = -\frac{5000}{3}x + 5000$$
$$G(x) = -\frac{5000}{3}x + 5000$$
Now find $G(2.34)$:
$$G(2.34) = -\frac{5000}{3}(2.34) + 5000 = 1100$$
gallons.

55. a. $C(x) = mx + b$

$b = \$1500$ (fixed costs)
Total cost of producing 100 rods is $2200.
$C(100) = m(100) + 1500 = \$2200 \Rightarrow m = 7$
Thus, $C(x) = 7x + 1500$.

b. Marginal cost at $x = 100$ is $m = \$7/\text{rod}$

c. Since the marginal cost = \$7, the cost of raising the daily production level form 100 to 101 rods is \$7. Alternatively,
$C(101) - C(100) = 2207 - 2200 = \$7$.

57. If the monopolist wants to sell one more unit of goods, then the price per unit must be lowered by 2 cents. No one will pay 7 dollars or more for a unit of goods.

59. The point (0, 1.5) is on the line and the slope is 6 (ml/min). Let y be the amount of drug in the body x minutes from the start of the infusion. Then $y - 1.5 = 6(x - 0) \Rightarrow y = 6x + 1.5$.

61. The diver starts at a depth of 212 ft, which is represented as −212. Thus, the function is $y(t) = 2t - 212$.

63. a. $C(x) = 7x + 230$

b. $R(x) = 12x$

65. Using $\dfrac{f(x_2) - f(x_1)}{x_2 - x_1} = m$ and the hint,

$\dfrac{f(x) - f(x_1)}{x - x_1} = m \Rightarrow f(x) - f(x_1) = m(x - x_1)$

$f(x) = m(x - x_1) + f(x_1)$
$\quad = mx + (-mx_1 + f(x_1))$
Let $b = -mx_1 + f(x_1)$. Then $f(x) = mx + b$.

67. a. The points (0, 54) and (36, 66) lie on the line.

slope $= \dfrac{y_2 - y_1}{x_2 - x_1} = \dfrac{66 - 54}{36 - 0} = \dfrac{1}{3}$

$y - 54 = \dfrac{1}{3}(x - 0) \Rightarrow y = \dfrac{1}{3}x + 54$

b. Every year since 2014, $\frac{1}{3}\% = .33\%$ more of the world population becomes urban.

c. The year 2020 is represented by $x = 6$.

$f(6) = \dfrac{1}{3}(6) + 54 = 56$

Thus, in 2020, 56% of the world's population will be urban.

d. $72 = \dfrac{1}{3}x + 54 \Rightarrow 18 = \dfrac{1}{3}x \Rightarrow x = 54$

72% of the world's population will be urban 54 years after 2014, or in 2068.

1.2 The Slope of a Curve at a Point

1. $-\dfrac{4}{3}$

3. 1

5. 1

7. –2

9. Small positive slope; large positive slope

11. Zero slope; small negative slope

For 13–23, note that the slope of the line tangent to the graph of $y = x^2$ at the point (x, y) is $2x$.

13. The slope at (−.4, .16) is $2(-.4) = -.8$.
Let $(x_1, y_1) = (-.4, .16)$, $m = -.8$.

$y - .16 = -.8(x - (-.4))$
$y - .16 = -.8(x + .4)$
$\quad y = -.8x - .16$

15. The slope at $\left(\dfrac{1}{3}, \dfrac{1}{9}\right)$ is $m = 2x = 2\left(\dfrac{1}{3}\right) = \dfrac{2}{3}$.

Let $(x_1, y_1) = \left(\dfrac{1}{3}, \dfrac{1}{9}\right)$.

$y - \dfrac{1}{9} = \dfrac{2}{3}\left(x - \dfrac{1}{3}\right) \Rightarrow y - \dfrac{1}{9} = \dfrac{2}{3}x - \dfrac{2}{9} \Rightarrow$

$y = \dfrac{2}{3}x - \dfrac{1}{9}$

17. When $x = -\dfrac{1}{4}$, slope $= 2\left(-\dfrac{1}{4}\right) = -\dfrac{1}{2}$.

19. When $x = 2.5$, slope $= 2(2.5) = 5$ and $y = (2.5)^2 = 6.25$. Let $(x_1, y_1) = (2.5, 6.25)$, $m = 5$.
$y - 6.25 = 5(x - 2.5) \Rightarrow y - 6.25 = 5x - 12.5 \Rightarrow$
$y = 5x - 6.25$

21. The slope of the tangent is $2x$, so solve

$2x = \dfrac{7}{2} \Rightarrow x = \dfrac{7}{4}$. The point is

$\left(\dfrac{7}{4}, \left(\dfrac{7}{4}\right)^2\right) = \left(\dfrac{7}{4}, \dfrac{49}{16}\right)$.

23. The slope of the line $2x + 3y = 4$ is $-\dfrac{2}{3}$, so

the slope of the tangent line is also $-\dfrac{2}{3}$. Now

solve $2x = -\dfrac{2}{3} \Rightarrow x = -\dfrac{1}{3}$. The point is

$\left(-\dfrac{1}{3}, \left(-\dfrac{1}{3} \right)^2 \right) = \left(-\dfrac{1}{3}, \dfrac{1}{9} \right)$.

25. March 1, 2015: about \$52.00
January 1, 2016: about \$27.00
The price decreased about \$25.00
The price was rising on both days.

27. The price of a barrel of oil was about \$27.25.
It was rising at a rate of about

$\dfrac{27.50 - 27.25}{5} \approx \$.05$ per day.

For 29–31, note that the slope of the line tangent to
the graph of $y = x^3$ at the point (x, y) is $3x^2$.

29. Slope $= 3x^2$

When $x = 2$, slope $= 3(2)^2 = 12$.

31. Slope $= 3x^2$

When $x = -\dfrac{1}{2}$, slope $= 3\left(-\dfrac{1}{2} \right)^2 = \dfrac{3}{4}$.

33. The slope of the line tangent to $y = x^2$ at
$x = a$ is $2a$. The slope of $y = 2x - 1$ is 2.
Equating these gives $2a = 2 \Rightarrow a = 1$.
So, $f(a) = (1)^2 = 1$, $f'(1) = 2(1) = 2$

35. The slope of the curve $y = x^3$ at any point is

$3x^2$. Solve $3x^2 = \dfrac{3}{2} \Rightarrow x^2 = \dfrac{1}{2} \Rightarrow x = \pm\dfrac{1}{\sqrt{2}}$.

$x = \dfrac{1}{\sqrt{2}} \Rightarrow y = \left(\dfrac{1}{\sqrt{2}} \right)^3 = \dfrac{1}{2\sqrt{2}}$.

$x = -\dfrac{1}{\sqrt{2}} \Rightarrow y = \left(-\dfrac{1}{\sqrt{2}} \right)^3 = -\dfrac{1}{2\sqrt{2}}$. The

points are $\left(\dfrac{1}{\sqrt{2}}, \dfrac{1}{2\sqrt{2}} \right)$ and $\left(-\dfrac{1}{\sqrt{2}}, -\dfrac{1}{2\sqrt{2}} \right)$.

37. a. $m = \dfrac{13 - 4}{5 - 2} = 3$
length of d is $13 - 4 = 9$

b. The slope of line l increases.

39.

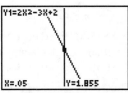

$[-.078125, .078125]$ by $[1.927923, 2.084173]$

When $x = 0$, $y = 2$. Find a second point on the
line using **VALUE**: $x = .05$, $y = 1.855$

$m = \dfrac{1.855 - 2}{.05 - 0} = -2.9$

The actual value of m is -3.

41.

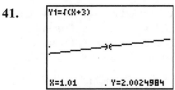

$[.6463, 1.3963]$ by $[1.5, 2.5]$

When $x = 1$, $y = 2$
Find a second point on the line using **VALUE**:
$x = 1.01$, $y = 2.0024984$

$m = \dfrac{2.0024984 - 2}{1.01 - 1} \approx .25$

The actual value of m is $\dfrac{1}{2\sqrt{3} + 1} = .25$.

1.3   The Derivative and Limits

For exercises $1 - 15$, refer to equations (1) and (2)
section 1.3 in the text along with the Power Rule
$f'(x) = rx^{r-1}$ for $f(x) = x^r$.

1. $f(x) = 3x + 7$, $f'(x) = 3$

3. $f(x) = \dfrac{3x}{4} - 2$, $f'(x) = \dfrac{3}{4}$

5. $f(x) = x^7$, $f'(x) = 7x^6$

7. $f(x) = x^{2/3}$, $f'(x) = \dfrac{2}{3} x^{-1/3} = \dfrac{2}{3\sqrt[3]{x}}$

9. $f(x) = \dfrac{1}{\sqrt{x^5}} = \dfrac{1}{x^{5/2}} = x^{-5/2}$,

$f'(x) = -\dfrac{5}{2} x^{-7/2} = -\dfrac{5}{2x^{7/2}}$

11. $f(x) = \sqrt[3]{x} = x^{1/3}$, $f'(x) = \dfrac{1}{3} x^{-2/3} = \dfrac{1}{3x^{2/3}}$

13. $f(x) = \dfrac{1}{x^{-2}} = x^2$, $f'(x) = 2x$

15. $f(x) = 4^2 = 16,\ f'(x) = 0$

In exercises 17–23, first find the derivative of the function, then evaluate the derivative for the given value of x.

17. $f(x) = x^3$ at $x = \dfrac{1}{2}$

$f'(x) = 3x^2$

$f'\left(\dfrac{1}{2}\right) = 3\left(\dfrac{1}{2}\right)^2 = \dfrac{3}{4}$

19. $f(x) = \dfrac{1}{x}$ at $x = \dfrac{2}{3}$

$f(x) = x^{-1};\ f'(x) = -x^{-2} = -\dfrac{1}{x^2}$

$f'\left(\dfrac{2}{3}\right) = -\dfrac{1}{(2/3)^2} = -\dfrac{1}{4/9} = -\dfrac{9}{4}$

21. $f(x) = x + 11$ at $x = 0$

$f'(x) = 1 \Rightarrow f'(0) = 1$

23. $f(x) = \sqrt{x}$ at $x = \dfrac{1}{16}$

$f(x) = x^{1/2}$

$f'(x) = \dfrac{1}{2}x^{-1/2} = \dfrac{1}{2\sqrt{x}}$

$f'(16) = \dfrac{1}{2\sqrt{\dfrac{1}{16}}} = \dfrac{1}{2 \cdot \dfrac{1}{4}} = 2$

25. Recall that the slope of a curve at a given point is the value of the derivative evaluated at that point.

$y = x^4$

slope $= y' = 4x^3$

at $x = 2,\ y' = 4(2)^3 = 32$

27. $f(x) = x^3;\ f(-5) = (-5)^3 = -125$

$f'(x) = 3x^2$

$f'(-5) = 3(-5)^2 = 75$

29. $f(x) = x^{1/3};\ f(8) = 8^{1/3} = 2$

$f'(x) = \dfrac{1}{3}x^{-2/3}$

$f'(8) = \dfrac{1}{3} \cdot 8^{-2/3} = \dfrac{1}{3} \cdot \dfrac{1}{4} = \dfrac{1}{12}$

31. $f(x) = \dfrac{1}{x^5} = x^{-5}$

$f(-2) = \dfrac{1}{(-2)^5} = -\dfrac{1}{32}$

$f'(x) = -5x^{-6} = -\dfrac{5}{x^6}$

$f'(-2) = -\dfrac{5}{(-2)^6} = -\dfrac{5}{64}$

For exercises 33–39, refer to Example 4 on page 77 in the text.

33. $f(x) = x^3 \Rightarrow f'(x) = 3x^2$

When $x = -2,\ f(x) = (-2)^3 = -8$.
The slope of the tangent at $x = -2$ is
$f'(-2) = 3(-2)^2 = 12$. Thus, the equation of the tangent at $(-2, -8)$ in point-slope form is
$y + 8 = 12(x + 2)$.

35. $f(x) = 3x + 1 \Rightarrow f'(x) = 3$

When $x = 4,\ f(x) = 3 \cdot 4 + 1 = 13$. The slope of the tangent at $x = 4$ is $f'(4) = 3$. Thus, the equation of the tangent at $(4, 13)$ in point-slope form is $y - 13 = 3(x - 4)$ or $y = 3x + 1$ in slope-intercept form.

37. $f(x) = \sqrt{x} = x^{1/2} \Rightarrow f'(x) = \dfrac{1}{2}x^{-1/2} = \dfrac{1}{2\sqrt{x}}$

When $x = \dfrac{1}{9},\ f(x) = \sqrt{\dfrac{1}{9}} = \dfrac{1}{3}$. The slope of the tangent at $x = \dfrac{1}{9}$ is

$f'\left(\dfrac{1}{9}\right) = \dfrac{1}{2}\left(\dfrac{1}{9}\right)^{-1/2} = \dfrac{1}{2}(9)^{1/2} = \dfrac{3}{2}$. Thus, the equation of the tangent at $\left(\dfrac{1}{9}, \dfrac{1}{3}\right)$ in point-slope form is $y - \dfrac{1}{3} = \dfrac{3}{2}\left(x - \dfrac{1}{9}\right)$.

39. $f(x) = \dfrac{1}{\sqrt{x}} = x^{-1/2} \Rightarrow f'(x) = -\dfrac{1}{2}x^{-3/2}$

When $x = 1$, $f(x) = \dfrac{1}{\sqrt{1}} = 1$. The slope of the

tangent at $x = 1$ is $f'(1) = -\dfrac{1}{2}(1)^{-3/2} = -\dfrac{1}{2}$.

Thus, the equation of the tangent at $(1,1)$ in

point-slope form is $y - 1 = -\dfrac{1}{2}(x-1)$.

41. Equation 6 on page 76 states that
$y - f(a) = f'(a)(x-a)$.

$y = f(x) = x^4 \Rightarrow y' = f'(x) = 4x^3$
For $a = 1$, $f(a) = f(1) = 1$ and
$f'(a) = f'(1) = 4$. Thus, the equation of the
tangent at $(1,1)$ in point-slope form is
$y - 1 = 4(x-1)$.

43. The slope of the tangent is $m = 2$.

$f(x) = \sqrt{x} = x^{1/2} \Rightarrow f'(x) = \dfrac{1}{2}x^{-1/2}$. The

slope of the tangent at $x = a$ is

$f'(a) = \dfrac{1}{2}a^{-1/2}$. Therefore $P = \left(\dfrac{1}{16}, \dfrac{1}{4}\right)$.

$\dfrac{1}{4} = 2\left(\dfrac{1}{16}\right) + b \Rightarrow b = \dfrac{1}{8}$.

45. a.   The slope of the tangent line is $m = \dfrac{1}{8}$

because the slopes of parallel lines are
equal.

$f(x) = \sqrt{x} = x^{1/2} \Rightarrow f'(x) = \dfrac{1}{2}x^{-1/2}$. The

slope of the tangent is at $x = a$ is

$f'(a) = \dfrac{1}{2}a^{-1/2}$.

$\dfrac{1}{2}a^{-1/2} = \dfrac{1}{8} \Rightarrow a = 16$. $f(16) = \sqrt{16} = 4$.

Therefore, the point we are looking for is
$(16, 4)$.

b.

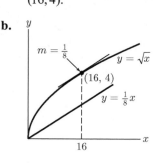

47. A tangent line perpendicular to $y = x$ has slope
-1. $f(x) = x^3 \Rightarrow f'(x) = 3x^2$. The slope of
the tangent at $x = a$ is $f'(a) = 3a^2$. Thus,

$3a^2 = -1$, which has no real solution.

Therefore, there is no point on $y = x^3$ where
the tangent line is perpendicular to $y = x$.

49. $\dfrac{d}{dx}\left(x^8\right) = 8x^7$

51. $\dfrac{d}{dx}\left(x^{3/4}\right) = \dfrac{3}{4}x^{-1/4}$

53. $y = 1 \Rightarrow \dfrac{dy}{dx} = \dfrac{d}{dx}(1) = 0$

55. $y = x^{1/5}, \dfrac{d}{dx}(x^{1/5}) = \dfrac{1}{5}x^{-4/5}$

57. The tangent line at $x = 6$ is $y = \dfrac{1}{3}x + 2$, so

$f(6) = \dfrac{1}{3}(6) + 2 = 4$.

The slope of $y = \dfrac{1}{3}x + 2$ is $\dfrac{1}{3}$, so $f'(6) = \dfrac{1}{3}$.

59. $y = f(x) = \sqrt{x} = x^{1/2}$
The slope of the tangent line at $x = a$ is

$f'(a) = \dfrac{1}{2}a^{-1/2} = \dfrac{1}{2\sqrt{a}}$.

The slope of the tangent line $y = \dfrac{1}{4}x + b$ is

$\dfrac{1}{4}$. First, find the value of a. Let $\dfrac{1}{4} = \dfrac{1}{2\sqrt{a}}$

and solve for a: $2\sqrt{a} = 4 \Rightarrow \sqrt{a} = 2 \Rightarrow a = 4$
When $x = 4$, $f(4) = \sqrt{4} = 2$.
Let $(x_1, y_1) = (4, 2)$. Then,

$y - 2 = \dfrac{1}{4}(x - 4) \Rightarrow y = \dfrac{1}{4}x - 1 + 2 \Rightarrow$

$y = \dfrac{1}{4}x + 1$, so $b = 1$.

61. At $x = a$, $y = 2.01a - .51$ or $y = 2.02a - .52$, so
$.01a = .01$, and $a = 1$
$y = f(1) = 2.01 - .51 = 1.5$.
$f'(a) = 2$ because the slope of the "smallest"
secant line is 2.01.

63. The coordinates of A are $(4, 5)$. From the graph of the derivative, we see that

$f'(4) = \dfrac{1}{2}$, so the slope of the tangent line is

$\dfrac{1}{2}$. By the point-slope formula, the equation of

the tangent line is $y - 5 = \dfrac{1}{2}(x - 4)$.

65. $f(x) = 2x^2$

$$\dfrac{f(x+h) - f(x)}{h} = \dfrac{2(x+h)^2 - 2x^2}{h}$$

$$= \dfrac{2x^2 + 4xh + 2h^2 - 2x^2}{h}$$

$$= \dfrac{h(4x+h)}{h} = 4x + h$$

67. $f(x) = -x^2 + 2x$

$$\dfrac{f(x+h) - f(x)}{h}$$

$$= \dfrac{\left[-(x+h)^2 + 2(x+h)\right] - \left(-x^2 + 2x\right)}{h}$$

$$= \dfrac{-x^2 - 2xh - h^2 + 2x + 2h + x^2 - 2x}{h}$$

$$= \dfrac{h(-2x + 2 - h)}{h} = -2x + 2 - h$$

69. $f(x) = x^3$

$$\dfrac{f(x+h) - f(x)}{h} = \dfrac{(x+h)^3 - x^3}{h}$$

$$= \dfrac{x^3 + 3x^2h + 3xh^2 + h^3 - x^3}{h}$$

$$= \dfrac{h\left(3x^2 + 3xh + h^2\right)}{h}$$

$$= 3x^2 + 3xh + h^2$$

For exercises 71–75, refer to the three-step method on page 78 in the text.

71. $f(x) = -x^2$

$$\dfrac{f(x+h) - f(x)}{h} = \dfrac{-(x+h)^2 - \left(-x^2\right)}{h}$$

$$= \dfrac{-x^2 - 2xh - h^2 + x^2}{h}$$

$$= \dfrac{h(-2x+h)}{h} = -2x + h$$

As h approaches 0, the quantity $-2x + h$ approaches $-2x$. Thus, $f'(x) = -2x$.

73. $f(x) = 7x^2 + x - 1$

$$\dfrac{f(x+h) - f(x)}{h}$$

$$= \dfrac{\left[7(x+h)^2 + (x+h) - 1\right] - \left(7x^2 + x - 1\right)}{h}$$

$$= \dfrac{7x^2 + 14xh + 7h^2 + x + h - 1 - 7x^2 - x + 1}{h}$$

$$= \dfrac{14xh + 7h^2 + h}{h} = \dfrac{h(14x + 7h + 1)}{h}$$

$$= 14x + 1 + 7h$$

As h approaches 0, the quantity $14x + 1 + 7h$ approaches $14x + 1$. Thus, $f'(x) = 14x + 1$.

75. $f(x) = x^3$

$$\dfrac{f(x+h) - f(x)}{h} = \dfrac{(x+h)^3 - x^3}{h}$$

$$= \dfrac{x^3 + 3x^2h + 3xh^2 + h^3 - x^3}{h}$$

$$= \dfrac{h\left(3x^2 + 3xh + h^2\right)}{h}$$

$$= 3x^2 + 3xh + h^2$$

As h approaches 0, the quantity $3x^2 + 3xh + h^2$ approaches $3x^2$. Thus, $f'(x) = 3x^2$.

77. a., b.

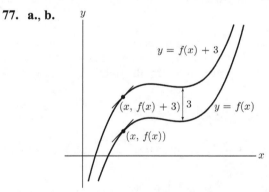

In both cases the tangent lines are parallel.

c. A vertical shift in a graph does not change its shape. Therefore, the slope at any point x remains the same for any shift in the y-direction. Since the slope at a given point is the value of the derivative at that point, $\dfrac{d}{dx} f(x) = \dfrac{d}{dx}\left(f(x) + 3\right)$.

79. $f'(0)$, where $f(x) = 2^x$

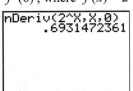

81. $f'(1)$, where $f(x) = \sqrt{1+x^2}$

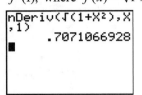

83. $f'(2)$, where $f(x) = \dfrac{x}{1+x}$

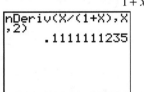

1.4   Limits and the Derivative

1. There is no limit because as x approaches 3 from the left, $g(x)$ approaches $-\infty$, while as x approaches 3 from the right, $g(x)$ approaches 2.

3. 1

5. There is no limit because as x approaches 3 from the left, $g(x)$ approaches 4, while as x approaches 3 from the right, $g(x)$ approaches 5.

7. $\displaystyle\lim_{x \to 1} (1 - 6x) = 1 - 6(1) = -5$

9. $\displaystyle\lim_{x \to 3} \sqrt{x^2 + 16} = \sqrt{(3)^2 + 16} = \sqrt{25} = 5$

11. $\displaystyle\lim_{x \to 5} \frac{x^2 + 1}{5 + x} = \frac{\displaystyle\lim_{x \to 5} x^2 + \lim_{x \to 5} 1}{\displaystyle\lim_{x \to 5} 5 + \lim_{x \to 5} x} = \frac{5^2 + 1}{5 + 5} = \frac{13}{5}$

13. $\displaystyle\lim_{x \to 7} \left(x + \sqrt{x - 6}\right)\left(x^2 - 2x + 1\right)$

$= \displaystyle\lim_{x \to 7} \left(x + \sqrt{x - 6}\right)(x - 1)^2$

$= \left(\displaystyle\lim_{x \to 7} x + \lim_{x \to 7} \sqrt{x - 6}\right)\left(\lim_{x \to 7} x - \lim_{x \to 7} 1\right)^2$

$= (7 + 1)(7 - 1)^2 = 8 \cdot 36 = 288$

15. $\displaystyle\lim_{x \to -5} \frac{\sqrt{x^2 - 5x - 36}}{8 - 3x}$

$= \dfrac{\left(\displaystyle\lim_{x \to -5} x^2 - \lim_{x \to -5} 5x - \lim_{x \to -5} 36\right)^{1/2}}{\displaystyle\lim_{x \to -5} 8 - \lim_{x \to -5} 3x}$

$= \dfrac{\left(25 - (-25) - 36\right)^{1/2}}{8 - (-15)} = \dfrac{\sqrt{14}}{23}$

17. $\displaystyle\lim_{x \to 0} \frac{x^2 + 3x}{x} = \lim_{x \to 0} \frac{x(x + 3)}{x} = \lim_{x \to 0} (x + 3) = 3$

19. $\displaystyle\lim_{x \to 2} \frac{-2x^2 + 4x}{x - 2} = \lim_{x \to 2} \frac{-2x(x - 2)}{(x - 2)}$

$= \displaystyle\lim_{x \to 2} (-2x) = -4$

21. $\displaystyle\lim_{x \to 4} \frac{x^2 - 16}{4 - x} = \lim_{x \to 4} \frac{(x - 4)(x + 4)}{-(x - 4)}$

$= \displaystyle\lim_{x \to 4} (-x - 4)$

$= \displaystyle\lim_{x \to 4} (-x) - \lim_{x \to 4} 4$

$= -4 - 4 = -8$

23. $\displaystyle\lim_{x \to 6} \frac{x^2 - 6x}{x^2 - 5x - 6} = \lim_{x \to 6} \frac{x(x - 6)}{(x - 6)(x + 1)}$

$= \dfrac{\displaystyle\lim_{x \to 6} x}{\displaystyle\lim_{x \to 6} (x + 1)} = \dfrac{6}{6 + 1} = \dfrac{6}{7}$

25. $\displaystyle\lim_{x \to 8} \frac{x^2 + 64}{x - 8}$ is undefined.

27. $\displaystyle\lim_{x \to 0} f(x) = -\frac{1}{2}$ and $\displaystyle\lim_{x \to 0} g(x) = \frac{1}{2}$.

a. $\displaystyle\lim_{x \to 0} (f(x) + g(x)) = \lim_{x \to 0} f(x) + \lim_{x \to 0} g(x)$

$= -\dfrac{1}{2} + \dfrac{1}{2} = 0$

b. $\displaystyle\lim_{x \to 0} (f(x) - 2g(x)) = \lim_{x \to 0} f(x) - 2 \cdot \lim_{x \to 0} g(x)$

$= -\dfrac{1}{2} - 2 \cdot \dfrac{1}{2} = -\dfrac{3}{2}$

c. $\displaystyle\lim_{x \to 0} (f(x) \cdot g(x)) = \left[\lim_{x \to 0} f(x)\right] \cdot \left[\lim_{x \to 0} g(x)\right]$

$= \left[-\dfrac{1}{2}\right] \cdot \left[\dfrac{1}{2}\right] = -\dfrac{1}{4}$

d. Since $\displaystyle\lim_{x \to 0} g(x) \neq 0$,

$\displaystyle\lim_{x \to 0} \frac{f(x)}{g(x)} = \dfrac{\displaystyle\lim_{x \to 0} f(x)}{\displaystyle\lim_{x \to 0} g(x)} = \dfrac{-\frac{1}{2}}{\frac{1}{2}} = -1.$

29. $f(x) = x^2 + 1$

$$f'(3) = \lim_{h \to 0} \frac{f(3+h) - f(3)}{h} = \lim_{h \to 0} \frac{(3+h)^2 + 1 - (3^2 + 1)}{h} = \lim_{h \to 0} \frac{9 + 6h + h^2 + 1 - 10}{h}$$

$$= \lim_{h \to 0} \frac{h^2 + 6h}{h} = \lim_{h \to 0} (h + 6) = 6$$

31. $f(x) = x^3 + 3x + 1$

$$f'(0) = \lim_{h \to 0} \frac{f(0+h) - f(0)}{h} = \lim_{h \to 0} \frac{h^3 + 3h + 1 - 1}{h} = \lim_{h \to 0} \frac{h(h^2 + 3)}{h} = \lim_{h \to 0} (h^2 + 3) = 3$$

For exercises 33–35, use the three-step method discussed on page 85 in section 1.4 and illustrated in Examples 6 and 7.

33. $f(x) = x^2 + 1$

Step 1: $\dfrac{f(x+h) - f(x)}{h} = \dfrac{\left[(x+h)^2 + 1\right] - \left(x^2 + 1\right)}{h}$

Step 2: $\dfrac{\left[(x+h)^2 + 1\right] - \left(x^2 + 1\right)}{h} = \dfrac{x^2 + 2xh + h^2 + 1 - x^2 - 1}{h} = \dfrac{2xh + h^2}{h} = \dfrac{h(2x+h)}{h} = 2x + h$

Step 3: $f'(x) = \lim_{h \to 0} (2x + h) = 2x$

35. $f(x) = x^3 - 1$

Step 1: $\dfrac{f(x+h) - f(x)}{h} = \dfrac{\left[(x+h)^3 - 1\right] - \left(x^3 - 1\right)}{h}$

Step 2: $\dfrac{\left[(x+h)^3 - 1\right] - \left(x^3 - 1\right)}{h} = \dfrac{x^3 + 3x^2 h + 3xh^2 + h^3 - 1 - x^3 + 1}{h} = \dfrac{3x^2 h + 3xh^2 + h^3}{h}$

$$= \dfrac{h\left(3x^2 + 3xh + h^2\right)}{h} = 3x^2 + 3xh + h^2$$

Step 3: $f'(x) = \lim_{h \to 0} \left(3x^2 + 3xh + h^2\right) = 3x^2$

37. $f(x) = 3x + 1$

$$f'(x) = \lim_{h \to 0} \frac{f(x+h) - f(x)}{h} = \lim_{h \to 0} \frac{3(x+h) + 1 - (3x+1)}{h} = \lim_{h \to 0} \frac{3x + 3h + 1 - 3x - 1}{h} = \lim_{h \to 0} \frac{3h}{h} = \lim_{h \to 0} 3 = 3$$

39. $f(x) = x + \dfrac{1}{x}$

$$f'(x) = \lim_{h \to 0} \frac{f(x+h) - f(x)}{h} = \lim_{h \to 0} \frac{(x+h) + \dfrac{1}{x+h} - \left(x + \dfrac{1}{x}\right)}{h} = \lim_{h \to 0} \frac{x + h + \dfrac{1}{x+h} - x - \dfrac{1}{x}}{h}$$

$$= \lim_{h \to 0} \frac{h + \dfrac{1}{x+h} - \dfrac{1}{x}}{h} = \lim_{h \to 0} \frac{\dfrac{h(x+h)(x) + x - (x+h)}{(x+h)(x)}}{h} = \lim_{h \to 0} \frac{\dfrac{hx^2 + h^2 x - h}{(x+h)(x)}}{h}$$

$$= \lim_{h \to 0} \frac{h(x^2 + hx - 1)}{(x+h)(x)} \left(\frac{1}{h}\right) = \lim_{h \to 0} \frac{(x^2 + hx - 1)}{(x+h)(x)} = \frac{x^2 - 1}{x^2} = 1 - \frac{1}{x^2}$$

41. $f(x) = \dfrac{x}{x+1}$

$$f'(x) = \lim_{h \to 0} \frac{f(x+h) - f(x)}{h} = \lim_{h \to 0} \frac{\dfrac{x+h}{x+h+1} - \dfrac{x}{x+1}}{h} = \lim_{h \to 0} \frac{\dfrac{(x+h)(x+1) - (x)(x+h+1)}{(x+h+1)(x+1)}}{h}$$

$$= \lim_{h \to 0} \frac{\dfrac{x^2 + xh + x + h - x^2 - xh - x}{(x+h+1)(x+1)}}{h} = \lim_{h \to 0} \frac{h}{(x+h+1)(x+1)}\left(\frac{1}{h}\right)$$

$$= \lim_{h \to 0} \frac{1}{(x+h+1)(x+1)} = \frac{1}{(x+1)(x+1)} = \frac{1}{(x+1)^2}$$

43. $f(x) = \dfrac{1}{x^2 + 1}$

$$f'(x) = \lim_{h \to 0} \frac{f(x+h) - f(x)}{h} = \lim_{h \to 0} \frac{\dfrac{1}{(x+h)^2 + 1} - \dfrac{1}{x^2 + 1}}{h} = \lim_{h \to 0} \frac{\dfrac{(x^2 + 1) - \left((x+h)^2 + 1\right)}{\left((x+h)^2 + 1\right)(x^2 + 1)}}{h}$$

$$= \lim_{h \to 0} \frac{\dfrac{x^2 + 1 - x^2 - 2xh - h^2 - 1}{\left((x+h)^2 + 1\right)(x^2 + 1)}}{h} = \lim_{h \to 0} \frac{h(-2x - h)}{\left((x+h)^2 + 1\right)(x^2 + 1)}\left(\frac{1}{h}\right)$$

$$= \lim_{h \to 0} \frac{(-2x - h)}{\left((x+h)^2 + 1\right)(x^2 + 1)} = \frac{-2x}{(x^2 + 1)(x^2 + 1)} = \frac{-2x}{(x^2 + 1)^2}$$

45. $f(x) = \sqrt{x + 2}$

$$f'(x) = \lim_{h \to 0} \frac{f(x+h) - f(x)}{h} = \lim_{h \to 0} \frac{\sqrt{x+h+2} - \sqrt{x+2}}{h}$$

$$= \lim_{h \to 0} \frac{\sqrt{x+h+2} - \sqrt{x+2}}{h}\left(\frac{\sqrt{x+h+2} + \sqrt{x+2}}{\sqrt{x+h+2} + \sqrt{x+2}}\right)$$

$$= \lim_{h \to 0} \frac{(x+h+2) - (x+2)}{h\left(\sqrt{x+h+2} + \sqrt{x+2}\right)} = \lim_{h \to 0} \frac{h}{h\left(\sqrt{x+h+2} + \sqrt{x+2}\right)}$$

$$= \lim_{h \to 0} \frac{1}{\sqrt{x+h+2} + \sqrt{x+2}} = \frac{1}{\sqrt{x+2} + \sqrt{x+2}} = \frac{1}{2\sqrt{x+2}}$$

47. $f(x) = \dfrac{1}{\sqrt{x}}$

$$f'(x) = \lim_{h \to 0} \frac{f(x+h) - f(x)}{h} = \lim_{h \to 0} \frac{\dfrac{1}{\sqrt{x+h}} - \dfrac{1}{\sqrt{x}}}{h} = \lim_{h \to 0} \frac{\dfrac{\sqrt{x} - \sqrt{x+h}}{\sqrt{x}\sqrt{x+h}}}{h}$$

$$= \lim_{h \to 0} \left(\frac{1}{h}\right) \frac{\sqrt{x} - \sqrt{x+h}}{\sqrt{x}\sqrt{x+h}}\left(\frac{\sqrt{x} + \sqrt{x+h}}{\sqrt{x} + \sqrt{x+h}}\right) = \lim_{h \to 0} \left(\frac{1}{h}\right) \frac{x - x - h}{\sqrt{x}\sqrt{x+h}\left(\sqrt{x} + \sqrt{x+h}\right)}$$

$$= \lim_{h \to 0} \frac{-1}{\sqrt{x}\sqrt{x+h}\left(\sqrt{x} + \sqrt{x+h}\right)} = \frac{-1}{\sqrt{x}\sqrt{x}\left(\sqrt{x} + \sqrt{x}\right)} = \frac{-1}{2x\sqrt{x}} = -\frac{1}{2x^{3/2}}$$

49. We want to find $f(x)$ such that

$\displaystyle \lim_{h \to 0} \frac{(1+h)^2 - 1}{h}$ has the same form as

$\displaystyle f'(a) = \lim_{h \to 0} \frac{f(a+h) - f(a)}{h}$. So,

$f(a+h) = (1+h)^2$ and $f(a) = 1$. Thus,

$f(x) = x^2$ and $a = 1$.

51. We want to find $f(x)$ such that

$\displaystyle \lim_{h \to 0} \frac{\frac{1}{10+h} - .1}{h}$ has the same form as

$\displaystyle f'(a) = \lim_{h \to 0} \frac{f(a+h) - f(a)}{h}$. So,

$f(a+h) = \dfrac{1}{10+h}$ and $f(a) = .1$. Thus,

$f(x) = x^{-1} = \dfrac{1}{x}$ and $a = 10$.

53. We want to find $f(x)$ such that

$\displaystyle \lim_{h \to 0} \frac{\sqrt{9+h} - 3}{h}$ has the same form as

$\displaystyle f'(a) = \lim_{h \to 0} \frac{f(a+h) - f(a)}{h}$. So,

$f(a+h) = \sqrt{9+h}$ and $f(a) = 3$. Thus,

$f(x) = \sqrt{x}$ and $a = 9$.

55. $\displaystyle \lim_{x \to \infty} \frac{1}{x^2} = 0$

57. $\displaystyle \lim_{x \to \infty} \frac{5x+3}{3x-2} = \lim_{x \to \infty} \frac{5 + \frac{3}{x}}{3 - \frac{2}{x}} = \frac{5}{3}$

59. $\displaystyle \lim_{x \to \infty} \frac{10x + 100}{x^2 - 30} = \lim_{x \to \infty} \frac{10 + \frac{100}{x}}{x - \frac{30}{x}} = 0$

61. $\displaystyle \lim_{x \to 0} f(x)$

As x approaches 0 from either side, $f(x)$

approaches $\dfrac{3}{4}$. So $\displaystyle \lim_{x \to 0} f(x) = \frac{3}{4}$.

63. $\displaystyle \lim_{x \to 0} xf(x) = \left[\lim_{x \to 0} x \right] \cdot \left[\lim_{x \to 0} f(x) \right]$

$\displaystyle = 0 \cdot \frac{3}{4} = 0$

65. $\displaystyle \lim_{x \to \infty} (1 - f(x)) = \lim_{x \to \infty} 1 - \lim_{x \to \infty} f(x) = 1 - 1 = 0$

67. $\displaystyle \lim_{x \to \infty} \sqrt{25 + x} - \sqrt{x}$

At large values of x the function goes to 0.

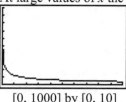

[0, 1000] by [0, 10]

69. $\displaystyle \lim_{x \to \infty} \frac{x^2 - 2x + 3}{2x^2 + 1}$

At large values of x the function goes to .5.

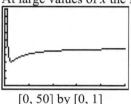

[0, 50] by [0, 1]

1.5 Differentiability and Continuity

1. No   **3.** Yes

5. No   **7.** No

9. Yes   **11.** No

13. $f(x) = x^2$

$\displaystyle \lim_{x \to 1} f(x) = \lim_{x \to 1} x^2 = 1$

$f(1) = 1^2 = 1$

Since $\displaystyle \lim_{x \to 1} f(x) = 1 = f(1)$, $f(x)$ is continuous

at $x = 1$.

$\displaystyle f'(1) = \lim_{h \to 0} \frac{f(1+h) - f(1)}{h} = \lim_{h \to 0} \frac{(1+h)^2 - (1)^2}{h}$

$\displaystyle = \lim_{h \to 0} \frac{1 + 2h + h^2 - 1}{h} = \lim_{h \to 0} \frac{h(2+h)}{h} = 2$

Therefore, $f(x)$ is continuous and differentiable at $x = 1$.

15. $f(x) = \begin{cases} x+2 & \text{for } -1 \le x \le 1 \\ 3x & \text{for } 1 < x \le 5 \end{cases}$

$\displaystyle \lim_{x \to 1} 3x = 3$

$\displaystyle \lim_{x \to 1} (x+2) = 3$

$f(1) = 1 + 2 = 3$

Since $\displaystyle \lim_{x \to 1} f(x) = 3 = f(1)$, $f(x)$ is continuous

at $x = 1$. The graph of $f(x)$ at $x = 1$ does not have a tangent line, so $f(x)$ is not differentiable at $x = 1$. Therefore, $f(x)$ is continuous but not differentiable at $x = 1$.

17. $f(x) = \begin{cases} 2x - 1 & \text{for } 0 \le x \le 1 \\ 1 & \text{for } 1 < x \end{cases}$

$\lim\limits_{x \to 1} 1 = 1$

$\lim\limits_{x \to 1} (2x - 1) = 1$

$f(1) = 2(1) - 1 = 1$

Since $\lim\limits_{x \to 1} f(x) = 1 = f(1)$, $f(x)$ is continuous at $x = 1$. The graph of $f(x)$ at $x = 1$ does not have a tangent line,

so $f(x)$ is not differentiable at $x = 1$. Therefore, $f(x)$ is continuous but not differentiable at $x = 1$.

19. $f(x) = \begin{cases} \dfrac{1}{x - 1} & \text{for } x \ne 1 \\ 0 & \text{for } x = 1 \end{cases}$

$\lim\limits_{x \to 1} f(x) = \lim\limits_{x \to 1} \dfrac{1}{x - 1}$ is undefined. Since $\lim\limits_{x \to 1} f(x)$ does not exist, $f(x)$ is not continuous at $x = 1$. By
Theorem 1, since $f(x)$ is not continuous at $x = 1$, it is not differentiable.

21. $\dfrac{x^2 - 7x + 10}{x - 5} = \dfrac{(x - 5)(x - 2)}{x - 5} = x - 2$, so define $f(5) = 5 - 2 = 3$.

23. $\dfrac{x^3 - 5x^2 + 4}{x^2}$

It is not possible to define $f(x)$ at $x = 0$ and make $f(x)$ continuous.

25. $\dfrac{(6 + x)^2 - 36}{x} = \dfrac{\left(x^2 + 12x + 36\right) - 36}{x} = \dfrac{x^2 + 12x}{x} = x + 12$

So, define $f(0) = 12$.

27. a. The function $T(x)$ is a piecewise-defined function.
For $0 \le x \le 27{,}050$, $T(x) = .15x$.
For $27{,}050 < x \le 65{,}550$, we have
$T(x) = .15 \cdot 27{,}050 + .275 \cdot (x - 27{,}050) = .275x - 3381.25$
For $65{,}550 < x \le 136{,}750$, we have
$T(x) = .275 \cdot 65{,}550 - 3381.25 + .305(x - 65{,}550) = .305x - 5347.75$

All together, the function is

$T(x) = \begin{cases} .15x & \text{for } 0 \le x \le 27{,}050 \\ .275x - 3381.25 & \text{for } 27{,}050 < x \le 65{,}550 \\ .305x - 5347.75 & \text{for } 65{,}550 < x \le 136{,}750 \end{cases}$

b.

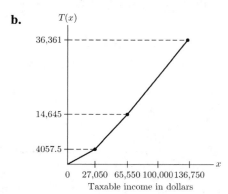

Taxable income in dollars

29. a. The function $R(x)$ is a piecewise function
For $0 \le x \le 100$, $R(x) = 2.50 + .07x$.
For $x > 100$, we have
$$R(x) = 2.50 + .07 \cdot 100 + .04(x - 100)$$
$$= 5.50 + .04x$$
All together, we have
$$R(x) = \begin{cases} .07x + 2.50 \text{ for } 0 \le x \le 100 \\ .04x + 5.50 \text{ for } x > 100 \end{cases}$$

b. Let $P(x)$ be the profit on x copies.
For $0 \le x \le 100$,
$$P(x) = 2.50 + .07x - .03x = 2.50 + .04x$$
For $x > 100$,
$$P(x) = 5.50 + .04x - .03x = 5.50 + .01x$$
All together, we have
$$P(x) = \begin{cases} .04x + 2.50 \text{ for } 0 \le x \le 100 \\ .01x + 5.50 \text{ for } x > 100 \end{cases}$$

31. a. The rate of sales is the slope of the line
connecting the points $(8, 4)$ and $(10, 10)$.
$$m = \frac{y_2 - y_1}{x_2 - x_1} = \frac{10 - 4}{10 - 8} = 3$$
The rate of sales between 8 a.m. and 10 a.m. is about \$3000 per hour.

b. We need to find the 2-hour period with the greatest slope. Looking at the graph gives 3 possibilities:
8 a.m. $-$ 10 a.m., $m = 3$ (from part a)
12 p.m. $-$ 2 p.m.,
$$m = \frac{y_2 - y_1}{x_2 - x_1} = \frac{16 - 12}{14 - 12} = 2$$
6 p.m. $-$ 8 p.m., $m = \dfrac{22 - 18}{8 - 6} = 2$
The interval from 8 a.m. to 10 a.m. has the greatest rate, which is \$3000 per hour.

33. For the function to be continuous, $\lim\limits_{x \to 0} f(x)$
must exist and equal $f(0)$. Therefore
$\lim\limits_{x \to 0} (x + a) = \lim\limits_{x \to 0} 1$, so $a = 1$.

1.6   Some Rules for Differentiation

1. $y = 6x^3$
$$\frac{dy}{dx} = \frac{d}{dx}\left(6x^3\right) = 18x^2$$

3. $y = 3\sqrt[3]{x} = 3x^{1/3}$
$$\frac{dy}{dx} = \frac{d}{dx}\left(3x^{1/3}\right) = \frac{1}{3} \cdot 3x^{-2/3}$$
$$= x^{-2/3} = \frac{1}{x^{2/3}} = \frac{1}{\sqrt[3]{x^2}}$$

5. $y = \dfrac{x}{2} - \dfrac{2}{x} = \dfrac{1}{2}x - 2x^{-1}$
$$\frac{dy}{dx} = \frac{d}{dx}\left(\frac{1}{2}x - 2x^{-1}\right) = \frac{d}{dx}\left(\frac{1}{2}x\right) - \frac{d}{dx}\left(2x^{-1}\right)$$
$$= \frac{1}{2} + 2x^{-2} = \frac{1}{2} + \frac{2}{x^2}$$

7. $f(x) = x^4 + x^3 + x$
$$\frac{d}{dx}\left(x^4 + x^3 + x\right) = \frac{d}{dx}x^4 + \frac{d}{dx}x^3 + \frac{d}{dx}x$$
$$= 4x^3 + 3x^2 + 1$$

9. $y = (2x + 4)^3$
$$\frac{dy}{dx} = \frac{d}{dx}(2x + 4)^3 = 3(2x + 4)^2 \frac{d}{dx}(2x + 4)$$
$$= 3(2x + 4)^2 (2) = 6(2x + 4)^2$$

11. $y = \left(x^3 + x^2 + 1\right)^7$
$$\frac{dy}{dx} = \frac{d}{dx}\left(x^3 + x^2 + 1\right)^7$$
$$= 7\left(x^3 + x^2 + 1\right)^6 \frac{d}{dx}\left(x^3 + x^2 + 1\right)$$
$$= 7\left(x^3 + x^2 + 1\right)^6 \left(3x^2 + 2x\right)$$

13. $y = \dfrac{4}{x^2} = 4x^{-2}$
$$\frac{dy}{dx} = 4\frac{d}{dx}x^{-2} = -\frac{8}{x^3}$$

15. $y = 3\sqrt[3]{2x^2 + 1} = 3\left(2x^2 + 1\right)^{1/3}$
$$\frac{dy}{dx} = \frac{d}{dx}3\left(2x^2 + 1\right)^{1/3}$$
$$= 3 \cdot \frac{1}{3}\left(2x^2 + 1\right)^{-2/3} \frac{d}{dx}\left(2x^2 + 1\right)$$
$$= (2x^2 + 1)^{-2/3}(4x) = \left(4x\right)\left(2x^2 + 1\right)^{-2/3}$$

17. $y = 2x + (x + 2)^3$
$$\frac{dy}{dx} = \frac{d}{dx}\left(2x + (x + 2)^3\right) = \frac{d}{dx}2x + \frac{d}{dx}(x + 2)^3$$
$$= 2 + 3(x + 2)^2 \frac{d}{dx}(x + 2) = 2 + 3(x + 2)^2 (1)$$
$$= 2 + 3(x + 2)^2$$

19. $y = \dfrac{1}{5x^5} = \dfrac{1}{5}x^{-5}$
$$\frac{dy}{dx} = \frac{d}{dx}\left(\frac{1}{5}x^{-5}\right) = \frac{1}{5}(-5)x^{-6} = -\frac{1}{x^6}$$

21. $y = \dfrac{1}{x^3+1} = \left(x^3+1\right)^{-1}$

$\dfrac{dy}{dx} = \dfrac{d}{dx}\left(x^3+1\right)^{-1} = -\left(x^3+1\right)^{-2}\dfrac{d}{dx}\left(x^3+1\right)$

$\quad = -\left(x^3+1\right)^{-2}\left(3x^2\right) = -3x^2\left(x^3+1\right)^{-2}$

$\quad = -\dfrac{3x^2}{\left(x^3+1\right)^2}$

23. $y = x + \dfrac{1}{x+1} = x + (x+1)^{-1}$

$\dfrac{dy}{dx} = \dfrac{d}{dx}\left(x+(x+1)^{-1}\right) = \dfrac{d}{dx}x + \dfrac{d}{dx}(x+1)^{-1}$

$\quad = 1 + \left(-(x+1)^{-2}\dfrac{d}{dx}(x+1)\right)$

$\quad = 1 + \left(-(x+1)^{-2}(1)\right) = 1 - (x+1)^{-2}$

25. $f(x) = 5\sqrt{3x^3+x} = 5(3x^3+x)^{1/2}$

$\dfrac{dy}{dx} = \dfrac{d}{dx}\left[5\left(3x^3+x\right)^{1/2}\right] = 5\dfrac{d}{dx}\left(3x^3+x\right)^{1/2}$

$\quad = 5\cdot\dfrac{1}{2}\left(3x^3+x\right)^{-1/2}\cdot\dfrac{d}{dx}\left(3x^3+x\right)$

$\quad = \dfrac{5\left(9x^2+1\right)}{2\sqrt{3x^3+x}} = \dfrac{45x^2+5}{2\sqrt{3x^3+x}}$

27. $y = 3x + \pi^3$

$\dfrac{dy}{dx} = \dfrac{d}{dx}\left(3x+\pi^3\right) = \dfrac{d}{dx}3x + \dfrac{d}{dx}\pi^3 = 3$

29. $y = \sqrt{1+x+x^2} = \left(1+x+x^2\right)^{1/2}$

$\dfrac{dy}{dx} = \dfrac{d}{dx}\left(1+x+x^2\right)^{1/2}$

$\quad = \dfrac{1}{2}\left(1+x+x^2\right)^{-1/2}\cdot\dfrac{d}{dx}\left(1+x+x^2\right)$

$\quad = \dfrac{1}{2}\left(1+x+x^2\right)^{-1/2}\cdot(1+2x) = \dfrac{1+2x}{2\sqrt{1+x+x^2}}$

31. $y = \dfrac{2}{1-5x} = 2(1-5x)^{-1}$

$\dfrac{dy}{dx} = \dfrac{d}{dx}\left[2(1-5x)^{-1}\right] = 2\cdot\dfrac{d}{dx}\left[(1-5x)^{-1}\right]$

$\quad = 2\cdot(-1)(1-5x)^{-2}\cdot\dfrac{d}{dx}(1-5x)$

$\quad = -2(1-5x)^{-2}\cdot(-5) = 10(1-5x)^{-2}$

$\quad = \dfrac{10}{\left(1-5x\right)^2}$

33. $y = \dfrac{45}{1+x+\sqrt{x}} = 45\left(1+x+x^{1/2}\right)^{-1}$

$\dfrac{dy}{dx} = \dfrac{d}{dx}\left[45\left(1+x+x^{1/2}\right)^{-1}\right]$

$\quad = 45(-1)\left(1+x+x^{1/2}\right)^{-2}\dfrac{d}{dx}\left(1+x+x^{1/2}\right)$

$\quad = -45\left(1+x+\sqrt{x}\right)^{-2}\left(1+\dfrac{1}{2}x^{-1/2}\right)$

35. $y = x+1+\sqrt{x+1} = x+1+(x+1)^{1/2}$

$\dfrac{dy}{dx} = \dfrac{d}{dx}\left(x+1+(x+1)^{1/2}\right)$

$\quad = \dfrac{d}{dx}x + \dfrac{d}{dx}1 + \dfrac{d}{dx}(x+1)^{1/2}$

$\quad = 1 + \dfrac{1}{2}(x+1)^{-1/2}$

37. $f(x) = \left(\dfrac{\sqrt{x}}{2}+1\right)^{3/2}$

$\dfrac{d}{dx}\left(\dfrac{\sqrt{x}}{2}+1\right)^{3/2} = \dfrac{3}{2}\left(\dfrac{\sqrt{x}}{2}+1\right)^{1/2}\cdot\dfrac{d}{dx}\left(\dfrac{\sqrt{x}}{2}+1\right)$

$\quad = \dfrac{3}{2}\left(\dfrac{\sqrt{x}}{2}+1\right)^{1/2}\cdot\dfrac{1}{2}\cdot\dfrac{1}{2}\cdot x^{-1/2}$

$\quad = \dfrac{3}{2}\left(\dfrac{\sqrt{x}}{2}+1\right)^{1/2}\left(\dfrac{1}{4}x^{-1/2}\right)$

$\quad = \dfrac{3}{8\sqrt{x}}\left(\dfrac{\sqrt{x}}{2}+1\right)^{1/2}$

39. $f(x) = 3x^2-2x+1,\ (1,2)$

$f'(x) = \dfrac{d}{dx}(3x^2-2x+1) = 6x-2$

slope $= f'(1) = 6(1)-2 = 4$

41. $y = x^3+3x-8$

$y' = \dfrac{d}{dx}\left(x^3+3x-8\right) = 3x^2+3$

slope $= f'(2) = 3(2)^2+3 = 15$

43. $y = f(x) = (x^2 - 15)^6$

$$\frac{dy}{dx} = \frac{d}{dx}(x^2 - 15)^6$$

$$= 6(x^2 - 15)^5 \cdot \frac{d}{dx}(x^2 - 15)$$

$$= 6(x^2 - 15)^5 \cdot 2x = 12x(x^2 - 15)^5$$

$$f'(x) = 12x(x^2 - 15)^5$$

slope $= f'(4) = 12(4)(16 - 15)^5 = 48$

$f(4) = (4^2 - 15)^6 = 1$

Let $(x_1, y_1) = (4, 1)$, slope $= 48$.

$y - 1 = 48(x - 4) \Rightarrow y = 48x - 191$

45. $f(x) = (3x^2 + x - 2)^2$

a. $\dfrac{d}{dx}(3x^2 + x - 2)^2$

$$= 2(3x^2 + x - 2) \cdot \frac{d}{dx}(3x^2 + x - 2)$$

$$= 2(3x^2 + x - 2)(6x + 1)$$

$$= 36x^3 + 18x^2 - 22x - 4$$

b. $(3x^2 + x - 2)(3x^2 + x - 2)$

$$= 9x^4 + 3x^3 - 6x^2 + 3x^3$$

$$\qquad + x^2 - 2x - 6x^2 - 2x + 4$$

$$= 9x^4 + 6x^3 - 11x^2 - 4x + 4$$

$$\frac{d}{dx}(9x^4 + 6x^3 - 11x^2 - 4x + 4)$$

$$= 36x^3 + 18x^2 - 22x - 4$$

47. $f(1) = .6(1) + 1 = 1.6$, so $g(1) = 3f(1) = 4.8$.
$f'(1) = .6$ (slope of the line),
$g'(1) = 3f'(1) = 1.8$

49. $h(5) = 3f(5) + 2g(5) = 3(2) + 2(4) = 14$
$h'(5) = 3f'(5) + 2g'(5) = 3(3) + 2(1) = 11$

51. $f(x) = 5\sqrt{g(x)}$

$$f'(x) = \frac{5}{2\sqrt{g(x)}} g'(x)$$

$$f(1) = 5\sqrt{g(1)} = 5\sqrt{4} = 10$$

$$f'(1) = \frac{5}{2\sqrt{g(1)}} g'(1) = \frac{5}{2\sqrt{4}} \cdot 3 = \frac{15}{4}$$

53. $\dfrac{dy}{dx} = x^2 - 8x + 18 = 3$, since the slope of

$6x - 2y = 1$ is 3.

$x^2 - 8x + 15 = 0 \Rightarrow (x - 5)(x - 3) = 0 \Rightarrow$

$x = 5$ or $x = 3$

Fit 3 and 5 back into the equation to get the

points $(3, 49)$ and $\left(5, \dfrac{161}{3}\right)$.

55. $y = f(x)$

slope $= \dfrac{3 - 5}{0 - 4} = \dfrac{1}{2}$

Let $(x_1, y_1) = (4, 5)$.

$y - 5 = \dfrac{1}{2}(x - 4) \Rightarrow y = \dfrac{1}{2}x + 3$

$f(4) = \dfrac{1}{2}(4) + 3 = 2 + 3 = 5$

$f'(4) = \dfrac{1}{2}$

1.7 More About Derivatives

1. $f(t) = (t^2 + 1)^5$

$$\frac{d}{dt}(t^2 + 1)^5 = 5(t^2 + 1)^4 \cdot \frac{d}{dt}(t^2 + 1)$$

$$= 5(t^2 + 1)^4 (2t)$$

$$= 10t(t^2 + 1)^4$$

3. $v(t) = 4t^2 + 11\sqrt{t} + 1 = 4t^2 + 11t^{1/2} + 1$

$$\frac{d}{dt}\left(4t^2 + 11t^{1/2} + 1\right) = 8t + \frac{11}{2}t^{-1/2}$$

5. $y = T^5 - 4T^4 + 3T^2 - T - 1$

$$\frac{dy}{dT} = \frac{d}{dT}\left(T^5 - 4T^4 + 3T^2 - T - 1\right)$$

$$= 5T^4 - 16T^3 + 6T - 1$$

7. $\dfrac{d}{dP}\left(3P^2 - \dfrac{1}{2}P + 1\right) = 6P - \dfrac{1}{2}$

9. $\dfrac{d}{dt}(a^2t^2 + b^2t + c^2) = 2a^2t + b^2 + 0$

$$= 2a^2t + b^2$$

11. $y = x + 1$
$y' = 1$
$y'' = 0$

13. $y = \sqrt{x} = x^{1/2}$

$y' = \dfrac{1}{2}x^{-1/2}$

$y'' = -\dfrac{1}{4}x^{-3/2}$

15. $y = \sqrt{x+1} = (x+1)^{1/2}$

$y' = \dfrac{1}{2}(x+1)^{-1/2}$

$y'' = -\dfrac{1}{2}\left(\dfrac{1}{2}\right)(x+1)^{-3/2} = -\dfrac{1}{4}(x+1)^{-3/2}$

17. $f(r) = \pi r^2$

$f'(r) = 2\pi r$

$f''(r) = 2\pi$

19. $f(P) = (3P+1)^5$

$f'(P) = 5(3P+1)^4 \cdot \dfrac{d}{dP}(3P+1)$

$\quad = 5(3P+1)^4 \cdot 3 = 15(3P+1)^4$

$f''(P) = 60(3P+1)^3 \cdot \dfrac{d}{dP}(3P+1)$

$\quad = 60(3P+1)^3 \cdot 3 = 180(3P+1)^3$

21. $\dfrac{d}{dx}(2x+7)^2 \Big|_{x=1} = \left[2(2x+7)\dfrac{d}{dx}(2x+7)\right]_{x=1}$

$\quad = \left[4(2x+7)\right]\big|_{x=1}$

$\quad = 4(2(1)+7) = 36$

23. $\dfrac{d}{dz}(z^2+2z+1)^7 \Big|_{z=-1}$

$\quad = \left[7(z^2+2z+1)^6\dfrac{d}{dz}(z^2+2z+1)\right]_{z=-1}$

$\quad = 7(2z+2)(z^2+2z+1)^6\big|_{z=-1}$

$\quad = 7(2(-1)+2)\left((-1)^2+2(-1)+1\right)$

$\quad = 0$

25. $\dfrac{d^2}{dx^2}(3x^3 - x^2 + 7x - 1)\Big|_{x=2}$

$\dfrac{d}{dx}\left(3x^3 - x^2 + 7x - 1\right) = 9x^2 - 2x + 7$

$\dfrac{d}{dx}\left(9x^2 - 2x + 7\right) = 18x - 2$

$\dfrac{d^2}{dx^2}(3x^3 - x^2 + 7x - 1)\Big|_{x=2} = 18(2) - 2 = 34$

27. $f'(1)$ and $f''(1)$, when $f(t) = \dfrac{1}{2+t}$.

$f'(t) = (-1)(2+t)^{-2}, \; f'(1) = -(2+1)^{-2} = -\dfrac{1}{9}$

$f''(t) = (-2)(-1)(2+t)^{-3}$,

$f''(1) = 2(2+1)^{-3} = \dfrac{2}{3^3} = \dfrac{2}{27}$

29. $\dfrac{d}{dt}\left(\dfrac{dv}{dt}\right)\Big|_{t=32}$, where $v(t) = 3t^3 + \dfrac{4}{t}$

$\dfrac{dv}{dt} = \dfrac{d}{dt}\left(3t^3 + \dfrac{4}{t}\right) = 9t^2 - \dfrac{4}{t^2}$

$\dfrac{d}{dt}\left(9t^2 - \dfrac{4}{t^2}\right)\Big|_{t=2} = \left(18t + \dfrac{8}{t^3}\right)\Big|_{t=2}$

$\quad = 18 \cdot 2 + \dfrac{8}{2^3} = 36 + 1 = 37$

31. $R = 1000 + 80x - .02x^2$, for $0 \le x \le 2000$

$\dfrac{dR}{dx} = 80 - .04x$

$\dfrac{dR}{dx}\Big|_{x=1500} = 80 - .04(1500) = 20$

33. $s = PT$

  a. $\dfrac{ds}{dP} = \dfrac{d}{dP}(PT) = T$

  b. $\dfrac{ds}{dT} = \dfrac{d}{dT}(PT) = P$

35. $s = Tx^2 + 3xP + T^2$

  a. $\dfrac{ds}{dx} = \dfrac{d}{dx}\left(Tx^2 + 3xP + T^2\right) = 2Tx + 3P$

  b. $\dfrac{ds}{dP} = \dfrac{d}{dP}\left(Tx^2 + 3xP + T^2\right) = 3x$

  c. $\dfrac{ds}{dT} = \dfrac{d}{dT}\left(Tx^2 + 3xP + T^2\right) = x^2 + 2T$

37. $C(50) = 5000$ means that it costs $5000 to manufacture 50 bicycles in one day.

$C'(50) = 45$ means that it costs an additional $45 to make the 51st bicycle.

39. $R(x) = 3x - .01x^2$, $R'(x) = 3 - .02x$

  a. $R'(20) = 3 - .02(20) = \$2.60$ per unit

  b. $R(x) = 3x - .01x^2 = 200 \Rightarrow$
$x = 100$ or $x = 200$ units

41. a. When 1200 chips are produced per day, the revenue is $22,000 $\Rightarrow R(12) = 22$, and the marginal revenue is $.75 per chip $\Rightarrow R'(12) = \$.075$ thousand / unit ($.75 per chip = $75 per unit = $.075 thousand/unit)

b. Marginal Profit = Marginal Revenue − Marginal Cost
$P'(12) = R'(12) - C'(12) = .75 - 1.5$
$\qquad = -\$.75$ per chip

43. a. The sales at the end of January reached $120,560, so $S(1) = \$120,560$.
Sales rising at a rate of $1500/month means that $S'(1) = \$1500$.

b. At the end of March, the sales for the month dropped to $80,000, so $S(3) = \$80,000$. Sales falling by about $200/day means that
$S'(30) = -\$200(30) = -\$6000$.

45. a. $S(10) = 3 + \dfrac{9}{11^2} \approx \$3.074$ thousand

$S'(10) = \dfrac{-18}{11^3} \approx \$-.0135$ thousand/day

b. $S(11) \approx S(10) + S'(10)$
$\qquad \approx 3.07438 + (-.013524)$
$\qquad \approx \$3.061$ thousand

$S(11) = 3 + \dfrac{9}{12^2} = \$3.0625$ thousand

47. a. When $8000 is spent on advertising, 1200 computers were sold , so $A(8) = 12$, and sales rising at the rate of 50 computers for each $1000 spent on advertising means that $A'(8) = .5$.

b. $A(9) \approx A(8) + A'(8) = 12.5$ (hundred)
$\qquad = 1250$ computers
If the company spends $9000 on advertising, 1250 computers will be sold.

49. a. $f(x) = x^5 - x^4 + 3x$
$f'(x) = 5x^4 - 4x^3 + 3$
$f''(x) = 20x^3 - 12x^2$
$f'''(x) = 60x^2 - 24x$

b. $f(x) = 4x^{5/2}$
$f'(x) = 10x^{3/2}$
$f''(x) = 15x^{1/2}$
$f'''(x) = \dfrac{15}{2} x^{-1/2} = \dfrac{15}{2\sqrt{x}}$

51. $f(x) = \dfrac{x}{1 + x^2}$

$Y_1 = \dfrac{X}{1+X^2};\ Y_2 = \text{nDeriv}(Y_1, X, X)$

$Y_3 = \text{nDeriv}(Y_2, X, X)$

$[-4, 4]$ by $[-2, 2]$

1.8   The Derivative as a Rate of Change

1. $f(x) = x^2 + 3x$

a. Over $1 \le x \le 2$,
$$\dfrac{f(b) - f(a)}{b - a} = \dfrac{\left(2^2 + 3(2)\right) - \left(1^2 + 3(1)\right)}{2 - 1}$$
$$= \dfrac{10 - 4}{1} = 6$$

b. over $1 \le x \le 1.5$,
$$\dfrac{f(b) - f(a)}{b - a} = \dfrac{\left(1.5^2 + 3(1.5)\right) - \left(1^2 + 3(1)\right)}{1.5 - 1}$$
$$= \dfrac{6.75 - 4}{.5} = 5.5$$

c. over $1 \le x \le 1.1$,
$$\dfrac{f(b) - f(a)}{b - a} = \dfrac{\left(1.1^2 + 3(1.1)\right) - \left(1^2 + 3(1)\right)}{1.1 - 1}$$
$$= \dfrac{4.51 - 4}{.1} = 5.1$$

3. $f(x) = 4x^2$

   a. Over $1 \le x \le 2$,

$$\frac{f(b) - f(a)}{b - a} = \frac{4(2)^2 - 4(1)^2}{2 - 1} = \frac{16 - 4}{1} = 12$$

   over $1 \le x \le 1.5$,

$$\frac{f(b) - f(a)}{b - a} = \frac{4(1.5)^2 - 4(1)^2}{1.5 - 1} = \frac{9 - 4}{.5} = 10$$

   over $1 \le x \le 1.1$,

$$\frac{f(b) - f(a)}{b - a} = \frac{4(1.1)^2 - 4(1)^2}{1.1 - 1} = \frac{4.84 - 4}{.1}$$
$$= 8.4$$

   b. $f'(x) = 8x \Rightarrow f'(1) = 8$

5. $f(t) = t^2 + 3t - 7$

   a. Over $5 \le x \le 6$,
$$\frac{f(b) - f(a)}{b - a}$$
$$= \frac{6^2 + 3(6) - 7 - (5^2 + 3(5) - 7)}{6 - 5}$$
$$= 36 + 18 - 7 - 25 - 15 + 7 = 14$$

   b. $f'(t) = 2t + 3$
$$f'(5) = 2(5) + 3 = 13$$

7. $s(t) = 2t^2 + 4t$

   a. $s'(t) = 4t + 4$
$$s'(6) = 4(6) + 4 = 28 \text{ km/hr}$$

   b. $s(6) = 2(6)^2 + 4(6) = 72 + 24 = 96 \text{ km}$

   c. When does $s'(t) = 6$?
$$s'(t) = 4t + 4$$
$$6 = 4t + 4 \Rightarrow t = \frac{1}{2}$$

   The object is traveling at the rate of 6 km/hr when $t = \dfrac{1}{2}$ hr.

9. $f(t) = 60t + t^2 - \dfrac{1}{12}t^3$

$$f'(t) + 60 + 2t - \frac{1}{4}t^2$$

$$f'(2) = 60 + 2(2) - \frac{1}{4}(2)^2 = 60 + 4 - 1$$
$$= 63 \text{ units/hour}$$

11. $s(t) = -6t^2 + 72t$

   a. $v(t) = s'(t) = -12t + 72 \text{ ft per sec}$

   b. $a(t) = v'(t) = -12 \text{ ft per sec per sec}$

   c. The velocity is zero when the rocket reaches its maximum height, so find the maximum height by solving $v(t) = 0$.
$$-12t + 72 = 0 \Rightarrow t = 6$$
The rocket reaches its maximum height at 6 seconds.

   d. The maximum height is
$$s(6) = -6 \cdot 6^2 + 72 \cdot 6 = 216 \text{ ft.}$$

13. $s(t) = 160t - 16t^2$

   a. $s'(t) = 160 - 32t$
$$s'(0) = 160 - 32(0) = 160 \text{ ft/sec}$$

   b. $s'(2) = 160 - 32(2) = 160 - 64 = 96 \text{ ft/sec}$

   c. $s''(t) = -32$
$$s''(3) = -32 \text{ ft/sec}^2$$

   d. When will $s(t) = 0$?
$$160t - 16t^2 = 0 \Rightarrow 16t(10 - t) = 0 \Rightarrow$$
$$t = 0 \text{ sec or } t = 10 \text{ sec}$$
The rocket will hit the ground after 10 sec.

   e. What is $s'(t)$ when $t = 10$?
$$s'(10) = 160 - 32(10) = 160 - 320$$
$$= -160 \text{ ft/sec}$$

15. A. The velocity of the ball after 3 seconds is the first derivative evaluated at $t = 3$, or $s'(3)$. The solution is **b.**

   B. To find when the velocity will be 3 feet per second, set $s'(t) = 3$ and solve for t. The solution is **d.**

   C. The average velocity during the first 3 seconds can be found from:
$$\frac{f(b) - f(a)}{b - a} = \frac{s(3) - s(0)}{3}$$
The solution is **f.**

   D. The ball will be 3 feet above the ground when, for some value a, $s(a) = 3$. The solution is **e.**

   E. The ball will hit the ground when $s(t) = 0$. Solve for t. The solution is **a.**

   F. The ball will be $s(3)$ feet high after 3 seconds. The solution is **c.**

   G. The ball travels $s(3) - s(0)$ feet during the first 3 seconds. The solution is **g.**

17. $s(t) = t^2 + 3t + 2$

    a. $s'(t) = 2t + 3$

        $s'(6) = 2(6) + 3 = 12 + 3 = 15$ feet/second

    b. No; the positive velocity indicates the object is moving away from the reference point.

    c. The object is 6 feet from the reference point when $s(t) = 6$.

        $s(t) = t^2 + 3t + 2 = 6$

        $t^2 + 3t - 4 = 0 \Rightarrow (t + 4)(t - 1) = 0 \Rightarrow$
        $t = -4$ or $t = 1$

        Time is positive, so $t = 1$ second. The velocity at this time is:

        $s'(1) = 2(1) + 3 = 5$ feet/second

19. $f(100) = 5000$
    $f'(100) = 10$
    $f(a + h) - f(a) \approx f'(a) \cdot h$
    $f(a + h) \approx f'(a) \cdot h + f(a)$

    a. $101 = 100 + 1$

        $f(100 + 1) \approx f'(100) \cdot 1 + f(100)$
                $\approx 10 + 5000 \approx 5010$

    b. $100.5 = 100 + .5$

        $f(100 + .5) \approx f'(100) \cdot .5 + f(100)$
                $\approx 10 \cdot .5 + 5000 \approx 5005$

    c. $99 = 100 + (-1)$

        $f(100 + (-1)) \approx f'(100) \cdot (-1) + f(100)$
                $\approx 10 \cdot (-1) + 5000 \approx 4990$

    d. $98 = 100 + (-2)$

        $f(100 + (-2)) \approx f'(100) \cdot (-2) + f(100)$
                $\approx 10(-2) + 5000 \approx 4980$

    e. $99.75 = 100 + (-.25)$

        $f(100 + (-.25))$
            $\approx f'(100) \cdot (-.25) + f(100)$
            $\approx 10(.25) + 5000 \approx 4997.5$

21. $f(4) = 120; \ f'(4) = -5$

    Four minutes after it has been poured, the temperature of the coffee is $120°$. At that time, its temperature is decreasing by $5°$ per minute.
    At 4.1 minutes: $4.1 = 4 + .1$
    $f(4 + .1) \approx f'(4) \cdot .1 + f(4)$
             $\approx -5 \cdot .1 + 120 \approx 119.5°$

23. $f(10,000) = 200,000; \ f'(10,000) = -3$

    When the price of a car is \$10,000, 200,000 cars are sold. At that price, the number of cars sold decreases by 3 for each dollar increase in the price.

25. $f(12) = 60; \ f'(12) = -2$

    When the price of a computer is \$1200, 60,000 computers will be sold. At that price, the number of computers sold decreases by 2000 for every \$100 increase in price.

    $f(12.5) \approx f(12) + .5f'(12)$
            $= 60 + .5(-2) = 59$

    About 59,000 computers will be sold if the price increases to \$1250.

27. $P(100) = 90,000; \ P'(100) = 1200$

    The profit from manufacturing and selling 100 luxury cars is \$90,000. Each additional car made and sold creates an additional profit of \$1200.
    At 99 cars: $99 = 100 + (-1)$
    $f(100 + (-1)) \approx f'(100) \cdot (-1) + f(100)$
               $\approx 1200(-1) + 90,000$
               $\approx \$88,800$

29. $C(x) = 6x^2 + 14x + 18$

    a. $C'(x) = 12x + 14$

        $C'(5) = 12(5) + 14 = \$74$ thousand/unit

    b. $C(5.25) \approx C'(5)(.25) + C(5)$
                $= 74(.25) + 238$
                $= \$256.5$ thousand

    c. Solve
        $6x^2 + 14x + 18 = -x^2 + 37x + 38$
        $7x^2 - 23x - 20 = 0$
        $(x - 4)(7x + 5) = 0 \Rightarrow x = 4$ or $x = -\dfrac{5}{7}$
        The break even point is $x = 4$ items.

    d. $R'(x) = -2x + 37$
        $R'(4) = \$29$ thousand/unit
        $C'(x) = 12x + 14$
        $C'(4) = \$62$ thousand/unit

        No, the company should not increase production beyond $x = 4$ items. The additional cost is greater than the additional revenue generated and the company will lose money.

31. a. $f(7) \approx \$500$ billion

b. $f'(7) \approx \$50$ billion/year

c. $f(t) = 1000$ at $t = 14$, or 1994.

d. $f'(t) = 100$ at $t = 14$, or 1994.

33. $f(t) = .36 + .77(t - .5)^{-.36}$

a.   Graph:

$Y_1 = .36 + .77(X - .5)^{-.36}$

$Y_2 = \text{nDeriv}(Y_1, X, X)$

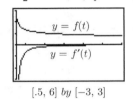

[.5, 6] *by* [−3, 3]

b.   Evaluate at $t = 4$.
$f(4) \approx .85$ seconds

c.   Graphing the line $y = .8$ and using the
INTERSECT command, the point
(5.23, .8) is on both graphs. The judgment
time was about .8 seconds after 5 days.

d.   Evaluate $f'(t)$ at $t = 4$.
$f'(4) \approx -.05$ seconds/day

e.   Graphing the line $y = -.08$ and $f'(t)$, and
using the **INTERSECT** command, the
point (2.994, −.08) is on both graphs. The
judgment time was changing at the rate of
−.08 seconds per day after 3 days.

Chapter 1 Review Exercises

1.   Let $(x_1, y_1) = (0, 3)$.
$y - 3 = -2(x - 0)$
$y = 3 - 2x$

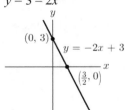

2.   Let $(x_1, y_1) = (0, -1)$.

$y - (-1) = \dfrac{3}{4}(x - 0) \Rightarrow y = \dfrac{3}{4}x - 1$

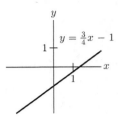

3.   Let $(x_1, y_1) = (2, 0)$.
$y - 0 = 5(x - 2)$
$y = 5x - 10$

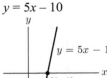

4.   Let $(x_1, y_1) = (1, 4)$.

$y - 4 = -\dfrac{1}{3}(x - 1)$

$y = -\dfrac{x}{3} + \dfrac{13}{3}$

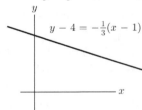

5. $y = -2x$, slope $= -2$
Let $(x_1, y_1) = (3, 5)$.
$y - 5 = -2(x - 3)$
$y = 11 - 2x$ or
$y = -2x + 11$

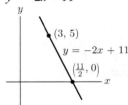

6. $-2x + 3y = 6$

$y = 2 + \dfrac{2}{3}x$, slope $= \dfrac{2}{3}$

Let $(x_1, y_1) = (0, 1)$.

$y - 1 = \dfrac{2}{3}(x - 0)$

$y = \dfrac{2}{3}x + 1$

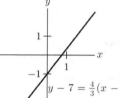

7. slope $= \dfrac{7 - 4}{3 - (-1)} = \dfrac{3}{4}$

Let $(x_1, y_1) = (3, 7)$.

$y - 7 = \dfrac{3}{4}(x - 3)$

$y = \dfrac{3}{4}x + \dfrac{19}{4}$

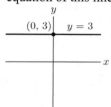

8. slope $= \dfrac{1 - 1}{5 - 2} = 0$

Let $(x_1, y_1) = (2, 1)$.

$y - 1 = 0(x - 2)$

$y = 1$

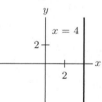

9. Slope of $y = 3x + 4$ is 3, thus a perpendicular line has slope of $-\dfrac{1}{3}$. The perpendicular line through $(1, 2)$ is

$y - 2 = \left(-\dfrac{1}{3}\right)(x - 1)$

$y = -\dfrac{1}{3}x + \dfrac{7}{3}$

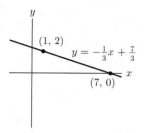

10. Slope of $3x + 4y = 5$ is $-\dfrac{3}{4}$ since

$y = -\dfrac{3}{4}x + \dfrac{5}{4}$, thus a perpendicular line has

slope of $\dfrac{4}{3}$. The perpendicular line through

$(6, 7)$ is $y - 7 = \dfrac{4}{3}(x - 6)$ or $y = \dfrac{4}{3}x - 1$.

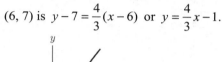

11. The equation of the x-axis is $y = 0$, so the equation of this line is $y = 3$.

12. The equation of the y-axis is $x = 0$, so 4 units to the right is $x = 4$.

13.

14.

15. $y = x^7 + x^3$; $y' = 7x^6 + 3x^2$

16. $y = 5x^8$; $y' = 40x^7$

17. $y = 6\sqrt{x} = 6x^{1/2}$; $y' = 3x^{-1/2} = \dfrac{3}{\sqrt{x}}$

18. $y = x^7 + 3x^5 + 1$; $y' = 7x^6 + 15x^4$

19. $y = \dfrac{3}{x} = 3x^{-1}$; $y' = -3x^{-2} = -\dfrac{3}{x^2}$

20. $y = x^4 - \dfrac{4}{x} = x^4 - 4x^{-1}$

$y' = 4x^3 + 4x^{-2} = 4x^3 + \dfrac{4}{x^2}$

21. $y = (3x^2 - 1)^8$

$y' = 8(3x^2 - 1)^7 (6x) = 48x(3x^2 - 1)^7$

22. $y = \dfrac{3}{4}x^{4/3} + \dfrac{4}{3}x^{3/4}$

$y' = x^{1/3} + x^{-1/4}$

23. $y = \dfrac{1}{5x - 1} = (5x - 1)^{-1}$

$\dfrac{dy}{dx} = -(5x - 1)^{-2}(5) = -\dfrac{5}{(5x - 1)^2}$

24. $y = (x^3 + x^2 + 1)^5$

$y' = 5(x^3 + x^2 + 1)^4 (3x^2 + 2x)$

25. $y = \sqrt{x^2 + 1} = (x^2 + 1)^{1/2}$

$y' = \dfrac{1}{2}(x^2 + 1)^{-1/2}(2x)$

$= \dfrac{1}{2}x(x^2 + 1)^{-1/2} = \dfrac{x}{\sqrt{x^2 + 1}}$

26. $y = \dfrac{5}{7x^2 + 1} = 5(7x^2 + 1)^{-1}$

$\dfrac{dy}{dx} = -5(7x^2 + 1)^{-2}(14x) = -\dfrac{70x}{(7x^2 + 1)^2}$

27. $f(x) = \dfrac{1}{\sqrt[4]{x}} = x^{-1/4}$

$f'(x) = -\dfrac{1}{4}x^{-5/4} = -\dfrac{1}{4x^{5/4}}$

28. $f(x) = (2x + 1)^3$

$f'(x) = 3(2x + 1)^2(2) = 6(2x + 1)^2$

29. $f(x) = 5$; $f'(x) = 0$

30. $f(x) = \dfrac{5x}{2} - \dfrac{2}{5x} = \dfrac{5}{2}x - \dfrac{2}{5}x^{-1}$

$f'(x) = \dfrac{5}{2} + \dfrac{2}{5}x^{-2} = \dfrac{5}{2} + \dfrac{2}{5x^2}$

31. $f(x) = [x^5 - (x - 1)^5]^{10}$

$f'(x) = 10[x^5 - (x - 1)^5]^9[5x^4 - 5(x - 1)^4]$

32. $f(t) = t^{10} - 10t^9$; $f'(t) = 10t^9 - 90t^8$

33. $g(t) = 3\sqrt{t} - \dfrac{3}{\sqrt{t}} = 3t^{1/2} - 3t^{-1/2}$

$g'(t) = \dfrac{3}{2}t^{-1/2} + \dfrac{3}{2}t^{-3/2}$

34. $h(t) = 3\sqrt{2}$; $h'(t) = 0$

35. $f(t) = \dfrac{2}{t - 3t^3} = 2(t - 3t^3)^{-1}$

$f'(t) = -2(t - 3t^3)^{-2}(1 - 9t^2) = \dfrac{-2(1 - 9t^2)}{(t - 3t^3)^2}$

$= \dfrac{2(9t^2 - 1)}{(t - 3t^3)^2}$

36. $g(P) = 4P^{.7}$; $g'(P) = 2.8P^{-0.3}$

37. $h(x) = \dfrac{3}{2}x^{3/2} - 6x^{2/3}$; $h'(x) = \dfrac{9}{4}x^{1/2} - 4x^{-1/3}$

38. $f(x) = \sqrt{x + \sqrt{x}} = (x + x^{1/2})^{1/2}$

$f'(x) = \dfrac{1}{2}(x + x^{1/2})^{-1/2}\left(1 + \dfrac{1}{2}x^{-1/2}\right)$

$= \dfrac{1}{2\sqrt{x + \sqrt{x}}}\left(1 + \dfrac{1}{2\sqrt{x}}\right)$

39. $f(t) = 3t^3 - 2t^2$

$f'(t) = 9t^2 - 4t$

$f'(2) = 36 - 8 = 28$

40. $V(r) = 15\pi r^2$

$V'(r) = 30\pi r$

$V'\left(\dfrac{1}{3}\right) = 10\pi$

41. $g(u) = 3u - 1$

$g(5) = 15 - 1 = 14$

$g'(u) = 3$

$g'(5) = 3$

42. $h(x) = -\dfrac{1}{2}; \quad h(-2) = -\dfrac{1}{2}$

$h'(x) = 0; \quad h'(-2) = 0$

43. $f(x) = x^{5/2}; \quad f'(x) = \dfrac{5}{2}x^{3/2}$

$f''(x) = \dfrac{15}{4}x^{1/2}; \quad f''(4) = \dfrac{15}{2}$

44. $g(t) = \dfrac{1}{4}(2t - 7)^4$

$g'(t) = (2t - 7)^3(2) = 2(2t - 7)^3$

$g''(t) = 6(2t - 7)^2(2) = 12(2t - 7)^2$

$g''(3) = 12[2(3) - 7]^2 = 12$

45. $y = (3x - 1)^3 - 4(3x - 1)^2$

slope $= y' = 3(3x - 1)^2(3) - 8(3x - 1)(3)$

$= 9(3x - 1)^2 - 24(3x - 1)$

When $x = 0$, slope $= 9 + 24 = 33$.

46. $y = (4 - x)^5$

slope $= y' = 5(4 - x)^4(-1) = -5(4 - x)^4$

When $x = 5$, slope $= -5$.

47. $\dfrac{d}{dx}(x^4 - 2x^2) = 4x^3 - 4x$

48. $\dfrac{d}{dt}(t^{5/2} + 2t^{3/2} - t^{1/2})$

$= \dfrac{5}{2}t^{3/2} + 3t^{1/2} - \dfrac{1}{2}t^{-1/2}$

49. $\dfrac{d}{dP}\left(\sqrt{1 - 3P}\right) = \dfrac{d}{dP}(1 - 3P)^{1/2}$

$= \dfrac{1}{2}(1 - 3P)^{-1/2}(-3)$

$= -\dfrac{3}{2}(1 - 3P)^{-1/2}$

50. $\dfrac{d}{dn}(n^{-5}) = -5n^{-6}$

51. $\dfrac{d}{dz}(z^3 - 4z^2 + z - 3)\Big|_{z=-2} = (3z^2 - 8z + 1)\Big|_{z=-2}$

$= 12 + 16 + 1 = 29$

52. $\dfrac{d}{dx}(4x - 10)^5\Big|_{x=3} = [5(4x - 10)^4(4)]\Big|_{x=3}$

$= [20(4x - 10)^4]\Big|_{x=3} = 320$

53. $\dfrac{d^2}{dx^2}(5x + 1)^4 = \dfrac{d}{dx}[4(5x + 1)^3(5)]$

$= 60(5x + 1)^2(5) = 300(5x + 1)^2$

54. $\dfrac{d^2}{dt^2}\left(2\sqrt{t}\right) = \dfrac{d^2}{dt^2}2t^{1/2} = \dfrac{d}{dt}t^{-1/2} = -\dfrac{1}{2}t^{-3/2}$

55. $\dfrac{d^2}{dt^2}(t^3 + 2t^2 - t)\Big|_{t=-1} = \dfrac{d}{dt}(3t^2 + 4t - 1)\Big|_{t=-1}$

$= (6t + 4)\Big|_{t=-1} = -2$

56. $\dfrac{d^2}{dP^2}(3P + 2)\Big|_{P=4} = \dfrac{d}{dP}3\Big|_{P=4} = 0\Big|_{P=4} = 0$

57. $\dfrac{d^2y}{dx^2}(4x^{3/2}) = \dfrac{dy}{dx}(6x^{1/2}) = 3x^{-1/2}$

58. $\dfrac{d}{dt}\left(\dfrac{1}{3t}\right) = \dfrac{d}{dt}\left(\dfrac{1}{3}t^{-1}\right) = -\dfrac{1}{3}t^{-2}$ or $-\dfrac{1}{3t^2}$

$\dfrac{d}{dt}\left(-\dfrac{1}{3}t^{-2}\right) = \dfrac{2}{3}t^{-3}$ or $\dfrac{2}{3t^3}$

59. $f(x) = x^3 - 4x^2 + 6$

slope $= f'(x) = 3x^2 - 8x$

When $x = 2$, slope $= 3(2)^2 - 8(2) = -4$.

When $x = 2$, $y = 2^3 - 4(2)^2 + 6 = -2$.

Let $(x_1, y_1) = (2, -2)$.

$y - (-2) = -4(x - 2) \Rightarrow y = -4x + 6$

60. $y = \dfrac{1}{3x-5} = (3x-5)^{-1}$

$y' = -(3x-5)^{-2}(3) = -\dfrac{3}{(3x-5)^2}$

When $x = 1$, slope $= -\dfrac{3}{(3(1)-5)^2} = -\dfrac{3}{4}$.

When $x = 1$, $y = \dfrac{1}{3(1)-5} = -\dfrac{1}{2}$.

Let $(x_1, y_1) = \left(1, -\dfrac{1}{2}\right)$.

$y - \left(-\dfrac{1}{2}\right) = -\dfrac{3}{4}(x-1) \Rightarrow y = -\dfrac{3}{4}x + \dfrac{1}{4}$

61. $y = x^2$

slope $= y' = 2x$

When $x = \dfrac{3}{2}$, slope $= 2\left(\dfrac{3}{2}\right) = 3$.

Let $(x_1, y_1) = \left(\dfrac{3}{2}, \dfrac{9}{4}\right)$.

$y - \dfrac{9}{4} = 3\left(x - \dfrac{3}{2}\right) \Rightarrow y = 3x - \dfrac{9}{4}$

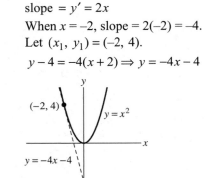

62. $y = x^2$

slope $= y' = 2x$

When $x = -2$, slope $= 2(-2) = -4$.

Let $(x_1, y_1) = (-2, 4)$.

$y - 4 = -4(x+2) \Rightarrow y = -4x - 4$

63. $y = 3x^3 - 5x^2 + x + 3$

slope $= y' = 9x^2 - 10x + 1$

When $x = 1$, slope $= 9(1)^2 - 10(1) + 1 = 0$.

When $x = 1$, $y = 3(1)^3 - 5(1)^2 + 1 + 3 = 2$.

Let $(x_1, y_1) = (1, 2)$.

$y - 2 = 0(x-1) \Rightarrow y = 2$

64. $y = (2x^2 - 3x)^3$

slope $= y' = 3(2x^2 - 3x)^2(4x - 3)$

When $x = 2$,

slope $= 3\left(2(2)^2 - 3(2)\right)^2\left(4(2) - 3\right) = 60$.

When $x = 2$, $y = \left(2(2)^2 - 3(2)\right)^3 = 8$.

Let $(x_1, y_1) = (2, 8)$.

$y - 8 = 60(x-2) \Rightarrow y = 60x - 112$

65. The line has slope -1 and contains the point $(5, 0)$.

$y - 0 = -1(x - 5) \Rightarrow y = -x + 5$

$f(2) = -2 + 5 = 3$

$f'(2) = -1$

66. The tangent line contains the points $(0, 2)$ and (a, a^3) and has slope $= 3a^2$. Thus,

$\dfrac{a^3 - 2}{a} = 3a^2 \Rightarrow a^3 - 2 = 3a^3 \Rightarrow -2 = 2a^3 \Rightarrow$

$a = -1$

67. $s'(t) = -32t + 32$

The binoculars will hit the ground when $s(t) = 0$, i.e.,

$s(t) = -16t^2 + 32t + 128 = 0$

$-16(t^2 - 2t - 8) = 0$

$-16(t - 4)(t + 2) = 0$

$t = 4$ or $t = -2$

$s'(4) = -32(4) + 32 = -96$ feet/sec.

Therefore, when the binoculars hit the ground, they will be falling at the rate of 96 feet/sec.

68. $40t + t^2 - \dfrac{1}{15}t^3$ tons is the total output of a coal mine after t hours. The rate of output is

$40 + 2t - \dfrac{1}{5}t^2$ tons per hour. At $t = 5$, the rate

of output is $40 + 2(5) - \dfrac{1}{5}(5)^2 = 45$ tons/hour.

69. 11 feet

70. $\dfrac{s(4)-s(1)}{4-1}=\dfrac{6-1}{4-1}=\dfrac{5}{3}$ ft/sec

71. Slope of the tangent line is $\dfrac{5}{3}$ so $\dfrac{5}{3}$ ft/sec.

72. $t=6$, since $s(t)$ is steeper at $t=6$ than at $t=5$.

73. $C(x)=.1x^3-6x^2+136x+200$

   a. $C(21)-C(20)$
$$=.1(21)^3-6(21)^2+136(21)+200$$
$$-(.1(20)^3-6(20)^2+136(20)+200)$$
$$=1336.1-1320=\$16.10$$

   b. $C'(x)=.3x^2-12x+136$

$$C'(20)=.3(20)^2-12(20)+136=\$16$$

74. $f(235)=4600$

$f'(235)=-100$

$f(a+h)\approx f'(a)\cdot h+f(a)$

   a. $237=235+2$
$$f(235+2)\approx f'(235)\cdot 2+f(235)$$
$$\approx -100\cdot 2+4600$$
$$\approx 4400 \text{ riders}$$

   b. $234=235+(-1)$
$$f(235+(-1))\approx f'(235)\cdot(-1)+f(235)$$
$$\approx -100\cdot(-1)+4600$$
$$\approx 4700 \text{ riders}$$

   c. $240=235+5$
$$f(235+5)\approx f'(235)\cdot 5+f(235)$$
$$\approx -100\cdot 5+4600$$
$$\approx 4100 \text{ riders}$$

   d. $232=235+(-3)$
$$f(235+(-3))\approx f'(235)\cdot(-3)+f(235)$$
$$\approx -100\cdot(-3)+4600$$
$$\approx 4900 \text{ riders}$$

75. $h(12.5)-h(12)\approx h'(12)(.5)$
$$=(1.5)(.5)=.75 \text{ in.}$$

76. $f\left(7+\dfrac{1}{2}\right)-f(7)\approx f'(7)\cdot\dfrac{1}{2}$

$$=(25.06)\cdot\dfrac{1}{2}=12.53$$

$12.53 is the additional money earned if the bank paid $7\dfrac{1}{2}\%$ interest.

77. $\displaystyle\lim_{x\to 2}\dfrac{x^2-4}{x-2}=\lim_{x\to 2}\dfrac{(x+2)(x-2)}{x-2}$
$$=\lim_{x\to 2}(x+2)=2+2=4$$

78. The limit does not exist.

79. The limit does not exist.

80. $\displaystyle\lim_{x\to 5}\dfrac{x-5}{x^2-7x+2}=\dfrac{5-5}{25-35+2}=0$

81. $f'(5)=\displaystyle\lim_{h\to 0}\dfrac{f(5+h)-f(5)}{h}$

If $f(x)=\dfrac{1}{2x}$, then

$$f(5+h)-f(5)=\dfrac{1}{2(5+h)}-\dfrac{1}{2(5)}$$
$$=\dfrac{1}{2(5+h)}\cdot\dfrac{5}{5}-\dfrac{1}{2(5)}\cdot\left(\dfrac{5+h}{5+h}\right)$$
$$=\dfrac{5-(5+h)}{10(5+h)}=\dfrac{-h}{10(5+h)}$$

Thus,

$$f'(5)=\lim_{h\to 0}[f(5+h)-f(5)]\cdot\dfrac{1}{h}$$
$$=\lim_{h\to 0}\dfrac{-h}{10(5+h)}\cdot\dfrac{1}{h}$$
$$=\lim_{h\to 0}\dfrac{-1}{10(5+h)}=-\dfrac{1}{50}$$

82. $f'(3)=\displaystyle\lim_{h\to 0}\dfrac{f(3+h)-f(3)}{h}$

If $f(x)=x^2-2x+1$, then
$f(3+h)-f(3)$
$$=(3+h)^2-2(3+h)+1-(9-6+1)$$
$$=h^2+4h.$$
Thus,

$$f'(3)=\lim_{h\to 0}\dfrac{f(3+h)-f(3)}{h}=\lim_{h\to 0}\dfrac{h^2+4h}{h}$$
$$=\lim_{h\to 0}(h+4)=4.$$

83. The slope of a secant line at $(3, 9)$

84. $\dfrac{\dfrac{1}{2+h}-\dfrac{1}{2}}{h}=\dfrac{\dfrac{2-2-h}{2(2+h)}}{h}=\dfrac{-1}{2(2+h)}$

As $h\to 0$, $\dfrac{-1}{2(2+h)}\to -\dfrac{1}{4}$.

Chapter 2 Applications of the Derivative

2.1 Describing Graphs of Functions

1. (a), (e), (f)

3. (b), (c), (d)

5. Increasing for $x < .5$, relative maximum point at $x = .5$, maximum value $= 1$, decreasing for $x > .5$, concave down, y-intercept $(0, 0)$, x-intercepts $(0, 0)$ and $(1, 0)$.

7. Decreasing for $x < 0$, relative minimum point at $x = 0$, relative minimum value $= 2$, increasing for $0 < x < 2$, relative maximum point at $x = 2$, relative maximum value $= 4$, decreasing for $x > 2$, concave up for $x < 1$, inflection point at $(1, 3)$, concave down for $x > 1$, y-intercept at $(0, 2)$, x-intercept $(3.6, 0)$.

9. Decreasing for $x < 2$, relative minimum at $x = 2$, minimum value $= 3$, increasing for $x > 2$, concave up for all x, no inflection point, defined for $x > 0$, the line $y = x$ is an asymptote, the y-axis is an asymptote.

11. Decreasing for $1 \le x < 3$, relative minimum at $x = 3$, relative minimum value $= .9$, increasing for $x > 3$, maximum value $= 6$ (at $x = 1$), minimum value $= .9$ (at $x = 3$), concave up for $1 \le x < 4$, inflection point at $(4, 1.5)$, concave down for $x > 4$; the line $y = 4$ is an asymptote.

13. The slope decreases for all x.

15. Slope decreases for $x < 1$, increases for $x > 1$. Minimum slope occurs at $x = 1$.

17. **a.** C, F **b.** A, B, F

    c. C

19.

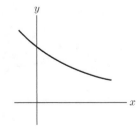

21.

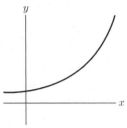

23.

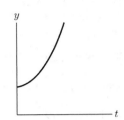

25.

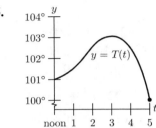

27.

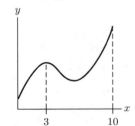

29. 1960

31. The parachutist's speed levels off to 15 ft/sec.

33.

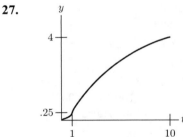

35.

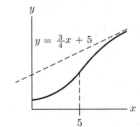

37. **a.** Yes; there is a relative minimum point between the two relative maximum points.

    b. Yes; there is an inflection point between the two relative extreme points.

39.

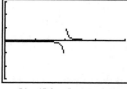

[0, 4] by [−15, 15]
Vertical asymptote: $x = 2$

41.

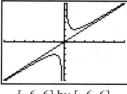

[−6, 6] by [−6, 6]
The line $y = x$ is the asymptote of the first

function, $y = \dfrac{1}{x} + x$.

2.2 The First and Second Derivative Rules

1. (e)

3. (a), (b), (d), (e)

5. (d)

7.

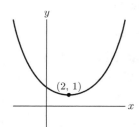

9.

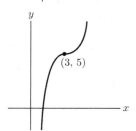

11.

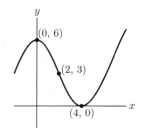

13.

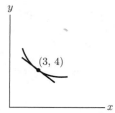

15.

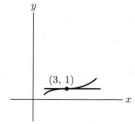

17.

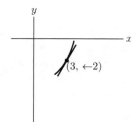

19.

	f	f'	f''
A	POS	POS	NEG
B	0	NEG	0
C	NEG	0	POS

21. $t = 1$ because the slope is more positive at $t = 1$.

23. a. $f'(9) < 0$, so $f(x)$ is decreasing at $x = 9$.

b. The function $f(x)$ is increasing for $1 \le x < 2$ because the values of $f'(x)$ are positive. The function $f(x)$ is decreasing for $2 < x \le 3$ because the values of $f'(x)$ are negative. Therefore, $f(x)$ has a relative maximum at $x = 2$. Since $f(2) = 9$, the coordinates of the relative maximum point are (2, 9).

c. The function $f(x)$ is decreasing for $9 \le x < 10$ because the values of $f'(x)$ are negative. The function $f(x)$ is increasing for $10 < x \le 11$ because the values of $f'(x)$ are positive. Therefore, $f(x)$ has a relative minimum at $x = 10$.

d. $f''(2) < 0$, so the graph is concave down.

e. $f''(x) = 0,$ so the inflection point is at $x = 6.$ Since $f(6) = 5,$ the coordinates of the inflection point are $(6, 5).$

f. The x-coordinate where $f'(x) = 6$ is $x = 15.$

25. The slope is positive because $f'(6) = 2.$

27. The slope is 0 because $f'(3) = 0.$ Also $f'(x)$ is positive for x slightly less than 3, and $f'(x)$ is negative for x slightly greater than 3. Hence $f(x)$ changes from increasing to decreasing at $x = 3.$

29. $f'(x)$ is increasing at $x = 0,$ so the graph of $f(x)$ is concave up.

31. At $x = 1,$ $f'(x)$ changes from increasing to decreasing, so the slope of the graph of $f(x)$ changes from increasing to decreasing. The concavity of the graph of $f(x)$ changes from concave up to concave down.

33. $f'(x) = 2,$ so $m = 2 \Rightarrow y - 3 = 2(x - 6) \Rightarrow$ $y = 2x - 9$

35. $f(0.25) \approx f(0) + f'(0)(.25)$
$= 3 + (1)(.25) = 3.25$

37. a. $h(100.5) \approx h(100) + h'(100)(.5)$
The change $= h(100.5) - h(100)$
$\approx h'(100)(.5) = \dfrac{1}{3} \cdot \dfrac{1}{2} = \dfrac{1}{6}$ inch.

b. (ii) because the water level is falling.

39. $f'(x) = 4(3x^2 + 1)^3(6x) = 24x(3x^2 + 1)^3$
Graph II cannot be the graph of $f(x)$ because $f'(x)$ is always positive for $x > 0.$

41. $f'(x) = \dfrac{5}{2}x^{3/2}; \ f''(x) = \dfrac{15}{4}x^{1/2}$
Graph I could be the graph of $f(x)$ since $f''(x) > 0$
for $x > 0.$

43. a. Since $f(65) \approx 2,$ there were about 2 million farms.

b. Since $f'(65) \approx -.03,$ the rate of change was -0.03 million farms per year. The number of farms was declining at the rate of about 30,000 farms per year.

c. The solution of $f(t) = 6$ is $t \approx 15,$ so there were 6 million farms in 1940.

d. The solutions of $f'(t) = -.06$ are $t \approx 20$ and $t \approx 53,$ so the number of farms was declining at the rate of 60,000 farms per year in 1945 and in 1978.

e. The graph of $f'(t)$ reaches its minimum at $t \approx 35.$ Confirm this by observing that the graph of $y = f''(t)$ crosses the t-axis at $t \approx 35.$ The number of farms was decreasing fastest in 1960.

45. $f(x) = 3x^5 - 20x^3 - 120x$

$y = f'(x)$ $y = f(x)$

$[-4, 4]$ by $[-325, 325]$

Note that $f'(x) = 15x^4 - 60x^2 - 120,$ or use the calculator's ability to graph numerical derivatives.
Relative maximum: $x \approx -2.34$
Relative minimum: $x \approx 2.34$
Inflection point: $x \approx \pm 1.41, x = 0$

2.3   The First and Second Derivative Tests and Curve Sketching

1. $f(x) = x^3 - 27x$
$f'(x) = 3x^2 - 27 = 3(x^2 - 9) = 3(x - 3)(x + 3)$
$f'(x) = 0$ if $x = -3$ or $x = 3$
$f(-3) = 54, f(3) = -54$
Critical points: $(-3, 54), (3, -54)$

Critical Points, Intervals	$x < -3$	$-3 < x < 3$	$3 < x$
$x - 3$	−	−	+
$x + 3$	−	+	+
$f'(x)$	+	−	+
$f(x)$	Increasing on $(-\infty, -3)$	Decreasing on $(-3, 3)$	Increasing on $(3, \infty)$

Relative maximum at $(-3, 54),$ relative minimum at $(3, -54).$

3. $f(x) = -x^3 + 6x^2 - 9x + 1$

$f'(x) = -3x^2 + 12x - 9$
$= -3(x^2 - 4x + 3) = -3(x-1)(x-3)$
$f'(x) = 0$ if $x = 1$ or $x = 3$
$f(1) = -3, f(3) = 1$
Critical points: $(1, -3), (3, 1)$

Critical Points, Intervals	$x < 1$	$1 < x < 3$	$3 < x$
$-3(x-1)$	+	−	−
$x - 3$	−	−	+
$f'(x)$	−	+	−
$f(x)$	Decreasing on $(-\infty, 1)$	Increasing on $(1, 3)$	Decreasing on $(3, \infty)$

Relative maximum at $(3, 1)$, relative minimum at $(1, -3)$.

5. $f(x) = \dfrac{1}{3}x^3 - x^2 + 1$

$f'(x) = x^2 - 2x = x(x-2)$
$f'(x) = 0$ if $x = 0$ or $x = 2$

$f(0) = 1; \ f(2) = -\dfrac{1}{3}$

Critical points: $(0, 1), \left(2, -\dfrac{1}{3}\right)$

Critical Points, Intervals	$x < 0$	$0 < x < 2$	$2 < x$
x	−	+	+
$x - 2$	−	−	+
$f'(x)$	+	−	+
$f(x)$	Increasing on $(-\infty, 0)$	Decreasing on $(0, 2)$	Increasing on $(2, \infty)$

Relative maximum at $(0, 1)$, relative minimum at $\left(2, -\dfrac{1}{3}\right)$.

7. $f(x) = -x^3 - 12x^2 - 2$

$f'(x) = -3x^2 - 24x = -3x(x+8)$
$f'(x) = 0$ if $x = -8$ or $x = 0$
$f(-8) = -258, f(0) = -2$
Critical points: $(-8, -258), (0, -2)$

Critical Points, Intervals	$x < -8$	$-8 < x < 0$	$0 < x$
$x + 8$	−	+	+
$-3x$	+	+	−
$f'(x)$	−	+	−
$f(x)$	Decreasing on $(-\infty, -8)$	Increasing on $(-8, 0)$	Decreasing on $(0, \infty)$

Relative maximum at $(0, -2)$, relative minimum at $(-8, -258)$.

9. $f(x) = 2x^3 - 8$
$f'(x) = 6x^2$
$f'(x) = 0$ if $x = 0$
$f(0) = -8$
Critical point: $(0, -8)$

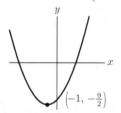

11. $f(x) = \dfrac{1}{2}x^2 + x - 4$
$f'(x) = x + 1$
$f'(x) = 0$ if $x = -1$

$f(-1) = -\dfrac{9}{2}$

Critical point: $\left(-1, -\dfrac{9}{2}\right)$

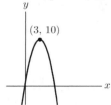

13. $f(x) = 1 + 6x - x^2$
$f'(x) = 6 - 2x$
$f'(x) = 0$ if $x = 3$
$f(3) = 10$
Critical point: $(3, 10)$

15. $f(x) = -x^2 - 8x - 10$
$f'(x) = -2x - 8$
$f'(x) = 0$ if $x = -4$
$f(-4) = 6$
Critical point: $(-4, 6)$

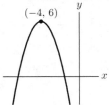

17. $f(x) = x^3 + 6x^2 + 9x$
$f'(x) = 3x^2 + 12x + 9$
$f''(x) = 6x + 12$
$f'(x) = 0$ if $x = -3$ or $x = -1$
$f(-3) = 0 \Rightarrow (-3, 0)$ is a critical pt.
$f(-1) = -4 \Rightarrow (-1, -4)$ is a critical pt.
$f''(-3) = -6 < 0 \Rightarrow (-3, 0)$ is a local max.
$f''(-1) = 18 > 0 \Rightarrow (-1, -4)$ is a local min.

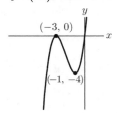

19. $f(x) = x^3 - 12x$
$f'(x) = 3x^2 - 12$
$f''(x) = 6x$
$f'(x) = 0$ if $x = -2$ or $x = 2$
$f(-2) = 16 \Rightarrow (-2, 16)$ is a critical pt.
$f(2) = -4 \Rightarrow (2, -16)$ is a critical pt.
$f''(-2) = -12 < 0 \Rightarrow (-2, 16)$ is a local max.
$f''(2) = 12 > 0 \Rightarrow (2, -16)$ is a local min.

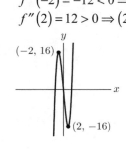

21. $f(x) = -\frac{1}{9}x^3 + x^2 + 9x$
$f'(x) = -\frac{1}{3}x^2 + 2x + 9$
$f''(x) = -\frac{2}{3}x + 2$
$f'(x) = 0$ if $x = -3$ or $x = 9$
$f(-3) = 0 \Rightarrow (-3, -15)$ is a critical pt.
$f(9) = 81 \Rightarrow (9, 81)$ is a critical pt.
$f''(-3) = 4 > 0 \Rightarrow (-3, -15)$ is a local min.
$f''(9) = -4 < 0 \Rightarrow (9, 81)$ is a local max.

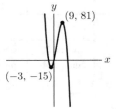

23. $f(x) = -\frac{1}{3}x^3 + 2x^2 - 12$
$f'(x) = -x^2 + 4x$
$f''(x) = -2x + 4$
$f'(x) = 0$ if $x = 0$ or $x = 4$
$f(0) = -12 \Rightarrow (0, -12)$ is a critical pt.
$f(4) = -\frac{4}{3} \Rightarrow \left(4, -\frac{4}{3}\right)$ is a critical pt.
$f''(0) = 4 > 0 \Rightarrow (-3, 0)$ is a local min.
$f''(4) = -4 < 0 \Rightarrow \left(4, -\frac{4}{3}\right)$ is a local max.

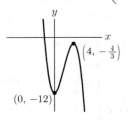

25. $y = x^3 - 3x + 2$
$y' = 3x^2 - 3$
$y'' = 6x$
$y' = 0$ if $x = -1$ or $x = 1$
$y(-1) = 4 \Rightarrow (-1, 4)$ is a critical pt.
$y(1) = 0 \Rightarrow (0, 0)$ is a critical pt.
$y''(-1) = -6 < 0 \Rightarrow (-1, 4)$ is a local max.
$y''(1) = 6 > 0 \Rightarrow (1, 0)$ is a local min.

Concavity reverses between $x = -1$ and $x = 1$, so there must be an inflection point.

(continued on next page)

(*continued*)

$y'' = 0$ when $x = 0$.
$y(0) = 2 \Rightarrow (0, 2)$ is an inflection pt.

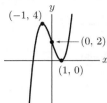

27. $y = 1 + 3x^2 - x^3$
$y' = 6x - 3x^2$
$y'' = 6 - 6x$
$y' = 0$ if $x = 0$ or $x = 2$
$y(0) = 1 \Rightarrow (0, 1)$ is a critical pt.
$y(2) = 5 \Rightarrow (2, 5)$ is a critical pt.
$y''(0) = 6 > 0 \Rightarrow (0, 1)$ is a local min.
$y''(2) = -6 < 0 \Rightarrow (2, 5)$ is a local max.

Concavity reverses between $x = 0$ and $x = 2$, so there must be an inflection point.
$y'' = 0$ when $x = 1$.
$y(1) = 3 \Rightarrow (1, 3)$ is an inflection pt.

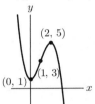

29. $y = \dfrac{1}{3}x^3 - x^2 - 3x + 5$
$y' = x^2 - 2x - 3$
$y'' = 2x - 2$
$y' = 0$ if $x = -1$ or $x = 3$
$y(-1) = \dfrac{20}{3} \Rightarrow \left(-1, \dfrac{20}{3}\right)$ is a critical pt.
$y(3) = -4 \Rightarrow (3, -4)$ is a critical pt.
$y''(-1) = -4 < 0 \Rightarrow \left(1, \dfrac{20}{3}\right)$ is a local max.
$y''(3) = 4 > 0 \Rightarrow (3, -4)$ is a local min.

Concavity reverses between $x = -1$ and $x = 3$, so there must be an inflection point.
$y'' = 0$ when $x = 1$.
$y(1) = \dfrac{4}{3} \Rightarrow \left(1, \dfrac{4}{3}\right)$ is an inflection pt.

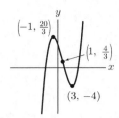

31. $y = 2x^3 - 3x^2 - 36x + 20$
$y' = 6x^2 - 6x - 36$
$y'' = 12x - 6$
$y' = 0$ if $x = -2$ or $x = 3$
$y(-2) = 64 \Rightarrow (-2, 64)$ is a critical pt.
$y(3) = -61 \Rightarrow (3, -61)$ is a critical pt.
$y''(-2) = -30 < 0 \Rightarrow (-2, 64)$ is a local max.
$y''(3) = 30 > 0 \Rightarrow (3, -61)$ is a local min.

Concavity reverses between $x = -2$ and $x = 3$, so there must be an inflection point.
$y'' = 0$ when $x = \dfrac{1}{2}$.
$y\left(\dfrac{1}{2}\right) = \dfrac{3}{2} \Rightarrow \left(\dfrac{1}{2}, \dfrac{3}{2}\right)$ is an inflection pt.

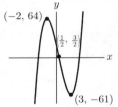

33. $f'(x) = 2ax + b$; $f''(x) = 2a$
It is not possible for the graph of $f(x)$ to have an inflection point because $f''(x) = 2a \neq 0$.

35. $f(x) = \dfrac{1}{4}x^2 - 2x + 7$; $f'(x) = \dfrac{1}{2}x - 2$;
$f''(x) = \dfrac{1}{2}$
Set $f'(x) = 0$ and solve for x,
$\dfrac{1}{2}x - 2 = 0, x = 4$;
$f(4) = \dfrac{1}{4}(4)^2 - 2(4) + 7 = 3$; $f''(4) = \dfrac{1}{2}$
Since $f''(4)$ is positive, the graph is concave up at $x = 3$ and therefore $(4, 3)$ is a relative minimum point.

37. $g(x) = 3 + 4x - 2x^2; g'(x) = 4 - 4x$
$g''(x) = -4$

Set $g'(x) = 0$ and solve for x.

$4 - 4x = 0 \Rightarrow x = 1$

$g(1) = 3 + 4(1) - 2(1)^2 = 5; g''(1) = -4$

Since $g''(1)$ is negative, the graph is concave down at $x = 1$ and therefore $(1, 5)$ is a relative maximum point.

39. $f(x) = 5x^2 + x - 3; f'(x) = 10x + 1; f''(x) = 10$

Set $f'(x) = 0$ and solve for x.

$$10x + 1 = 0 \Rightarrow x = -\frac{1}{10} = -.1$$

$$f(-.1) = 5(-.1)^2 + (-.1) - 3 = -3.05$$

$$f''(-.1) = 10$$

Since $f''(-.1)$ is positive, the graph is concave up at $x = -.1$ and therefore $(-.1, -3.05)$ is a relative minimum point.

41. $y = g(x)$ is the derivative of $y = f(x)$ because the zero of $g(x)$ corresponds to the extreme point of $f(x)$.

43. a. f has a relative minimum.

b. f has an inflection point.

45. a. $(.47, 41), (.18, 300); \ m = \dfrac{300 - 41}{.18 - .47} = -\dfrac{259}{.29};$

$$y - 41 = -\frac{259}{.29}(x - .47) \Rightarrow$$
$$y = -\frac{259}{.29}x + \frac{133.62}{.29}$$
$$\approx -893.103x + 460.759$$

$A(x) = -893.103x + 460.759$ billion dollars

b. $R(x) = \dfrac{x}{100} \cdot A(x)$

$$= \frac{x}{100}(-893.103x + 460.759)$$

$R(.3) \approx \$.578484$ billion or $\$578.484$ million

$R(.1) \approx \$.371449$ billion or $\$371.449$ million

c. $R'(x) = -17.8621x + 4.6076$

$R'(x) = 0$ when $x = .258$, The fee that maximizes revenue is $.258\%$ and the maximum revenue is
$R(.258) = \$.594273$ billion or
$\$594.273$ million

47.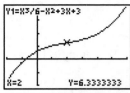

$[-2, 6]$ by $[-10, 20]$

Since $f(x)$ is always increasing, $f'(x)$ is always nonnegative.

49.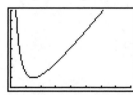

$[0, 16]$ by $[0, 16]$

This graph is like the graph of a parabola that opens upward because (for $x > 0$) the entire graph is concave up and it has a minimum value. Unlike a parabola, it is not symmetric. Also, this graph has a vertical asymptote ($x = 0$), while a parabola does not have an asymptote.

2.4 Curve Sketching (Conclusion)

1. $y = x^2 - 3x + 1$

$$x = \frac{-(-3) \pm \sqrt{(-3)^2 - 4(1)(1)}}{2(1)} = \frac{3 \pm \sqrt{5}}{2}$$

The x-intercepts are $\left(\dfrac{3 + \sqrt{5}}{2}, 0\right)$ and

$\left(\dfrac{3 - \sqrt{5}}{2}, 0\right)$

3. $y = 2x^2 + 5x + 2$

$$x = \frac{-5 \pm \sqrt{5^2 - 4(2)(2)}}{2(2)} = \frac{-5 \pm 3}{4} = -\frac{1}{2}, -2$$

The x-intercepts are $\left(-\dfrac{1}{2}, 0\right)$ and $(-2, 0)$.

5. $y = 4x - 4x^2 - 1$

$$x = \frac{-4 \pm \sqrt{4^2 - 4(-4)(-1)}}{2(-4)} = \frac{-4 \pm 0}{-8}$$

The x-intercept is $\left(\dfrac{1}{2}, 0\right)$.

7. $f(x) = \frac{1}{3}x^3 - 2x^2 + 5x;\ f'(x) = x^2 - 4x + 5$

$$x = \frac{-(-4) \pm \sqrt{(-4)^2 - 4(1)(5)}}{2(1)} = \frac{4 \pm \sqrt{-4}}{2}$$

Since $f'(x)$ has no real zeros, $f(x)$ has no relative extreme points.

9. $f(x) = x^3 - 6x^2 + 12x - 6$

$f'(x) = 3x^2 - 12x + 12$

$f''(x) = 6x - 12$

To find possible extrema, set $f'(x) = 0$ and solve for x.

$3x^2 - 12x + 12 = 0$

$3(x^2 - 4x + 4) = 0$

$(x-2)^2 = 0 \Rightarrow x = 2$

$f(2) = 2^3 - 6 \cdot 2^2 + 12 \cdot 2 - 6 = 2$

Thus, $(2, 2)$ is a critical point.

Critical Points, Intervals	$x < 2$	$2 < x$
$x - 2$	$-$	$+$
$f'(x)$	$+$	$+$
$f(x)$	Increasing on $(-\infty, 2]$	Increasing on $[2, \infty)$.

No relative maximum or relative minimum.
Since $f'(x) \geq 0$ for all x, the graph is always increasing.
To find possible inflection points, set $f''(x) = 0$ and solve for x.
$6x - 12 = 0 \Rightarrow x = 2$
Since $f''(x) < 0$ for $x < 2$ (meaning the graph is concave down) and $f''(x) > 0$ for $x > 2$ (meaning the graph is concave up), the point $(2, 2)$ is an inflection point.
$f(0) = -6$, so the y-intercept is $(0, -6)$.

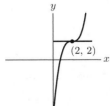

11. $f(x) = x^3 + 3x + 1$

$f'(x) = 3x^2 + 3$

$f''(x) = 6x$

To find possible extrema, set $f'(x) = 0$ and solve for x.

$3x^2 + 3 = 0 \Rightarrow$ no real solution
Thus, there are no extrema.
Since $f'(x) \geq 0$ for all x, the graph is always increasing.
To find possible inflection points, set $f''(x) = 0$ and solve for x.
$6x = 0 \Rightarrow x = 0$
$f(0) = 1$
Since $f''(x) < 0$ for $x < 0$ (meaning the graph is concave down) and $f''(x) > 0$ for $x > 0$ (meaning the graph is concave up), the point $(0, 1)$ is an inflection point. This is also the y-intercept.

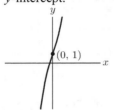

13. $f(x) = 5 - 13x + 6x^2 - x^3$

$f'(x) = -13 + 12x - 3x^2$

$f''(x) = 12 - 6x$

To find possible extrema, set $f'(x) = 0$ and solve for x.

$-13 + 12x - 3x^2 = 0 \Rightarrow$ no real solution
Thus, there are no extrema.
Since $f'(x) \leq 0$ for all x, the graph is always decreasing.
To find possible inflection points, set $f''(x) = 0$ and solve for x.
$12 - 6x = 0 \Rightarrow x = 2$
$f(2) = -5$
Since $f''(x) > 0$ for $x < 2$ (meaning the graph is concave up) and $f''(x) < 0$ for $x > 2$ (meaning the graph is concave down), the point $(2, -5)$ is an inflection point.
$f(0) = 5$, so the y-intercept is $(0, 5)$.

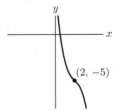

15. $f(x) = \dfrac{4}{3}x^3 - 2x^2 + x$

$f'(x) = 4x^2 - 4x = 4x(x-1)$

$f''(x) = 8x - 4$

To find possible extrema, set $f'(x) = 0$ and solve for x.

$4x^2 - 4x = 0 \Rightarrow x = 0,\ 1$

$f(0) = 0 \Rightarrow (0, 0)$ is a critical point

$f(1) = \dfrac{1}{3} \Rightarrow \left(1, \dfrac{1}{3}\right)$ is a critical point

Critical Points, Intervals	$x < 0$	$0 < x < 1$	$1 < x$
$4x$	$-$	$+$	$+$
$(x-1)$	$-$	$-$	$+$
$f'(x)$	$+$	$-$	$+$
$f(x)$	Decreasing on $(-\infty, 0)$	Decreasing on $(0, 1)$	Increasing on $(1, \infty)$

We have identified $(0, 0)$ and $\left(1, \dfrac{1}{3}\right)$ as critical points. However, neither is a local maximum, nor a local minimum. Therefore, they may be inflection points. However, $f''(0) \neq 0$ and $f'' \neq 0$, so neither is an inflection point. Since $f'(x) \geq 0$ for all x, the graph is always increasing.

To find possible inflection points, set $f''(x) = 0$ and solve for x.

$8x - 4 = 0 \Rightarrow x = \dfrac{1}{2}$

$f\left(\dfrac{1}{2}\right) = \dfrac{1}{6}$

Since $f''(x) < 0$ for $x < \tfrac{1}{2}$ (meaning the graph is concave down) and $f''(x) > 0$ for $x > \tfrac{1}{2}$ (meaning the graph is concave up), the point $\left(\tfrac{1}{2}, \tfrac{1}{6}\right)$ is an inflection point.

$f(0) = 0 \Rightarrow (0, 0)$ is the y-intercept.

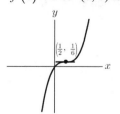

17. $f(x) = 1 - 3x + 3x^2 - x^3$

$f'(x) = -3 + 6x - 3x^2 = -3\left(x^2 - 2x + 1\right)$

$f''(x) = 6 - 6x$

To find possible extrema, set $f'(x) = 0$ and solve for x.

$-3 + 6x - 3x^2 = 0 \Rightarrow x = 1$

$f(1) = 0$

Since $f'(x) \leq 0$ for all x, the graph is always decreasing, and thus, there are no extrema. Therefore, $(1, 0)$ may be an inflection point. Set $f''(x) = 0$ and solve for x.

$6 - 6x = 0 \Rightarrow x = 1$

Since $f''(x) > 0$ for $x < 1$ (meaning the graph is concave up) and $f''(x) < 0$ for $x > 1$ (meaning the graph is concave down), the point $(1, 0)$ is an inflection point.

$f(0) = 1$, so the y-intercept is $(0, 1)$.

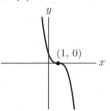

19. $f(x) = x^4 - 6x^2$

$f'(x) = 4x^3 - 12x$

$f''(x) = 12x^2 - 12$

To find possible extrema, set $f'(x) = 0$ and solve for x.

$4x^3 - 12x = 0$

$4x\left(x^2 - 3\right) = 0 \Rightarrow x = 0,\ x = -\sqrt{3},\ x = \sqrt{3}$

$f(0) = 0^4 - 6 \cdot 0^2 = 0$

$f\left(-\sqrt{3}\right) = \left(-\sqrt{3}\right)^4 - 6 \cdot \left(-\sqrt{3}\right)^2 = 9 - 18 = -9$

$f\left(\sqrt{3}\right) = \left(\sqrt{3}\right)^4 - 6 \cdot \left(\sqrt{3}\right)^2 = 9 - 18 = -9$

Thus, $(0, 0)$, $\left(-\sqrt{3}, -9\right)$, and $\left(\sqrt{3}, -9\right)$ are critical points.

$f''(0) = -12$, so the graph is concave down at $x = 0$, and $(0, 0)$ is a relative maximum.

$f''\left(-\sqrt{3}\right) = 12\left(-\sqrt{3}\right)^2 - 12 = 24$ so the graph is concave up at $x = -\sqrt{3}$, and $\left(-\sqrt{3}, -9\right)$, is a relative minimum.

(continued on next page)

(*continued*)

$f''\left(\sqrt{3}\right)=12\left(\sqrt{3}\right)^2-12=24$ so the graph is concave up at $x=\sqrt{3}$, and $\left(\sqrt{3},-9\right)$, is a relative minimum.

The concavity of this function reverses twice, so there must be at least two inflection points. Set $f''(x)=0$ and solve for x:

$12x^2-12=0\Rightarrow 12\left(x^2-1\right)=0\Rightarrow$
$(x-1)(x+1)=0\Rightarrow x=\pm1$
$f(-1)=(-1)^4-6(-1)^2=-5$
$f(1)=1^4-6(1)^2=-5$

Thus, the inflection points are $(-1,-5)$ and $(1,-5)$.

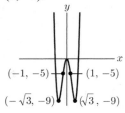

21. $f(x)=(x-3)^4$
$f'(x)=4(x-3)^3$
$f''(x)=12(x-3)$

To find possible extrema, set $f'(x)=0$ and solve for x.

$4(x-3)^3=0\Rightarrow x=3$
$f(3)=0$

Thus, $(3, 0)$ is a critical point.

$f''(3)=0$, so we must use the first derivative rule to determine if $(3, 0)$ is a local maximum or minimum.

Critical Points, Intervals	$x<3$	$x>3$
$x-3$	$-$	$+$
$f'(x)$	$-$	$+$
$f(x)$	Decreasing on $(-\infty,3)$	Increasing on $(3,\infty)$

Thus, $(3, 0)$ is local minimum.
Since $f''(x)=0,$ when $x=3$, $(3, 0)$ is also an inflection point.
The y-intercept is $(0, 81)$.

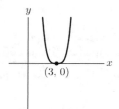

23. $y=\dfrac{1}{x}+\dfrac{1}{4}x,\ x>0$

$y'=-\dfrac{1}{x^2}+\dfrac{1}{4}$

$y''=\dfrac{2}{x^3}$

To find possible extrema, set $y'=0$ and solve for x:

$-\dfrac{1}{x^2}+\dfrac{1}{4}=0\Rightarrow x=2$

Note that we need to consider the positive solution only because the function is defined only for $x>0$. When $x=2$, $y=1$, and

$y''=\dfrac{1}{4}>0,$ so the graph is concave up, and $(2, 1)$ is a relative minimum.

Since y'' can never be zero, there are no inflection points. The term $\dfrac{1}{x}$ tells us that the y-axis is an asymptote. As $x\to\infty$, the graph approaches $y=\frac{1}{4}x,$ so this is also an asymptote of the graph.

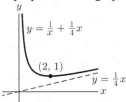

25. $y=\dfrac{9}{x}+x+1,\ x>0$

$y'=-\dfrac{9}{x^2}+1$

$y''=\dfrac{18}{x^3}$

To find possible extrema, set $y'=0$ and solve for x:

$-\dfrac{9}{x^2}+1=0\Rightarrow -\dfrac{9}{x^2}=-1\Rightarrow 9=x^2\Rightarrow x=3$

(*continued on next page*)

(*continued*)

Note that we need to consider the positive solution only because the function is defined only for $x > 0$. When $x = 3$, $y = \dfrac{9}{3} + 3 + 1 = 7$,

and $y'' = \dfrac{18}{3^3} > 0$, so the graph is concave up,

and $(3, 7)$ is a relative minimum.

Since y'' can never be zero, there are no

inflection points. The term $\dfrac{9}{x}$ tells us that the

y-axis is an asymptote. As $x \to \infty$, the graph approaches $y = x + 1$, so this is an asymptote of the graph.

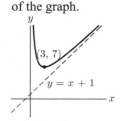

27. $y = \dfrac{2}{x} + \dfrac{x}{2} + 2, \; x > 0$

$y' = -\dfrac{2}{x^2} + \dfrac{1}{2}$

$y'' = \dfrac{4}{x^3}$

To find possible extrema, set $y' = 0$ and solve for x:

$-\dfrac{2}{x^2} + \dfrac{1}{2} = 0 \Rightarrow -\dfrac{2}{x^2} = -\dfrac{1}{2} \Rightarrow x = 2$

Note that we need to consider the positive solution only because the function is defined only for $x > 0$. When $x = 2$, $y = 4$ and

$y'' = \dfrac{4}{2^3} > 0$, so the graph is concave up, and

$(2, 4)$ is a relative minimum.

Since y'' can never be zero, there are no

inflection points. The term $\dfrac{2}{x}$ tells us that the

y-axis is an asymptote. As $x \to \infty$, the graph

approaches $y = \dfrac{x}{2} + 2$, so this is also an

asymptote of the graph.

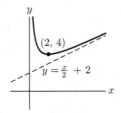

29. $y = 6\sqrt{x} - x, \; x > 0$

$y' = 3x^{-1/2} - 1 = \dfrac{3}{\sqrt{x}} - 1$

$y'' = -\dfrac{3}{2}x^{-3/2}$

To find possible extrema, set $y' = 0$ and solve for x:

$\dfrac{3}{\sqrt{x}} - 1 = 0 \Rightarrow x = 9$

Note that we need to consider the positive solution only because the function is defined only for $x > 0$. When $x = 9$, $y = 9$, and $y'' < 0$, so the graph is concave down, and $(9, 9)$ is a relative maximum. Since y'' can never be zero, there are no inflection points.

When $x = 0$, $y = 0$, so $(0, 0)$ is the y-intercept.

$y = 6\sqrt{x} - x = 0 \Rightarrow 36x = x^2 \Rightarrow x = 36$, so $(36, 0)$ is an x-intercept

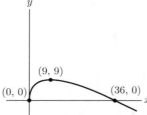

31. $g(x) = f'(x)$. The 3 zeros of $g(x)$ correspond to the 3 extreme points of $f(x)$. $f(x) \neq g(x)$, the zeros of $f(x)$ do not correspond with the extreme points of $g(x)$.

33. $f(x) = ax^2 + bx + c; \; f'(x) = 2ax + b$

$f'(0) = b = 0$ (There is a local maximum at $x = 0 \Rightarrow f'(0) = 0$).

Therefore, $f(x) = ax^2 + c; \; f(0) = c = 1;$

$f(2) = 0 \Rightarrow 4a + 2b + c = 0 \Rightarrow$

$4a + 1 = 0 \Rightarrow a = -\dfrac{1}{4}:$

Thus, $f(x) = -\dfrac{1}{4}x^2 + 1.$

35. Since $f'(a) = 0$ and $f'(x)$ is increasing at $x = a$, $f' < 0$ for $x < a$ and $f' > 0$ for $x > a$. According to the first derivative test, f has a local minimum at $x = a$.

37. a.

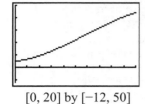

[0, 20] by [−12, 50]

b. Since $f(7) = 15.0036$, the rat weighed about 15.0 grams.

c. Using graphing calculator techniques, solve $f(t) = 27$ to obtain $t \approx 12.0380$. The rat's weight reached 27 grams after about 12.0 days.

d.

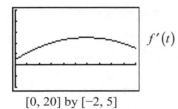

$f'(t)$

[0, 20] by [−2, 5]

Note that $f'(t) = .48 + .34t - .0144t^2$.

Since $f'(4) = 1.6096$, the rat was gaining weight at the rate of about 1.6 grams per day.

e. Using graphing calculator techniques, solve $f'(t) = 2$ to obtain $t \approx 5.990$ or $t \approx 17.6207$. The rat was gaining weight at the rate of 2 grams per day after about 6.0 days and after about 17.6 days.

f. The maximum value of $f'(t)$ appears to occur at $t \approx 11.8$. To confirm, note that $f''(t) = .34 - .0288x$, so the solution of $f''(t) = 0$ is $t \approx 11.8056$. The rat was growing at the fastest rate after about 11.8 days.

2.5 Optimization Problems

1. $g(x) = 10 + 40x - x^2 \Rightarrow g'(x) = 40 - 2x \Rightarrow g''(x) = -2$

The maximum value of $g(x)$ occurs at $x = 20$; $g(20) = 410$.

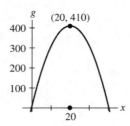

3. $f(t) = t^3 - 6t^2 + 40 \Rightarrow f'(t) = 3t^2 - 12t \Rightarrow$
$f''(t) = 6t - 12$

The minimum value for $t \geq 0$ occurs at $t = 4$; $f(4) = 8$.

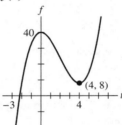

5. Solving $x + y = 2$ for y gives $y = 2 - x$. Substituting into $Q = xy$ gives
$$Q(x) = x(2 - x) = 2x - x^2.$$
$$\frac{dQ}{dx} = 2 - 2x$$
$$\frac{dQ}{dx} = 0 \Rightarrow 2 - 2x = 0 \Rightarrow x = 1$$
$$\frac{d^2Q}{dx^2} = -2$$
The maximum value of $Q(x)$ occurs at $x = 1$, $y = 1$. $Q(1) = 2(1) - (1)^2 = 1$.

7. $x + y = 6 \Rightarrow y = 6 - x$
$$Q(x) = x^2 + (6 - x)^2 = 2x^2 - 12x + 36$$
$$\frac{dQ}{dx} = 4x - 12; \quad \frac{d^2Q}{dx^2} = 4$$
$$\frac{dQ}{dx} = 0 \Rightarrow 4x - 12 = 0 \Rightarrow x = 3$$
The minimum of $Q(x)$ occurs at $x = 3$. The minimum is $Q(3) = 3^2 + (6 - 3)^2 = 18$

9. $xy = 36 \rightarrow y = \dfrac{36}{x}$

$$S(x) = x + \frac{36}{x}$$

$$S'(x) = 1 - \frac{36}{x^2}$$

$$S'(x) = 0 \rightarrow 1 - \frac{36}{x^2} = 0 \rightarrow x = 6 \text{ or } -6$$

$$S''(x) = 1 + \frac{72}{x^3}, S''(6) = \frac{4}{3}$$

The positive value $x = 6$ minimizes $S(x)$, and

$y = \dfrac{36}{6} = 6$. $S(6,6) = 6 + 6 = 12$

11. Let A = area.

   a. Objective equation: $A = xy$
Constraint equation: $8x + 4y = 320$

   b. Solving constraint equation for y in terms of x gives $y = 80 - 2x$. Substituting into objective equation yields
$A = x(80 - 2x) = -2x^2 + 80x$.

   c. $\dfrac{dA}{dx} = -4x + 80 \Rightarrow \dfrac{d^2 A}{dx^2} = -4$
The maximum value of A occurs at $x = 20$. Substituting this value into the equation for y in part b gives $y = 80 - 40 = 40$.
Answer: $x = 20$ ft, $y = 40$ ft

13. **a.**

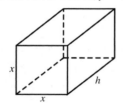

   b. length + girth = $h + 4x$

   c. Objective equation: $V = x^2 h$
Constraint equation: $h + 4x = 84$ or $h = 84 - 4x$

   d. Substituting $h = 84 - 4x$ into the objective equation, we have
$V = x^2(84 - 4x) = -4x^3 + 84x^2$.

   e. $V' = -12x^2 + 168x$
$V'' = -24x + 168$
The maximum value of V for $x > 0$ occurs at $x = 14$ in. Solving for h gives
$h = 84 - 4(14) = 28$ in.

15.

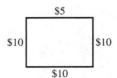

Let C = cost of materials.
Objective: $C = 15x + 20y$
Constraint: $xy = 75$
Solving the constraint for y and substituting
gives $C = 15x + 20\left(\dfrac{75}{x}\right) = 15x + \dfrac{1500}{x}$;

$$\frac{dC}{dx} = 15 - \frac{1500}{x^2}; \frac{d^2 C}{dx^2} = \frac{3000}{x^3}$$

The minimum value for $x > 0$ occurs at $x = 10$.
Answer: $x = 10$ ft, $y = 7.5$ ft

17. Let x = length of base, h = height,
M = surface area.

Constraint: $x^2 h = 8000 \Rightarrow h = \dfrac{8000}{x^2}$

Objective: $M = 2x^2 + 4xh$
Solving the constraint for y and substituting
gives $M = 2x^2 + 4x\left(\dfrac{8000}{x^2}\right) = 2x^2 + \dfrac{32,000}{x}$

$$\frac{dM}{dx} = 4x - \frac{32,000}{x^2}; \frac{d^2 M}{dx^2} = 4 + \frac{64,000}{x^2}$$

The minimum value of M for $x > 0$ occurs at
$x = 20$. Answer: 20 cm × 20 cm × 20 cm

19. Let x = length of side parallel to river,
y = length of side perpendicular to river.
Constraint: $6x + 15y = 1500$
Objective: $A = xy$
Solving the constraint for y and substituting
gives $A = x\left[-\dfrac{2}{5}x + 100\right] = -\dfrac{2}{5}x^2 + 100x$

$$\frac{dA}{dx} = -\frac{4}{5}x + 100; \frac{d^2 A}{dx^2} = -\frac{4}{5}$$

The minimum value of A for $x > 0$ occurs at
$x = 125$. Answer: $x = 125$ ft, $y = 50$ ft

21. Constraint: $x + y = 100$
Objective: $P = xy$
Solving the constraint for y and substituting
gives $P = x(100 - x) = -x^2 + 100x$

$$\frac{dP}{dx} = -2x + 100; \frac{d^2 P}{dx^2} = -2$$

The maximum value of P occurs at $x = 50$.
Answer: $x = 50$, $y = 50$

23.

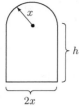

Constraint: $2x + 2h + \pi x = 14$ or
$(2 + \pi)x + 2h = 14$

Objective: $A = 2xh + \dfrac{\pi}{2}x^2$

Solving the constraint for h and substituting gives

$$A = 2x\left(7 - \frac{2+\pi}{2}x\right) + \frac{\pi}{2}x^2$$

$$= 14x - \left(\frac{\pi}{2} + 2\right)x^2$$

$$\frac{dA}{dx} = 14 - (4 + \pi)x; \quad \frac{d^2A}{dx^2} = -4 - \pi$$

The maximum value of A occurs at $x = \dfrac{14}{4 + \pi}$.

Answer: $x = \dfrac{14}{4 + \pi}$ ft

25. $A = 20w - \dfrac{1}{2}w^2; \quad \dfrac{dA}{dw} = 20 - w; \quad \dfrac{d^2A}{dw^2} = -1$

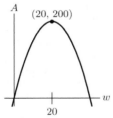

The maximum value of A occurs at $w = 20$.

$$x = 20 - \frac{1}{2}w = 20 - \frac{1}{2}(20) = 10$$

Answer: $w = 20$ ft, $x = 10$ ft

27.

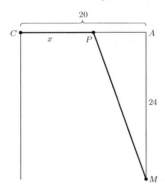

If $x =$ distance from C to P, let $y =$ be the distance from P to M. Then cost is the objective: $C = 6x + 10y$ and the constraint

$$y^2 = (20 - x)^2 + 24^2 = 976 - 40x + x^2.$$

Solving the constraint for y and substituting

gives $C = 6x + 10\left(976 - 40x + x^2\right)^{1/2}$.

$$\frac{dC}{dx} = 6 + 5(976 - 40x + x^2)^{-1/2}(-40 + 2x).$$

$$= 6 + \frac{5(-40 + 2x)}{\sqrt{976 - 40x + x^2}}$$

Solve $\dfrac{dC}{dx} = 0$:

$$6 + \frac{5(-40 + 2x)}{\sqrt{976 - 40x + x^2}} = 0$$

$$\frac{5(-40 + 2x)}{\sqrt{976 - 40x + x^2}} = -6$$

$$-200 + 10x = -6\sqrt{976 - 40x + x^2}$$

$$40000 - 4000x + 100x^2 = 36x^2 - 1440x + 35136$$

$$64x^2 - 2560x + 4864 = 0$$

$$x^2 - 40x + 76 = 0 \Rightarrow x = 2, \ x = 38.$$

But $x < 20 \Rightarrow x \neq 38$ and $\left.\dfrac{d^2C}{dx^2}\right|_{x=2} > 0.$

Therefore, the value of x that minimizes the cost of installing the cable is $x = 2$ meters and the minimum cost is $C = \$312$.

29. Distance $= \sqrt{(x - 2)^2 + y^2}$

By the hint we minimize

$$D = (x - 2)^2 + y^2 = (x - 2)^2 + x, \text{ since}$$

$$y = \sqrt{x}\ .$$

$$\frac{dD}{dx} = 2(x - 2) + 1$$

Set $\dfrac{dD}{dx} = 0$ to give: $2x = 3$, or

$$x = \frac{3}{2}, \ y = \sqrt{\frac{3}{2}}\ . \text{ So the point is } \left(\frac{3}{2}, \sqrt{\frac{3}{2}}\right).$$

31. Distance $= \sqrt{x^2 + y^2} = \sqrt{x^2 + (-2x+5)^2}$
$= \sqrt{5x^2 - 20x + 25}$

The distance has its smallest value when
$5x^2 - 20x + 25$ does, so we minimize
$D(x) = 5x^2 - 20x + 25 \Rightarrow D'(x) = 10x - 20$
Now set $D'(x) = 0$ and solve for x:
$10x - 20 = 0 \Rightarrow x = 2$
$y = -2(2) + 5 = 1$
The point is (2, 1).

2.6 Further Optimization Problems

1. a. At any given time during the order-reorder period, the inventory is between 180 pounds and 0 pounds. The average is
$\dfrac{180}{2} = 90$ pounds.

b. The maximum is 180 pounds.

c. The number of orders placed during the year can be found by counting the peaks in the figure.

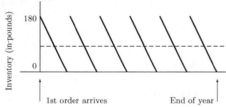

There were 6 orders placed during the year.

d. There were 180 pounds of cherries sold in each order-reorder period, and there were 6-order-reorder periods in the year. So there were $6 \cdot 180 = 1080$ pounds sold in one year.

3. a. The order cost is $16r$, and the carrying cost is $4 \cdot \dfrac{x}{2} = 2x$. The inventory cost C is
$C = 2x + 16r$.

b. The order quantity multiplied by the number of orders per year gives the total number of packages ordered per year. The constraint function is then $rx = 800$.

c. Solving the constraint function for r gives $r = \dfrac{800}{x}$. Substituting into the cost equation yields $C(x) = 2x + \dfrac{12,800}{x}$.

$C'(x) = 2 - \dfrac{12,800}{x^2} \Rightarrow 2 - \dfrac{12,800}{x^2} = 0 \Rightarrow$
$x^2 = \dfrac{12,800}{2} = 6400 \Rightarrow x = 80, \ r = 10$
The minimum inventory cost is
$C(80) = \$320$.

5. Let x be the order quantity and r the number of orders placed in the year. Then the inventory cost is $C = 80r + 5x$. The constraint is
$rx = 10,000$, so $r = \dfrac{10,000}{x}$ and we can write
$C(x) = \dfrac{800,000}{x} + 5x$.

a. $C(500) = \dfrac{800,000}{500} + 5(500) = \$4100$

b. $C'(x) = -\dfrac{800,000}{x^2} + 5 \Rightarrow$
$-\dfrac{800,000}{x^2} + 5 = 0 \Rightarrow$
$x^2 = \dfrac{800,000}{5} = 160,000 \Rightarrow x = 400$
The minimum value of $C(x)$ occurs at $x = 400$.

7. Let x be the number of microscopes produced in each run and let r be the number of runs. The objective function is
$C = 2500r + 15x + 20\left(\dfrac{x}{2}\right) = 2500r + 25x$.

The constraint is $xr = 1600$, $x = \dfrac{1600}{r}$, so
$C(r) = 2500r + \dfrac{40,000}{r}$.
$C'(r) = 2500 - \dfrac{40,000}{r^2} \Rightarrow$
$2500 - \dfrac{40,000}{r^2} = 0 \Rightarrow r^2 = \dfrac{40,000}{2500} \Rightarrow r = 4$
C has a minimum at $r = 4$. There should be 4 production runs.

9. The inventory cost is $C = hr + s\left(\dfrac{x}{2}\right)$ where r is the number of orders placed and x is the order size. The constraint is $rx = Q$, so $r = \dfrac{Q}{x}$ and we can write $C(x) = \dfrac{hQ}{x} + \dfrac{sx}{2}$.

$C'(x) = \dfrac{-hQ}{x^2} + \dfrac{s}{2}$. Setting $C'(x) = 0$ gives

$\dfrac{-hQ}{x^2} + \dfrac{s}{2} = 0$, $x^2 = \dfrac{2hQ}{s}$, $x = \pm\sqrt{\dfrac{2hQ}{s}}$. The

positive value $\sqrt{\dfrac{2hQ}{s}}$ gives the minimum

value for $C(x)$ for $x > 0$.

11.

The objective is $A = (x + 100)w$ and the constraint is
$x + (x + 100) + 2w = 2x + 2w + 100 = 400$; or
$x + w = 150$, $w = 150 - x$.
$A(x) = (x + 100)(150 - x)$
$\qquad = -x^2 + 50x + 15,000$
$A'(x) = -2x + 50$, $A'(25) = 0$

The maximum value of A occurs at $x = 25$.
Thus the optimal values are $x = 25$ ft,
$w = 150 - 25 = 125$ ft.

13.

The objective is $F = 2x + 3w$, and the

constraint is $xw = 54$, or $w = \dfrac{54}{x}$, so

$F(x) = 2x + \dfrac{162}{x}$,

$F'(x) = 2 - \dfrac{162}{x^2} \Rightarrow 2 - \dfrac{162}{x^2} = 0 \Rightarrow$

$x^2 = \dfrac{162}{2} \Rightarrow x = 9$

The minimum value of F for $x > 0$ is $x = 9$.
The optimal dimensions are thus $x = 9$ m,
$w = 6$ m.

15. **a.** $(0, 1000)$, $(5, 1500) \Rightarrow$

$m = \dfrac{1500 - 1000}{5 - 0} = 100$.

$y - 1500 = 100(x - 5)$; $y = 100x + 1000 \Rightarrow$

$A(x) = 100x + 1000$.

b. Let x be the discount per pizza. Then, for $0 \le x \le 18$,
revenue $= R(x) = (100x + 1000)(18 - x)$
$\qquad\qquad = 18000 + 800x - 100x^2$
$R'(x) = 800 - 200x \Rightarrow$
$800 - 200x = 0 \Rightarrow x = 4$
Therefore, revenue is maximized when the discount is $x = \$4$.

c. Let each pizza cost $\$9$ and let x be the discount per pizza. Then
$A(x) = 100x + 1000$ and, for $0 \le x \le 9$,
revenue $= R(x) = (100x + 1000)(9 - x)$.

$R(x) = 9000 - 100x - 100x^2$
$R'(x) = -100 - 200x \Rightarrow$
$-100 - 200x = 0 \Rightarrow x = -.5$
In this case, revenue is maximized when the discount is $x = -\$.50$. Since $0 \le x \le 9$, the revenue is maximized when $x = 0$.

17. Let x be the length and width of the base and let y be the height of the shed. The objective is
$C = 4x^2 + 2x^2 + 4 \cdot 2.5xy = 6x^2 + 10xy$. The

constraint is $x^2 y = 150 \Rightarrow y = \dfrac{150}{x^2}$.

$C(x) = 6x^2 + \dfrac{1500}{x}$, $C'(x) = 12x - \dfrac{1500}{x^2}$

$C'(x) = 0 \Rightarrow 12x - \dfrac{1500}{x^2} = 0 \Rightarrow x = 5$

The optimal dimensions are 5 ft $\times$ 5 ft $\times$ 6 ft.

19. Let x be the length of the square end and let h be the other dimension. The objective is
$V = x^2 h$ and the constraint is $2x + h = 120 \Rightarrow$
$h = 120 - 2x$.
$V(x) = 120x^2 - 2x^3$, $V'(x) = 240x - 6x^2 \Rightarrow$
$V'(x) = 0 \Rightarrow 240x - 6x^2 = 0 \Rightarrow$
$6x(40 - x) = 0 \Rightarrow x = 0$ or $x = 40$.

The maximum value of V for $x > 0$ occurs at
$x = 40$ cm, $h = 40$ cm.
The optimal dimensions are 40 cm $\times$ 40 cm $\times$ 40 cm.

21.

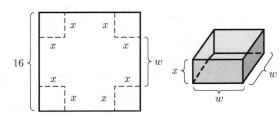

The objective equation is $V = w^2 x$ and the constraint is $w + 2x = 16 \Rightarrow w = 16 - 2x$.

$V(x) = (16 - 2x)^2 x = 4x^3 - 64x^2 + 256x$

$V'(x) = 12x^2 - 128x + 256$

$V'(x) = 0 \Rightarrow 12x^2 - 128x + 256 = 0 \Rightarrow$

$4(x - 8)(3x - 8) = 0 \Rightarrow x = 8 \text{ or } x = \dfrac{3}{8}$

$V''\left(\dfrac{8}{3}\right) < 0, V''(8) > 0$

The maximum value of V for x between 0 and 8 occurs at $x = \dfrac{8}{3}$ in.

23. We want to find the maximum value of $f'(t)$.

$f'(t) = \dfrac{10}{(t + 10)^2} - \dfrac{200}{(t + 10)^3}$;

$f''(t) = \dfrac{-20}{(t + 10)^3} + \dfrac{600}{(t + 10)^4}$. Setting $f''(t) = 0$ gives

$\dfrac{20}{(t + 10)^3} = \dfrac{600}{(t + 10)^4} \Rightarrow 20 = \dfrac{600}{(t + 10)} \Rightarrow t = 20$.

$f'''(t) = \dfrac{60}{(t + 10)^4} - \dfrac{2400}{(t + 10)^5}$; $f'''(20) < 0$, so $t = 20$ is the maximum value of $f'(t)$. Oxygen content is increasing fastest after 20 days.

25. Let (x, y) be the top right-hand corner of the window. The objective is $A = 2xy$ and the constraint is $y = 9 - x^2$. Thus,

$A(x) = 2x(9 - x^2) = 18x - 2x^3$,

$A'(x) = 18 - 6x^2$

$A'(x) = 0 \Rightarrow 18 - 6x^2 = 0 \Rightarrow x = \sqrt{3}$.

The maximum value of A for $x > 0$ occurs at $x = \sqrt{3}$. Thus, the window should be 6 units high and $2\sqrt{3}$ units wide.

27.

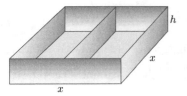

$A = x^2 + 5xh$ (Area: where x is the length of the square base and h is the height.)

$V = x^2 h = 400 \Rightarrow h = \dfrac{400}{x^2}$, so

$A = x^2 + \dfrac{2000}{x}$, and $\dfrac{dA}{dx} = 2x - \dfrac{2000}{x^2}$.

Setting $\dfrac{dA}{dx} = 0$ gives $2x^3 = 2000$ or $x = 10$ which in turn yields $h = 4$ in. The dimensions should be 10 in. $\times$ 10 in. $\times$ 4 in.

29.

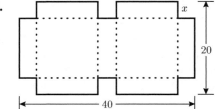

Let V = volume of box, and let l and w represent the dimensions of the base of the box.

Objective: $V = lwx$

Constraints: $l = \dfrac{40 - 3x}{2}$, $w = 20 - 2x$

Substituting, the volume of the box is given by

$V = \left(\dfrac{40 - 3x}{2}\right)(20 - 2x)x = 3x^3 - 70x^2 + 400x$.

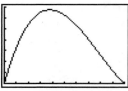

[0, 10] by [0, 700]

Since we require the dimensions of the box to be positive, the appropriate domain is $0 < x < 10$. Using graphing calculator techniques, the maximum function value on this domain occurs at $x \approx 3.7716$.

(continued on next page)

(continued)

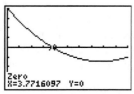

[0, 10] by [−400, 400]

To confirm this, use the calculator's numerical differentiation capability or the function

$$\frac{dV}{dx} = 9x^2 - 140x + 400 \text{ to graph the}$$

derivative, and observe that the solution of

$$\frac{dV}{dx} = 0 \text{ is } x \approx 3.7716.$$

The maximum volume occurs when $x \approx 3.77$ cm.

2.7 Applications of Derivatives to Business and Economics

1. The marginal cost function is

$$M(x) = C'(x) = 3x^2 - 12x + 13 .$$

$$M'(x) = 6x - 12$$

$$M'(x) = 0 \Rightarrow 6x - 12 = 0 \Rightarrow x = 2$$

The minimum value of $M(x)$ occurs at $x = 2$.
The minimum marginal cost is $M(2) = \$1$.

3. $R(x) = 200 - \dfrac{1600}{x+8} - x, \ R'(x) = \dfrac{1600}{(x+8)^2} - 1,$

$$R'(x) = 0 \Rightarrow \frac{1600}{(x+8)^2} - 1 = 0 \Rightarrow$$

$$1600 = (x+8)^2 \Rightarrow 40 = x + 8 \Rightarrow x = 32$$

The maximum value of $R(x)$ occurs at $x = 32$.

5. The profit function is

$$P(x) = R(x) - C(x)$$

$$= 28x - (x^3 - 6x^2 + 13x + 15)$$

$$= -x^3 + 6x^2 + 15x - 15$$

$$P'(x) = -3x^2 + 12x + 15$$

$$P'(x) = 0 \Rightarrow -3x^2 + 12x + 15 = 0 \Rightarrow ,$$

$$-3(x-5)(x+1) = 0 \Rightarrow x = 5 \text{ or } x = -1$$

The maximum value of $P(x)$ for $x > 0$ occurs at $x = 5$.

7. The revenue function is

$$R(x) = x\left(\frac{1}{12}x^2 - 10x + 300\right)$$

$$= \frac{1}{12}x^3 - 10x^2 + 300x$$

$$R'(x) = \frac{1}{4}x^2 - 20x + 300 \Rightarrow R'(x) = 0 \Rightarrow$$

$$\frac{1}{4}x^2 - 20x + 300 = 0 \Rightarrow x^2 - 80x + 1200 = 0 \Rightarrow$$

$$(x - 60)(x - 20) = 0 \Rightarrow x = 60 \text{ or } x = 20$$

$$R''(x) = \frac{1}{2}x - 20$$

$$R''(20) < 0, R''(60) > 0$$

The maximum value of $R(x)$ occurs at $x = 20$.

The corresponding price is $\$133\dfrac{1}{3}$ or $\$133.33$.

9. The revenue function is

$$R(x) = x(256 - 50x) = 256x - 50x^2 . \text{ Thus, the}$$

profit function is

$$P(x) = R(x) - C(x) = 256x - 50x^2 - 182 - 56x$$

$$= -50x^2 + 200x - 182$$

$$P'(x) = -100x + 200$$

$$P'(x) = 0 \Rightarrow -100x + 200 = 0 \Rightarrow x = 2$$

The maximum profit occurs at $x = 2$ (million tons). The corresponding price is $256 - 50(2) = 156$ dollars per ton.

11. a. Let p stand for the price of hamburgers and let x be the quantity. Using the point-slope equation,

$$p - 4 = \frac{4.4 - 4}{8000 - 10,000}(x - 10,000) \text{ or}$$

$p = -.0002x + 6$. Thus, the revenue function is

$$R(x) = x(-.0002x + 6) = -.0002x^2 + 6x.$$

$$R'(x) = -.0004x + 6$$

$$R'(x) = 0 \Rightarrow -.0004x + 6 = 0 \Rightarrow$$

$$x = 15,000$$

The maximum value of $R(x)$ occurs at $x = 15,000$. The optimal price is thus $-.0002(15,000) + 6 = \$3.00$

b. The cost function is $C(x) = 1000 + .6x$, so the profit function is $P(x) = R(x) - C(x)$

$$P(x) = R(x) - C(x)$$

$$= -.0002x^2 + 6x - (1000 + .6x)$$

$$= -.0002x^2 + 5.4x - 1000$$

$$P'(x) = -.0004x + 5.4$$

$$P'(x) = 0 \Rightarrow -.0004x + 5.4 = 0 \Rightarrow$$

$$x = 13,500$$

The maximum value of $P(x)$ occurs at $x = 13,500$. The optimal price is $-.0002(13,500) + 6 = \$3.30$.

13. Let x be the number of prints the artist sells.
Then his revenue = [price] · [quantity].

$$\begin{cases} (400 - 5(x - 50))x & \text{if } x > 50 \\ 400x & \text{if } x \leq 50 \end{cases}$$

For $x > 50$, $r(x) = -5x^2 + 650x$,

$r'(x) = -10x + 650$
$r'(x) = 0 \Rightarrow -10x + 650 = 0 \Rightarrow x = 65$

The maximum value of $r(x)$ occurs at $x = 65$.
The artist should sell 65 prints.

15. Let $P(x)$ be the profit from x tables.
Then $P(x) = (10 - (x - 12)(.5)x = -.5x^2 + 16x$
For $x \geq 12$, $P'(x) = 16 - x$
$P'(x) = 0 \Rightarrow 16 - x = 0 \Rightarrow x = 16$
The maximum value of $P(x)$ occurs at $x = 16$.
The cafe should provide 16 tables.

17. a. $R(x) = x\left(60 - 10^{-5}x\right) = 60x - 10^{-5}x^2$; so
the profit function is $P(x) = R(x) - C(x)$
$P(x) = R(x) - C(x)$
$\quad = \left(60x - 10^{-5}x^2\right) - \left(7 \cdot 10^6 + 30x\right)$
$\quad = -10^{-5}x^2 + 30x - 7 \cdot 10^6$
$P'(x) = -2 \cdot 10^{-5}x + 30$
$P'(x) = 0 \Rightarrow -2 \cdot 10^{-5}x + 30 = 0 \Rightarrow$
$x = 15 \cdot 10^5$
The maximum value of $P(x)$ occurs at
$x = 1.5 \cdot 10^5$ (thousand kilowatt-hours). The
corresponding price is
$p = 60 - 10^{-5}\left(15 \cdot 10^5\right) = 45.$
This represents $45/thousand kilowatt-hours.

b. The new profit function is
$P_1(x) = R(x) - C_1(x)$
$\quad = 60x - 10^{-5}x^2 - 7 \cdot 10^6 - 40x$
$\quad = -10^{-5}x^2 + 20x - 7 \cdot 10^6$
$P_1'(x) = -2 \cdot 10^{-5}x + 20$
$P_1'(x) = 0 \Rightarrow -2 \cdot 10^{-5}x + 20 = 0 \Rightarrow x = 10^6$.
The maximum value of $P_1(x)$ occurs at
$x = 10^6$ (thousand kilowatt-hours). The
corresponding price is
$p = 60 - 10^{-5}(10^6) = 50$, representing
$50/thousand kilowatt-hours.

The maximum profit will be obtained by
charging $50/thousand kilowatt-hours. Since
this represents an increase of only
$5/thousand kilowatt-hours over the answer
to part (a), the utility company should not
pass all of the increase on to consumers.

19. Let r be the percentage rate of interest ($r = 4$
represents a 4% interest rate).
Total deposit is $1,000,000r$. Total interest
paid out in one year is $10,000r^2$. Total
interest received on the loans of $1,000,000r$ is
$100,000r$.

$P = 100,000r - 10,000r^2$

$\dfrac{dP}{dr} = 100,000 - 20,000r$

Set $\dfrac{dP}{dr} = 0$ and solve for r:
$100,000 - 20,000r = 0 \Rightarrow r = 5$
An interest rate of 5% generates the greatest
profit.

21. a. Since $R(40) = 75$, the revenue is $75,000.

b. Since $R'(17.5) \approx 3.2$, the marginal
revenue is about $3200 per unit.

c. Since the solution of $R(x) = 45$ is $x = 15$,
the production level in 15 units.

d. Since the solution of $R'(x) = .8$ is
$x = 32.5$, the production level is 32.5
units.

e. Looking at the graph of $y = R(x)$, the
revenue appears to be greatest at $x \approx 35$.
To confirm, observe that the graph of
$y = R'(x)$ crosses the x-axis at $x = 35$.
The revenue is greatest at a production
level of 35 units.

Chapter 2 Review Exercises

1. a. The graph of $f(x)$ is increasing when
$f'(x) > 0$: $-3 < x < 1, x > 5$.
The graph of $f(x)$ is decreasing when
$f'(x) < 0$: $x < -3, 1 < x < 5$.

b. The graph of $f(x)$ is concave up when
$f'(x)$ is increasing: $x < -1, x > 3$.
The graph of $f(x)$ is concave down when
$f'(x)$ is decreasing: $-1 < x < 3$.

2. a. $f(3) = 2$

   b. The tangent line has slope $\frac{1}{2}$, so
   $$f'(3) = \frac{1}{2}.$$

   c. Since the point $(3, 2)$ appears to be an inflection point, $f''(3) = 0$.

3.

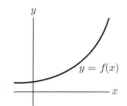

4.

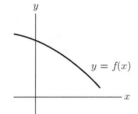

5.

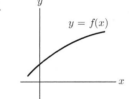

6.

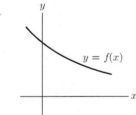

7. (d), (e)   **8.** (b)

9. (c), (d)   **10.** (a)

11. (e)   **12.** (b)

13. Graph goes through $(1, 2)$, increasing at $x = 1$.

14. Graph goes through $(1, 5)$, decreasing at $x = 1$.

15. Increasing and concave up at $x = 3$.

16. Decreasing and concave down at $x = 2$.

17. $(10, 2)$ is a relative minimum point.

18. Graph goes through $(4, -2)$, increasing and concave down at $x = 4$.

19. Graph goes through $(5, -1)$, decreasing at $x = 5$.

20. $(0, 0)$ is a relative minimum point.

21. a. $f(t) = 1$ at $t = 2$, after 2 hours.

   b. $f(5) = .8$

   c. $f'(t) = -.08$ at $t = 3$, after 3 hours.

   d. Since $f'(8) = -.02$, the rate of change is $-.02$ unit per hour.

22. a. Since $f(50) = 400$, the amount of energy produced was 400 trillion kilowatt-hours.

   b. Since $f'(50) = 35$, the rate of change was 35 trillion kilowatt-hours per year.

   c. Since $f(t) = 3000$ at $t = 95$, the production level reached 300 trillion kilowatt-hours in 1995.

   d. Since $f'(t) = 10$ at $t = 35$, the production level was rising at the rate of 10 trillion kilowatt-hours per year in 1935.

   e. Looking at the graph of $y = f'(t)$, the value of $f'(t)$ appears to be greatest at $t = 70$. To confirm, observe that the graph of $y = f''(t)$ crosses the t-axis at $t = 70$. Energy production was growing at the greatest rate in 1970. Since $f(70) = 1600$, the production level at that time was 1600 trillion kilowatt-hours.

23. $y = 3 - x^2$
$y' = -2x$
$y'' = -2$
$y' = 0$ if $x = 0$

If $x = 0$, $y = 3$, so $(0, 3)$ is a critical point and the y-intercept. $y'' < 0$, so $(0, 3)$ is a relative maximum.

$0 = 3 - x^2 \Rightarrow x = \pm\sqrt{3}$, so the x-intercepts are $\left(\pm\sqrt{3}, 0\right)$.

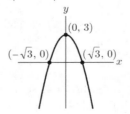

24. $y = 7 + 6x - x^2$
$y' = 6 - 2x$
$y'' = -2$
$y' = 0$ if $x = 3$
If $x = 3$, $y = 16$, so $(3, 16)$ is a critical point.
$y'' < 0$, so $(3, 16)$ is a relative maximum.
$0 = 7 + 6x - x^2 \Rightarrow x = -1$ or $x = 7$, so the
x-intercepts are $(-1, 0)$ and $(7, 0)$.
The y-intercept is $(0, 7)$.

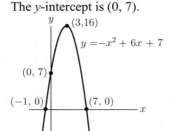

25. $y = x^2 + 3x - 10$
$y' = 2x + 3$
$y'' = 2$
$y' = 0$ if $x = -\dfrac{3}{2}$
If $x = -\dfrac{3}{2}$, $y = -\dfrac{49}{4}$ so $\left(-\dfrac{3}{2}, -\dfrac{49}{4}\right)$ is a critical
point. $y'' > 0$, so $\left(-\dfrac{3}{2}, -\dfrac{49}{4}\right)$ is a relative
minimum.
$x^2 + 3x - 10 = 0 \Rightarrow x = -5$ or $x = 2$, so the
x-intercepts are $(-5, 0)$ and $(2, 0)$.
The y-intercept is $(0, -10)$.

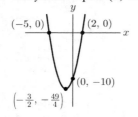

26. $y = 4 + 3x - x^2$
$y' = 3 - 2x$
$y'' = -2$
$y' = 0$ if $x = \dfrac{3}{2}$
If $x = \dfrac{3}{2}$, $y = -\dfrac{49}{4}$ so $\left(\dfrac{3}{2}, \dfrac{25}{4}\right)$ is a critical
point. $y'' < 0$, so $\left(\dfrac{3}{2}, \dfrac{25}{4}\right)$ is a relative
maximum.
$0 = 4 + 3x - x^2 \Rightarrow x = -1$ or $x = 4$, so the
x-intercepts are $(-1, 0)$ and $(4, 0)$.
The y-intercept is $(0, 4)$.

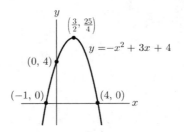

27. $y = -2x^2 + 10x - 10$
$y' = -4x + 10$
$y'' = -4$
$y' = 0$ if $x = \dfrac{5}{2}$
If $x = \dfrac{5}{2}$, $y = \dfrac{5}{2}$ so $\left(\dfrac{5}{2}, \dfrac{5}{2}\right)$ is a critical point.
$y'' < 0$, so $\left(\dfrac{5}{2}, \dfrac{5}{2}\right)$ is a relative maximum.
$0 = -2x^2 + 10x - 10 \Rightarrow x = \dfrac{5 \pm \sqrt{5}}{2}$, so the
x-intercepts are $\left(\dfrac{5-\sqrt{5}}{2}, 0\right)$ and $\left(\dfrac{5+\sqrt{5}}{2}, 0\right)$.
The y-intercept is $(0, -10)$.

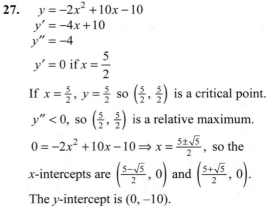

28. $y = x^2 - 9x + 19$
$y' = 2x - 9$
$y'' = 2$
$y' = 0$ if $x = \dfrac{9}{2}$
If $x = \dfrac{9}{2}$, $y = -\dfrac{5}{4}$ so $\left(\dfrac{9}{2}, -\dfrac{5}{4}\right)$ is a critical
point. $y'' > 0$, so $\left(\dfrac{9}{2}, -\dfrac{5}{4}\right)$ is a relative
minimum.
$0 = x^2 - 9x + 19 \Rightarrow x = \dfrac{9 \pm \sqrt{5}}{2}$, so the
x-intercepts are $\left(\dfrac{9-\sqrt{5}}{2}, 0\right)$ and $\left(\dfrac{9+\sqrt{5}}{2}, 0\right)$.
The y-intercept is $(0, 19)$.

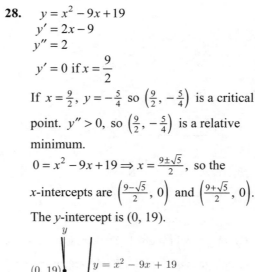

29. $y = x^2 + 3x + 2$
$y' = 2x + 3$
$y'' = 2$

$y' = 0$ if $x = -\dfrac{3}{2}$

If $x = -\frac{3}{2}$, $y = -\frac{1}{4}$ so $\left(-\frac{3}{2}, -\frac{1}{4}\right)$ is a critical point. $y'' > 0$, so $\left(-\frac{3}{2}, -\frac{1}{4}\right)$ is a relative minimum.

$0 = x^2 + 3x + 2 \Rightarrow x = -2$ or $x = -1$, so the x-intercepts are $(-2, 0)$ and $(-1, 0)$.
The y-intercept is $(0, 2)$.

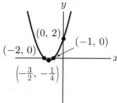

30. $y = -x^2 + 8x - 13$
$y' = -2x + 8$
$y'' = -2$
$y' = 0$ if $x = 4$

If $x = 4$, $y = 3$, so $(4, 3)$ is a critical point.
$y'' < 0$, so $(4, 3)$ is a relative maximum.

$0 = -x^2 + 8x - 13 \Rightarrow x = 4 \pm \sqrt{3}$, so the x-intercepts are $\left(4 - \sqrt{3}, 0\right)$ and $\left(4 + \sqrt{3}, 0\right)$.
The y-intercept is $(0, -13)$.

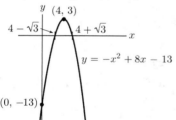

31. $y = -x^2 + 20x - 90$
$y' = -2x + 20$
$y'' = -2$
$y' = 0$ if $x = 10$

If $x = 10$, $y = 10$, so $(10, 10)$ is a critical point.
$y'' < 0$, so $(10, 10)$ is a relative maximum.

$0 = -x^2 + 20x - 90 \Rightarrow x = 10 \pm \sqrt{10}$, so the x-intercepts are $\left(10 - \sqrt{10}, 0\right)$ and $\left(10 + \sqrt{10}, 0\right)$.
The y-intercept is $(0, -13)$.

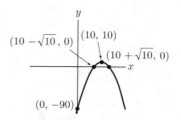

32. $y = 2x^2 + x - 1$
$y' = 4x + 1$
$y'' = 4$

$y' = 0$ if $x = -\dfrac{1}{4}$

If $x = -\frac{1}{4}$, $y = -\frac{9}{8}$ so $\left(-\frac{1}{4}, -\frac{9}{8}\right)$ is a critical point. $y'' > 0$, so $\left(-\frac{1}{4}, -\frac{9}{8}\right)$ is a relative minimum.

$0 = 2x^2 + x - 1 \Rightarrow x = -1$ or $x = \frac{1}{2}$, so the x-intercepts are $(-1, 0)$ and $\left(\frac{1}{2}, 0\right)$.
The y-intercept is $(0, -1)$.

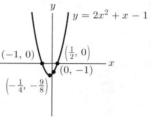

33. $f(x) = 2x^3 + 3x^2 + 1$
$f'(x) = 6x^2 + 6x$
$f''(x) = 12x + 6$
$f'(x) = 0$ if $x = 0$ or $x = -1$
$f(0) = 1 \Rightarrow (0, 1)$ is a critical pt.
$f(-1) = 2 \Rightarrow (-1, 2)$ is a critical pt.
$f''(0) = 6 > 0$, so the graph is concave up at $x = 0$, and $(0, 1)$ is a relative minimum.
$f''(-1) = -6 < 0$, so the graph is concave down at $x = -1$, and $(-1, 2)$ is a relative maximum.
$f''(x) = 0$ when $x = -\dfrac{1}{2}$.
$f\left(-\dfrac{1}{2}\right) = \dfrac{3}{2} \Rightarrow \left(-\dfrac{1}{2}, \dfrac{3}{2}\right)$ is an inflection pt.
The y-intercept is $(0, 1)$.

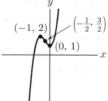

34. $f(x) = x^3 - \dfrac{3}{2}x^2 - 6x$

$f'(x) = 3x^2 - 3x - 6$

$f''(x) = 6x - 3$

$f'(x) = 0$ if $x = -1$ or $x = 2$

$f(-1) = \dfrac{7}{2} \Rightarrow \left(-1, \dfrac{7}{2}\right)$ is a critical pt.

$f(2) = -10 \Rightarrow (2, -10)$ is a critical pt.

$f''(-1) = -9 < 0$, so the graph is concave down

at $x = -1$, and $\left(-1, \frac{7}{2}\right)$ is a relative maximum.

$f''(2) = 9 > 0$, so the graph is concave up at

$x = 2$, and $(2, -10)$ is a relative minimum.

$f''(x) = 0$ when $x = \dfrac{1}{2}$.

$f\left(\dfrac{1}{2}\right) = -\dfrac{13}{4} \Rightarrow \left(\dfrac{1}{2}, -\dfrac{13}{4}\right)$ is an inflection pt.

The y-intercept is $(0, 1)$.

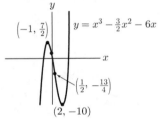

35. $f(x) = x^3 - 3x^2 + 3x - 2$

$f'(x) = 3x^2 - 6x + 3$

$f''(x) = 6x - 6$

To find possible extrema, set $f'(x) = 0$ and

solve for x.

$3x^2 - 6x + 3 = 0 \Rightarrow x = 1$

$f(1) = -1$, so $(1, -1)$ is a critical point.

Since $f'(x) \geq 0$ for all x, the graph is always

increasing, and $(1, -1)$ is neither a relative

maximum nor a relative minimum.

To find possible inflection points, set

$f''(x) = 0$ and solve for x.

$6x - 6 = 0 \Rightarrow x = 1$

Since $f''(x) < 0$ for $x < 1$ (meaning the graph

is concave down) and $f''(x) > 0$ for $x > 1$

(meaning the graph is concave up), the point

$(1, -1)$ is an inflection point. The y-intercept is

$(0, -2)$.

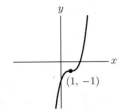

36. $f(x) = 100 + 36x - 6x^2 - x^3$

$f'(x) = 36 - 12x - 3x^2$

$f''(x) = -12 - 6x$

$f'(x) = 0$ if $x = -6$ or $x = 2$

$f(-6) = -116 \Rightarrow (-6, -116)$ is a critical pt.

$f(2) = 140 \Rightarrow (2, 140)$ is a critical pt.

$f''(-6) = 24 > 0$, so the graph is concave up at

$x = -6$, and $(-6, -116)$ is a relative minimum.

$f''(2) = -24 < 0$, so the graph is concave

down at $x = -1$, and $(2, 140)$ is a relative

maximum.

$f''(x) = 0$ when $x = -2$.

$f(-2) = 12 \Rightarrow (-2, 12)$ is an inflection pt.

The y-intercept is $(0, 100)$.

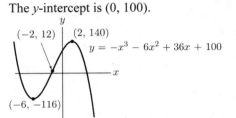

37. $f(x) = \dfrac{11}{3} + 3x - x^2 - \dfrac{1}{3}x^3$

$f'(x) = 3 - 2x - x^2$

$f''(x) = -2 - 2x$

$f'(x) = 0$ if $x = -3$ or $x = 1$

$f(-3) = -\dfrac{16}{3} \Rightarrow \left(-3, -\dfrac{16}{3}\right)$ is a critical pt.

$f(1) = \dfrac{16}{3} \Rightarrow \left(1, \dfrac{16}{3}\right)$ is a critical pt.

$f''(-3) = 4 > 0$, so the graph is concave up at

$x = -3$, and $\left(-3, -\frac{16}{3}\right)$ is a relative minimum.

$f''(1) = -4 < 0$, so the graph is concave down

at $x = 1$, and $\left(1, \frac{16}{3}\right)$ is a relative maximum.

$f''(x) = 0$ when $x = -1$.

$f(-1) = 0 \Rightarrow (-1, 0)$ is an inflection pt.

The y-intercept is $\left(0, \frac{11}{3}.\right)$

(*continued on next page*)

(*continued*)

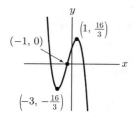

38. $f(x) = x^3 - 3x^2 - 9x + 7$

$f'(x) = 3x^2 - 6x - 9$

$f''(x) = 6x - 6$

$f'(x) = 0$ if $x = -1$ or $x = 3$

$f(-1) = 12 \Rightarrow (-1, 12)$ is a critical pt.

$f(3) = -10 \Rightarrow (3, -20)$ is a critical pt.

$f''(-1) = -12 < 0$, so the graph is concave down at $x = -1$, and $(-1, 12)$ is a relative maximum.

$f''(3) = 12 > 0$, so the graph is concave up at $x = 3$, and $(3, -20)$ is a relative minimum.

$f''(x) = 0$ when $x = 1$.

$f(1) = -4 \Rightarrow (1, -4)$ is an inflection pt.

The y-intercept is $(0, 7)$.

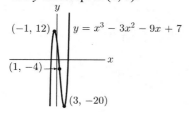

39. $f(x) = -\dfrac{1}{3}x^3 - 2x^2 - 5x$

$f'(x) = -x^2 - 4x - 5$

$f''(x) = -2x - 4$

To find possible extrema, set $f'(x) = 0$ and solve for x.

$-x^2 - 4x - 5 = 0 \Rightarrow$ no real solution

Thus, there are no extrema.

Since $f'(x) \leq 0$ for all x, the graph is always decreasing.

To find possible inflection points, set $f''(x) = 0$ and solve for x.

$-2x - 4 = 0 \Rightarrow x = -2$

$f(-2) = \dfrac{14}{3}$

Since $f''(x) > 0$ for $x < \frac{14}{3}$ (meaning the graph is concave up) and $f''(x) < 0$ for $x > \frac{14}{3}$ (meaning the graph is concave down), the point $\left(-2, \frac{14}{3}\right)$ is an inflection point.

$f(0) = -6$, so the y-intercept is $(0, 0)$.

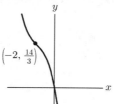

40. $y = x^3 - 6x^2 - 15x + 50$

$y' = 3x^2 - 12x - 15$

$y'' = 6x - 12$

$y' = 0$ if $x = -1$ or $x = 5$

If $x = -1$, $y = 58$. If $x = 5$, $y = -50$. So, $(-1, 58)$ and $(5, -50)$ are critical points.

If $x = -1$, $y'' = -18 < 0$, so the graph is concave down and $(-1, 58)$ is a relative maximum.

If $x = 5$, $y'' = 18 > 0$, so the graph is concave up and $(5, -50)$ is a relative minimum.

$y'' = 0$ when $x = 2$. If $x = 2$, $y = 4$, so $(2, 4)$ is an inflection point. The y-intercept is $(0, 50)$.

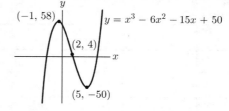

41. $y = x^4 - 2x^2$

$y' = 4x^3 - 4x$

$y'' = 12x^2 - 4$

$y' = 0$ if $x = 0$, $x = -1$, or $x = 1$

If $x = -1$, $y = -1$. If $x = 0$, $y = 0$. If $x = 1$, $y = -1$. So, $(-1, -1)$, $(0, 0)$, and $(1, -1)$ are critical points.

If $x = -1$, $y'' = 8 > 0$, so the graph is concave up and $(-1, -1)$ is a relative minimum.

If $x = 1$, $y = -1$, $y'' = 8 > 0$, so the graph is concave up and $(1, -1)$ is a relative minimum.

If $x = 0$, $y'' = 0$, so we must use the first derivative test. Since $y' < 0$ when $x < 0$ and also when $x > 0$, $(0, 0)$ is a relative maximum.

(*continued on next page*)

(*continued*)

$y'' = 0$ when $x = \pm\frac{1}{\sqrt{3}}$. If $x = -\frac{1}{\sqrt{3}}$, $y = -\frac{5}{9}$,

so $\left(-\frac{1}{\sqrt{3}}, -\frac{5}{9}\right)$ is an inflection point.

If $x = \frac{1}{\sqrt{3}}$, $y = -\frac{5}{9}$, so $\left(\frac{1}{\sqrt{3}}, -\frac{5}{9}\right)$ is an

inflection point. The y-intercept is $(0, 0)$.

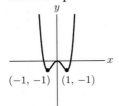

$(-1, -1)$ $(1, -1)$

42. $y = x^4 - 4x^3$

$y' = 4x^3 - 12x^2 = 4x^2(x - 3)$

$y'' = 12x^2 - 24x$

$y' = 0$ if $x = 0$ or $x = 3$

If $x = 0$, $y = 0$. If $x = 0$, $y = 0$. If $x = 3$,
$y = -27$. So, $(0, 0)$, and $(3, -27)$ are critical
points.

If $x = 3$, $y'' = 36 > 0$, so the graph is concave

up and $(3, -27)$ is a relative minimum.

If $x = 0$, $y'' = 0$, so we must use the first

derivative test.

Critical Points, Intervals	$x < 0$	$0 < x < 3$	$3 < x$
$4x^2$	+	+	+
$x - 3$	−	−	+
y'	−	−	+
y	Decreasing on $(-\infty, 0)$	Decreasing on $(0, 3)$	Increasing on $(3, \infty)$

Thus, $(0, 0)$ is neither a relative maximum nor
a relative minimum. It may be an inflection
point. Verify by using the second derivative
test.

$y'' = 0$ when $x = 0$ or $x = 2$.

Critical Points, Intervals	$x < 0$	$0 < x < 2$	$2 < x < 3$	$3 < x$
$12x$	−	+	+	+
$x - 2$	−	−	+	+
y''	+	−	+	+
Concavity	up	down	up	up

If $x = 0$, $y = 0$ so $(0, 0)$ is an inflection point.
If $x = 2$, $y = -16$ so $(2, -16)$ is an inflection
point. The y-intercept is $(0, 0)$.

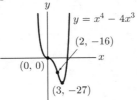

43. $y = \frac{x}{5} + \frac{20}{x} + 3$. $x > 0$

$y' = \frac{1}{5} - \frac{20}{x^2}$

$y'' = \frac{40}{x^3}$

$y' = 0$ if $x = \pm 10$

Note that we need to consider the positive
solution only because the function is defined
only for $x > 0$. When $x = 10$, $y = 7$, and
$y'' = \frac{1}{25} > 0$, so the graph is concave up and
$(10, 7)$ is a relative minimum.

Since y'' can never be zero, there are no

inflection points. The term $\frac{20}{x}$ tells us that the

y-axis is an asymptote. As $x \to \infty$, the graph

approaches $y = \frac{x}{4} + 3$, so this is also an

asymptote of the graph.

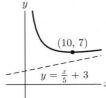

$(10, 7)$

$y = \frac{x}{5} + 3$

44. $y = \frac{1}{2x} + 2x + 1$, $x > 0$

$y' = -\frac{1}{2x^2} + 2$

$y'' = \frac{1}{x^3}$

$y' = 0$ if $x = \pm\frac{1}{2}$

Note that we need to consider the positive
solution only because the function is defined
only for $x > 0$. When $x = \frac{1}{2}$, $y = 3$, and
$y'' = 8 > 0$, so the graph is concave up and
$\left(\frac{1}{2}, 3\right)$ is a relative minimum.

(*continued on next page*)

(*continued*)

Since y'' can never be zero, there are no inflection points. The term $\dfrac{1}{2x}$ tells us that the y-axis is an asymptote. As $x \to \infty$, the graph approaches $y = 2x + 1$, so this is also an asymptote of the graph.

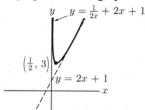

45. $f'(x) = \dfrac{3}{2}\left(x^2 + 2\right)^{1/2}(2x) = 3x\left(x^2 + 2\right)^{1/2}$

Since $f'(0) = 0$, f has a possible extreme value at $x = 0$.

46. $f'(x) = \dfrac{3}{2}\left(2x^2 + 3\right)^{1/2}(4x) = 6x\left(2x^2 + 3\right)^{1/2}$

$2x^2 + 3 > 0$ for all x, so the sign of $f'(x)$ is determined by the sign of $4x$. Therefore, $f'(x) > 0$ if $x > 0$, $f'(x) < 0$ if $x < 0$. This means that $f(x)$ is decreasing for $x < 0$ and increasing for $x > 0$.

47. $f''(x) = \dfrac{-2x}{(1+x^2)^2}$, so $f''(0) = 0$. Since $f'(x) > 0$ for all x, it follows that 0 must be an inflection point.

48. $f''(x) = \dfrac{1}{2}\left(5x^2 + 1\right)^{-1/2}(10x) = \dfrac{5x}{\sqrt{5x^2 + 1}}$, so $f''(0) = 0$. Since $f'(x) > 0$ for all x, $f''(x)$ is positive for $x > 0$ and negative for $x < 0$, and it follows that 0 must be an inflection point.

49. A – c, B – e, C – f, D – b, E – a, F – d

50. A – c, B – e, C – f, D – b, E – a, F – d

51. a. The number of people living between $10 + h$ and 10 miles from the center of the city.

   b. If so, $f(x)$ would be decreasing at $x = 10$, which is not possible.

52. $f(x) = \dfrac{1}{4}x^2 - x + 2 \ (0 \le x \le 8)$

$f'(x) = \dfrac{1}{2}x - 1$

$f'(x) = 0 \Rightarrow \dfrac{1}{2}x - 1 = 0 \Rightarrow x = 2$

$f''(x) = \dfrac{1}{2}$

Since $f'(2)$ is a relative minimum, the maximum value of $f(x)$ must occur at one of the endpoints. $f(0) = 2$, $f(8) = 10$. 10 is the maximum value, attained $x = 8$.

53. $f(x) = 2 - 6x - x^2 \ (0 \le x \le 5)$

$f'(x) = -6 - 2x$

Since $f'(x) < 0$ for all $x > 0$, $f(x)$ is decreasing on the interval [0, 5]. Thus, the maximum value occurs at $x = 0$. The maximum value is $f(0) = 2$.

54. $g(t) = t^2 - 6t + 9 \ (1 \le t \le 6)$

$g'(t) = 2t - 6$

$g'(t) = 0 \Rightarrow 2t - 6 = 0 \Rightarrow t = 3$

$g''(t) = 2$

The minimum value of $g(t)$ is $g(3) = 0$.

55. Let x be the width and h be the height. The objective is $S = 2xh + 4x + 8h$ and the constraint is $4xh = 200 \Rightarrow h = \dfrac{50}{x}$.

Thus,

$S(x) = 2x\left(\dfrac{50}{x}\right) + 4x + \dfrac{400}{x} = 100 + 4x + \dfrac{400}{x}$.

$S'(x) = 4 - \dfrac{400}{x^2}$

$S'(x) = 0 \Rightarrow 4 - \dfrac{400}{x^2} = 0 \Rightarrow x = 10$

$h = \dfrac{50}{x} = \dfrac{50}{10} = 5$

The minimum value of $S(x)$ for $x > 0$ occurs at $x = 10$. Thus, the dimensions of the box should be 10 ft × 4 ft × 5 ft.

56. Let x be the length of the base of the box and let y be the other dimension. The objective is
$V = x^2 y$ and the constraint is
$$3x^2 + x^2 + 4xy = 48$$
$$y = \frac{48 - 4x^2}{4x} = \frac{12 - x^2}{x}$$
$$V(x) = x^2 \cdot \frac{12 - x^2}{x} = 12x - x^3$$
$$V'(x) = 12 - 3x^2$$
$$V'(x) = 0 \Rightarrow 12 - 3x^2 = 0 \Rightarrow x = 2$$
$$V''(x) = -6x; \ V''(2) < 0.$$
The maximum value for x for $x > 0$ occurs at $x = 2$. The optimal dimensions are thus 2 ft $\times$ 2 ft $\times$ 4 ft.

57. Let x be the number of inches turned up on each side of the gutter. The objective is
$A(x) = (30 - 2x)x$
(A is the cross-sectional area of the gutter—maximizing this will maximize the volume).
$$A'(x) = 30 - 4x$$
$$A'(x) = 0 \Rightarrow 30 - 4x = 0 \Rightarrow x = \frac{15}{2}$$
$$A''(x) = -4, \ A''\left(\frac{15}{2}\right) < 0$$
$x = \frac{15}{2}$ inches gives the maximum value for A.

58. Let x be the number of trees planted. The objective is $f(x) = \left(25 - \frac{1}{2}(x - 40)\right)x \ (x \geq$
40). $f(x) = 45x - \frac{1}{2}x^2$
$$f'(x) = 45 - x$$
$$f'(x) = 0 \Rightarrow 45 - x = 0 \Rightarrow x = 45$$
$$f''(x) = -1; \ f''(45) < 0.$$
The maximum value of $f(x)$ occurs at $x = 45$. Thus, 45 trees should be planted.

59. Let r be the number of production runs and let x be the lot size. Then the objective is
$C = 1000r + .5\left(\frac{x}{2}\right)$ and the constraint is
$$rx = 400,000 \Rightarrow r = \frac{400,000}{x}$$
so $C(x) = \frac{4 \cdot 10^8}{x} + \frac{x}{4}$
$$C'(x) = \frac{-4 \cdot 10^8}{x^2} + \frac{1}{4}$$
$$C'(x) = 0 \Rightarrow \frac{-4 \cdot 10^8}{x^2} + \frac{1}{4} = 0 \Rightarrow x = 4 \cdot 10^4$$
$$C''(x) = \frac{8 \cdot 10^8}{x^3}; \ C''(4 \cdot 10^4) > 0.$$
The minimum value of $C(x)$ for $x > 0$ occurs at $x = 4 \cdot 10^4 = 40,000$. Thus the economic lot size is 40,000 books/run.

60. The revenue function is
$R(x) = (150 - .02x)x = 150x - .02x^2$.
Thus, the profit function is
$$P(x) = (150x - .02x^2) - (10x + 300)$$
$$= -.02x^2 + 140x - 300$$
$$P'(x) = -.04x + 140$$
$$P'(x) = 0 \Rightarrow -.04x + 140 = 0 \Rightarrow x = 3500$$
$$P''(x) = -.04; \ P''(3500) < 0.$$
The maximum value of $P(x)$ occurs at $x = 3500$.

61.

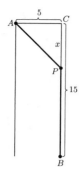

The distance from point A to point P is $\sqrt{25 + x^2}$ and the distance from point P to point B is $15 - x$. The time it takes to travel from point A to point P is $\dfrac{\sqrt{25 + x^2}}{8}$ and the time it takes to travel from point P to point B is $\dfrac{15 - x}{17}$. Therefore, the total trip takes $T(x) = \dfrac{1}{8}\left(25 + x^2\right)^{1/2} + \dfrac{1}{17}(15 - x)$ hours.

$$T'(x) = \frac{1}{16}(25 + x^2)^{-1/2}(2x) - \frac{1}{17}$$

$$T'(x) = 0 \Rightarrow \frac{1}{16}(25 + x^2)^{-1/2}(2x) - \frac{1}{17} = 0 \Rightarrow x = \frac{8}{3}$$

$$T''(x) = \frac{1}{8}(25 + x^2)^{-1/2} - \frac{1}{8}x^2(25 + x^2)^{-3/2}$$

$$T''\left(\frac{8}{3}\right) = \frac{1}{8}\left(25 + \left(\frac{8}{3}\right)^2\right)^{-1/2} - \frac{1}{8}\left(\frac{8}{3}\right)^2\left(25 + \left(\frac{8}{3}\right)^2\right)^{-3/2} = \frac{675}{39304} > 0$$

The minimum value for $T(x)$ occurs at $x = \frac{8}{3}$. Thus, Jane should drive from point A to point P, $\frac{8}{3}$ miles from point C, then down to point B.

62. Let $12 \le x \le 25$ be the size of the tour group. Then, the revenue generated from a group of x people, $R(x)$, is $R(x) = \left[800 - 20(x - 12)\right]x$. To maximize revenue:

$R'(x) = 1040 - 40x$
$R'(x) = 0 \Rightarrow 1040 - 40x = 0 \Rightarrow R = 26$
$R''(x) = -40; \ R''(26) < 0$

Revenue is maximized for a group of 26 people, which exceeds the maximum allowed. Although, $R(x)$ is an increasing function on [12, 25], therefore $R(x)$ reaches its maximum at $x = 25$ on the interval $12 \le x \le 25$. The tour group that produces the greatest revenue is size 25.

Chapter 3 Techniques of Differentiation

3.1   The Product and Quotient Rules

1. $\dfrac{d}{dx}\left[(x+1)\left(x^3+5x+2\right)\right]=(x+1)\left(3x^2+5\right)+\left(x^3+5x+2\right)=4x^3+3x^2+10x+7$

3. $\dfrac{d}{dx}\left[\left(2x^4-x+1\right)\left(-x^5+1\right)\right]=\left(2x^4-x+1\right)\left(-5x^4\right)+\left(8x^3-1\right)\left(-x^5+1\right)=-18x^8+6x^5-5x^4+8x^3-1$

5. $\dfrac{d}{dx}\left[x\left(x^2+1\right)^4\right]=x(4)\left(x^2+1\right)^3(2x)+(1)\left(x^2+1\right)^4=8x^2\left(x^2+1\right)^3+\left(x^2+1\right)^4=\left(x^2+1\right)^3\left(9x^2+1\right)$

7. $\dfrac{d}{dx}\left[\left(x^2+3\right)\left(x^2-3\right)^{10}\right]=\left(x^2+3\right)(10)\left(x^2-3\right)^9(2x)+\left(x^2-3\right)^{10}(2x)=2x\left(x^2-3\right)^9\left(11x^2+27\right)$

9. $\dfrac{d}{dx}\left[(5x+1)\left(x^2-1\right)+\dfrac{2x+1}{3}\right]=(5x+1)(2x)+5\left(x^2-1\right)+\dfrac{1}{3}(2)=15x^2+2x-\dfrac{13}{3}$

11. $\dfrac{d}{dx}\left[\dfrac{x-1}{x+1}\right]=\dfrac{(x+1)(1)-(x-1)(1)}{(x+1)^2}=\dfrac{2}{(x+1)^2}$

13. $\dfrac{d}{dx}\left[\dfrac{x^2-1}{x^2+1}\right]=\dfrac{\left(x^2+1\right)(2x)-\left(x^2-1\right)(2x)}{\left(x^2+1\right)^2}=\dfrac{4x}{\left(x^2+1\right)^2}$

15. $\dfrac{d}{dx}\left[\dfrac{x+3}{(2x+1)^2}\right]=\dfrac{(2x+1)^2(1)-(x+3)(2)(2x+1)(2)}{\left[(2x+1)^2\right]^2}=-\dfrac{4x^2+24x+11}{(2x+1)^4}=-\dfrac{(2x+1)(2x+11)}{(2x+1)^4}=-\dfrac{2x+11}{(2x+1)^3}$

17. $\dfrac{d}{dx}\left[\dfrac{1}{\pi}+\dfrac{2}{x^2+1}\right]=0+\dfrac{\left(x^2+1\right)(0)-2(2x)}{\left(x^2+1\right)^2}=\dfrac{-4x}{\left(x^2+1\right)^2}$

19. $\dfrac{d}{dx}\left[\dfrac{x^2}{\left(x^2+1\right)^2}\right]=\dfrac{\left(x^2+1\right)^2(2x)-x^2(2)\left(x^2+1\right)(2x)}{\left[\left(x^2+1\right)^2\right]^2}=\dfrac{2x^5+4x^3+2x-4x^5-4x^3}{\left(x^2+1\right)^4}=\dfrac{2x-2x^5}{\left(x^2+1\right)^4}$

Alternatively, we can factor $2x\left(x^2+1\right)$ in the numerator to obtain

$$\dfrac{\left(x^2+1\right)^2(2x)-x^2(2)\left(x^2+1\right)(2x)}{\left[\left(x^2+1\right)^2\right]^2}=\dfrac{2x\left(x^2+1\right)\left[\left(x^2+1\right)-2x^2\right]}{\left(x^2+1\right)^4}=\dfrac{2x\left(1-x^2\right)}{\left(x^2+1\right)^3}=\dfrac{2x-2x^3}{\left(x^2+1\right)^3}$$

21. $\dfrac{d}{dx}\left(\left[\left(3x^2+2x+2\right)(x-2)\right]^2\right)=2\left[\left(3x^2+2x+2\right)(x-2)\right]\left[\left(3x^2+2x+2\right)(1)+(6x+2)(x-2)\right]$

$$=2(x-2)\left(3x^2+2x+2\right)\left(9x^2-8x-2\right)$$

23. $\dfrac{d}{dx}\left[\dfrac{1}{\sqrt{x}+1}\right]=\dfrac{d}{dx}\left(x^{1/2}+1\right)^{-1}=-\left(\sqrt{x}+1\right)^{-2}\left(\dfrac{1}{2}x^{-1/2}\right)=-\dfrac{1}{2\sqrt{x}\left(\sqrt{x}+1\right)^2}$

25. $\dfrac{d}{dx}\left(\dfrac{x+11}{x-3}\right)^3 = 3\left(\dfrac{x+11}{x-3}\right)^2 \dfrac{d}{dx}\left(\dfrac{x+11}{x-3}\right) = 3\left(\dfrac{x+11}{x-3}\right)^2\left(\dfrac{(x-3)(1)-(x+11)(1)}{(x-3)^2}\right)$

$\qquad = 3\left(\dfrac{x+11}{x-3}\right)^2\left(\dfrac{-14}{(x-3)^2}\right) = -42\left(\dfrac{(x+11)^2}{(x-3)^4}\right)$

27. $\dfrac{d}{dx}\left[\sqrt{x+2}(2x+1)^2\right] = \dfrac{d}{dx}\left[(x+2)^{1/2}(2x+1)^2\right] = (x+2)^{1/2}(2)(2x+1)(2) + \dfrac{1}{2}(x+2)^{-1/2}(2x+1)^2$

$\qquad = \sqrt{x+2}\,(8x+4) + \dfrac{4x^2+4x+1}{2\sqrt{x+2}} = \dfrac{20x^2+44x+17}{2\sqrt{x+2}} = \dfrac{(2x+1)(10x+17)}{2\sqrt{x+2}}$

29. $y = (x-2)^5(x+1)^2$

$\dfrac{dy}{dx} = (x-2)^5(2)(x+1) + (x+1)^2(5)(x-2)^4 = (x-2)^4(x+1)(7x+1)$

$\left.\dfrac{dy}{dx}\right|_{x=3} = 88$

Equation of tangent line: $y - 16 = 88(x-3)$ or $y = 88x - 248$

31. $y = \dfrac{(x-2)^5}{(x-4)^3}$

$\dfrac{dy}{dx} = \dfrac{(x-4)^3(5)(x-2)^4 - (x-2)^5(3)(x-4)^2}{(x-4)^6} = \dfrac{(x-2)^4(2x-14)}{(x-4)^4}$

The tangent line is horizontal when $\dfrac{dy}{dx} = 0$.

$\dfrac{(x-2)^4(2x-14)}{(x-4)^4} = 0 \Rightarrow (x-2)^4(2x-14) = 0 \Rightarrow x = 2$ or $x = 7$

33. $y = \left(x^2-4\right)^3\left(2x^2+5\right)^5$

$\dfrac{dy}{dx} = \left(x^2-4\right)^3(5)\left(2x^2+5\right)^4(4x) + \left(2x^2+5\right)^5(3)\left(x^2-4\right)^2(2x) = (2x)\left(x^2-4\right)^2\left(2x^2+5\right)^4\left(16x^2-25\right)$

$\dfrac{dy}{dx} = 0$ for $x = 0, \pm 2, \pm\dfrac{5}{4}$

35. $y = \dfrac{x^2+3x-1}{x}$

$\dfrac{dy}{dx} = \dfrac{x(2x+3) - \left(x^2+3x-1\right)(1)}{x^2} = \dfrac{x^2+1}{x^2}$

$\dfrac{dy}{dx} = 5 \Rightarrow \dfrac{x^2+1}{x^2} = 5 \Rightarrow 4x^2 = 1 \Rightarrow x = \pm\dfrac{1}{2}$

Points on the graph are $\left(\dfrac{1}{2}, \dfrac{3}{2}\right), \left(-\dfrac{1}{2}, \dfrac{9}{2}\right)$.

37. $y = \left(x^2 + 1\right)^4$

$$\frac{dy}{dx} = 4\left(x^2 + 1\right)^3 (2x) = 8x\left(x^2 + 1\right)^3$$

$$\frac{d^2y}{dx^2} = 8x(3)\left(x^2 + 1\right)^2 (2x) + \left(x^2 + 1\right)^3 (8) = 8\left(x^2 + 1\right)^2 \left(7x^2 + 1\right)$$

39. $y = x\sqrt{x+1}$

$$\frac{dy}{dx} = x \cdot \frac{1}{2\sqrt{x+1}} + \sqrt{x+1}\,(1) = \frac{x}{2\sqrt{x+1}} + \sqrt{x+1} = \frac{3x+2}{2\sqrt{x+1}}$$

$$\frac{d^2y}{dx^2} = \frac{2\sqrt{x+1}\,(3) - (3x+2)\left[(2)\left(\dfrac{1}{2\sqrt{x+1}}\right) + \sqrt{x+1}\,(0)\right]}{\left(2\sqrt{x+1}\right)^2} = \frac{6\sqrt{x+1} - \dfrac{3x+2}{\sqrt{x+1}}}{\left(2\sqrt{x+1}\right)^2} = \frac{6(x+1) - (3x+2)}{4(x+1)^{3/2}} = \frac{3x+4}{4(x+1)^{3/2}}$$

41. $h(x) = xf(x)$

$$h'(x) = \frac{d}{dx}\big[xf(x)\big] = xf'(x) + (1)f(x) = xf'(x) + f(x)$$

43. $h(x) = \dfrac{f(x)}{x^2+1}$

$$h'(x) = \frac{d}{dx}\left[\frac{f(x)}{x^2+1}\right] = \frac{\left(x^2+1\right)f'(x) - f(x)(2x)}{\left(x^2+1\right)^2}$$

45. Let x be the width of the box and let h be its height. Constraint: $16 = 2xh + 2(3h) + 3x$

$$\left(h = \frac{16 - 3x}{2x + 6}\right)$$

Objective: $V = 3xh = 3x\left(\dfrac{16 - 3x}{2x + 6}\right) = \dfrac{48x - 9x^2}{2x + 6}$

$$\frac{dV}{dx} = \frac{(2x+6)(48 - 18x) - (48x - 9x^2)(2)}{(2x+6)^2} = \frac{-18(x+8)(x-2)}{(2x+6)^2} \text{ or } \frac{-18(x^2 + 6x - 16)}{(2x+6)^2}$$

$$\frac{d^2V}{dx^2} = -18\left[\frac{(2x+6)^2(2x+6) - (x^2 + 6x - 16)(2)(2x+6)(2)}{(2x+6)^4}\right] = -\frac{225}{(x+3)^3}$$

The maximum value of V occurs at $x = 2$.
Answer: $x = 2$, $h = 1$, i.e. optimal dimensions are
$2\,\text{ft} \times 3\,\text{ft} \times 1\,\text{ft}$

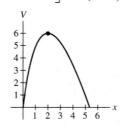

47. Let $AC(x)$ be the average cost of producing x units and let $C(x)$ be the total cost.

Then $AC(x) = \dfrac{C(x)}{x} = \dfrac{.1x^2 + 5x + 2250}{x}$.

$$AC'(x) = \frac{x(.2x+5) - (.1x^2 + 5x + 2250)(1)}{x^2} = \frac{.1x^2 - 2250}{x^2} = \frac{(.1)(x+150)(x-150)}{x^2}$$

$$AC''(x) = \frac{x^2(.2x) - (.1x^2 - 2250)(2x)}{x^4} = \frac{.2x^3 - .2x^3 + 4500x}{x^4} = \frac{4500}{x^3}$$

The minimum value of $A(x)$ occurs at $x = 150$.
The marginal cost function is $C'(x) = .2x + 5$.
At the level of 150 units, we have
$AC(150) = 35 = .2(150) + 5 = C'(150)$.

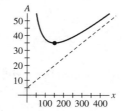

49. Recall that the marginal revenue, MR, is defined by $MR = R'(x)$. The average revenue is maximized when

$$\frac{d}{dx}(AR) = 0.$$

$$\frac{d}{dx}(AR) = \frac{d}{dx}\left(\frac{R(x)}{x}\right) = \frac{xR'(x) - R(x)(1)}{x^2}$$

If $\dfrac{xR'(x) - R(x)}{x^2} = 0$, then $xR'(x) - R(x) = 0 \Rightarrow xR'(x) = R(x) \Rightarrow R'(x) = \dfrac{R(x)}{x} \Rightarrow MR = AR$.

51. $A = W(t)L(t)$:

$$\frac{dA}{dt} = W(t)L'(t) + W'(t)L(t) = 5 \cdot 4 + 3 \cdot 6 = 38$$

The area of the rectangle is increasing by 38 square inches per second.

53. Let x be the number of years since the beginning of 1998, $P(x)$ the population, $c(x)$ the annual per capita consumption in gallons, and $t(x)$ the total annual consumption in gallons.
$P(x) = 1{,}856{,}000x + 268{,}924{,}000$
$c(x) = .2x + 52.3$
$t(x) = P(x)c(x)$
$t'(x) = P(x)c'(x) + P'(x)c(x) = (1{,}856{,}000x + 268{,}924{,}000)(.2) + (1{,}856{,}000)(.2x + 52.3)$
$t'(0) = (268{,}924{,}000)(.2) + (1{,}856{,}000)(52.3) = 150{,}853{,}600$

Total annual consumption was increasing at the beginning of 1998 at a rate of 150,853,600 gallons per year.

55. $$\frac{dy}{dx} = \frac{\left(1 + .25x^2\right)10 - 10x(.5x)}{\left(1 + .25x^2\right)^2} = \frac{10 - 2.5x^2}{\left(1 + .25x^2\right)^2}$$

$$\frac{dy}{dx} = 0 \Rightarrow \frac{10 - 2.5x^2}{\left(1 + .25x^2\right)^2} = 0 \Rightarrow 10 = 2.5x^2 \Rightarrow x = \pm 2$$

When $x = 2$, $y = 10$, so the maximum point is at $(2, 10)$.

57. $\dfrac{d}{dx}\Big[f(x)g(x)\Big]\Big|_{x=2} = g(2)f'(2) + f(2)g'(2) = 3(3) + 3\left(\dfrac{1}{3}\right) = 10$

59. $\dfrac{d}{dx}\Big[(f(x))^2\Big]\Big|_{x=2} = 2f(2)f'(2) = 2(3)(3) = 18$

61. $\dfrac{d}{dx}[xf(x)]\Big|_{x=2} = xf'(x) + f(x)(1) = 2(3) + 3 = 9$

63. $f(x) = \dfrac{1}{x}$, $g(x) = x^3$

    a. Using the product rule,

$$\frac{d}{dx}\left[\left(\frac{1}{x}\right)x^3\right] = \left(\frac{1}{x}\right)(3x^2) + x^3(-1)x^{-2} = 3x - x = 2x = \frac{d}{dx}\left[x^2\right]$$

    b. $f'(x) = -\dfrac{1}{x^2}$ and $g'(x) = 3x^2$, so $f'(x)g'(x) = -3$.

       Now $f(x)g(x) = \left(\dfrac{1}{x}\right)x^3 = x^2$ which has derivative $2x$. Thus, $f'(x)g'(x) \neq (f(x)g(x))'$.

65. $\dfrac{d}{dx}[f(x)g(x)h(x)] = \dfrac{d}{dx}\big[(f(x)g(x)) \cdot h(x)\big] = h(x)\dfrac{d}{dx}[f(x)g(x)] + [g(x)h(x)]h'(x)$

$$= h(x)[f'(x)g(x) + f(x)g'(x)] + g(x)h(x)h'(x)$$

$$= f'(x)g(x)h(x) + f(x)g'(x)h(x) + f(x)g(x)h'(x)$$

67. $b(t) = \dfrac{w(t)}{[h(t)]^2}$

$$b'(t) = \frac{d}{dt}\left[\frac{w(t)}{[h(t)]^2}\right]$$

$$= \frac{[h(t)]^2 \, w'(t) - w(t)(2)h(t)h'(t)}{[h(t)]^4}$$

$$= \frac{h(t)w'(t) - 2h'(t)w(t)}{[h(t)]^3}$$

69. a.

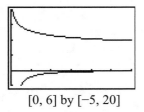

[0, 6] by [−5, 20]

    b. $f(3) \approx 10.8$ square millimeters

    c. Using a graphing calculator to find when $f(x) = 11$, $x \approx 2.61$ units of light.

    d. $f'(3) \approx -.55$ square millimeters per unit of light.

3.2 The Chain Rule and the General Power Rule

1. $f(x) = \dfrac{x}{x+1}$, $g(x) = x^3$; $f(g(x)) = \dfrac{x^3}{x^3+1}$

3. $f(x) = x(x^2 + 1)$, $g(x) = \sqrt{x}$

$$f(g(x)) = \sqrt{x}\left(\left(\sqrt{x}\right)^2 + 1\right) = \sqrt{x}\,(x+1)$$

5. $f(g(x)) = (x^3 + 8x - 2)^5$

$$f(x) = x^5, \; g(x) = x^3 + 8x - 2$$

7. $f(g(x)) = \sqrt{4 - x^2}$

$$f(x) = \sqrt{x}, \; g(x) = 4 - x^2$$

9. $f(g(x)) = \dfrac{1}{x^3 - 5x^2 + 1}$

$$f(x) = \frac{1}{x}, \; g(x) = x^3 - 5x^2 + 1$$

11. $\dfrac{d}{dx}(x^2 + 5)^{15} = 15(x^2 + 5)^{14}(2x)$

$$= 30x(x^2 + 5)^{14}$$

13. $\dfrac{d}{dx}6x^2(x-1)^3 = 6x^2(3)(x-1)^2(1) + 12x(x-1)^3$

$$= 18x^2(x-1)^2 + 12x(x-1)^3$$

$$= 6x(x-1)^2(5x-2)$$

15. $\dfrac{d}{dx}\Big[2(x^3-1)(3x^2+1)^4\Big]=2\Big[4(3x^2+1)^3(6x)(x^3-1)+3x^2(3x^2+1)^4\Big]=6x(3x^2+1)^3(11x^3+x-8)$

17. $\dfrac{d}{dx}\Big[f\big(g(x)\big)\Big]\Big|_{x=1}=f'\big(g(1)\big)g'(1)$

$\qquad\qquad\qquad\quad =\big(f'(3)\big)(4)$

$\qquad\qquad\qquad\quad =2(4)=8$

19. $\dfrac{d}{dx}\Big[f\big(f(x)\big)\Big]\Big|_{x=1}=f'\big(f(1)\big)f'(1)$

$\qquad\qquad\qquad\quad =\big(f'(1)\big)(5)$

$\qquad\qquad\qquad\quad =5(5)=25$

21. $h(x)=f(x^2)$

$\qquad h'(x)=\dfrac{d}{dx}\Big[f(x^2)\Big]$

$\qquad\qquad\; =f'(x^2)(2x)$

$\qquad\qquad\; =2xf'(x^2)$

23. $h(x)=-f(-x)$

$\qquad h'(x)=\dfrac{d}{dx}[-f(-x)]=-\dfrac{d}{dx}[f(-x)]$

$\qquad\qquad\; =-f'(-x)(-1)=f'(-x)$

25. $h(x)=\dfrac{f(x^2)}{x}$

$\qquad h'(x)=\dfrac{d}{dx}\left[\dfrac{f(x^2)}{x}\right]=\dfrac{xf'(x^2)(2x)-f(x^2)(1)}{x^2}$

$\qquad\qquad\; =\dfrac{2x^2f'(x^2)-f(x^2)}{x^2}$

27.

29. $f(x)=x^5,\ g(x)=6x-1$

$\qquad \dfrac{d}{dx}f(g(x))=\dfrac{d}{dx}(6x-1)^5=5(6x-1)^4(6)$

$\qquad\qquad\qquad\quad =30(6x-1)^4$

31. $f(x)=\dfrac{1}{x},\ g(x)=1-x^2$

$\qquad \dfrac{d}{dx}f(g(x))=\dfrac{d}{dx}\left[\dfrac{1}{1-x^2}\right]$

$\qquad\qquad\qquad\quad =\dfrac{0(1-x^2)-(-2x)(1)}{(1-x^2)^2}=\dfrac{2x}{(1-x^2)^2}$

33. $f(x)=x^4-x^2,\ g(x)=x^2-4$

$\qquad \dfrac{d}{dx}f(g(x))=\dfrac{d}{dx}\Big[(x^2-4)^4-(x^2-4)^2\Big]$

$\qquad\qquad\qquad\quad =4(x^2-4)^3(2x)-2(x^2-4)(2x)$

$\qquad\qquad\qquad\quad =8x(x^2-4)^3-4x(x^2-4)$

37. $y=u^{3/2},\ u=4x+1$

$\qquad \dfrac{dy}{du}=\dfrac{3}{2}u^{1/2};\ \dfrac{du}{dx}=4;$

$\qquad \dfrac{dy}{dx}=6u^{1/2}=6(4x+1)^{1/2}$

35. $f(x)=(x^3+1)^2,\ g(x)=x^2+5$

$\qquad \dfrac{d}{dx}f(g(x))=\dfrac{d}{dx}\Big[(x^2+5)^3+1\Big]^2$

$\qquad\qquad\qquad\quad =2\Big[(x^2+5)^3+1\Big]3(x^2+5)^2(2x)$

$\qquad\qquad\qquad\quad =12x\Big[(x^2+5)^3+1\Big](x^2+5)^2$

39. $y=\dfrac{u}{2}+\dfrac{2}{u},\ u=x-x^2,$

$\qquad \dfrac{dy}{du}=\dfrac{1}{2}-\dfrac{2}{u^2};\ \dfrac{du}{dx}=1-2x$

$\qquad \dfrac{dy}{dx}=\left[\dfrac{1}{2}-\dfrac{2}{(x-x^2)^2}\right](1-2x)$

$\qquad\qquad =\dfrac{\Big(\big(x-x^2\big)^2-4\Big)(1-2x)}{2(x-x^2)^2}$

41. $y=x^2-3x,\ x=t^2+3,\ t_0=0$

$\qquad \dfrac{dy}{dt}=\dfrac{dy}{dx}\dfrac{dx}{dt}=(2x-3)(2t)$

$\qquad\qquad =\Big(2\big(t^2+3\big)-3\Big)(2t)=(2t^2+6-3)(2t)$

$\qquad\qquad =4t^3+6t\Rightarrow\dfrac{dy}{dt}\Big|_{t=0}=0$

43. $y=\dfrac{x+1}{x-1},\ x=\dfrac{t^2}{4}=\dfrac{1}{4}t^2,\ t_0=3$

$\qquad \dfrac{dy}{dt}=\left(\dfrac{(x-1)(1)-(x+1)(1)}{(x-1)^2}\right)\left(\dfrac{1}{2}t\right)$

$\qquad\qquad =\dfrac{-t}{(x-1)^2}=\dfrac{-t}{\left(\frac{1}{4}t^2-1\right)^2}$

$\qquad \dfrac{dy}{dt}\Big|_{t=3}=\dfrac{-3}{\left(\frac{9}{4}-1\right)^2}=-\dfrac{48}{25}$

45. $\dfrac{dy}{dx} = 2(x-4)^6 + 12x(x-4)^5;\ \dfrac{dy}{dx}\Big|_{x=5} = 62$

The tangent line is $y - 10 = 62(x-5)$ or $y = 62x - 300$.

47. A horizontal tangent line has $\dfrac{dy}{dx} = 0$.

$\dfrac{dy}{dx} = 3(-x^2 + 4x - 3)^2(-2x + 4) = 0 \Rightarrow$

$-2x + 4 = 0 \Rightarrow x = 2$ or $-x^2 + 4x - 3 = 0 \Rightarrow$
$(x-3)(x-1) = 0 \Rightarrow x = 1, x = 3$
The x-coordinates are 1, 2, and 3.

49. a. $\dfrac{dV}{dt} = \dfrac{dV}{dx} \cdot \dfrac{dx}{dt}$

b. By the chain rule $\dfrac{dV}{dt} = 3x^2 \dfrac{dx}{dt}$.

We want $\dfrac{dV}{dt} = 12\dfrac{dx}{dt}$; so $3x^2 = 12$ or

$x = 2$.

51. a. $\dfrac{dy}{dt}, \dfrac{dP}{dy}, \dfrac{dP}{dt}$ **b.** $\dfrac{dP}{dt} = \dfrac{dP}{dy} \cdot \dfrac{dy}{dt}$

53. $P = \dfrac{200x}{100 + x^2},\ x = 4 + 2t$

a. $\dfrac{dP}{dx} = \dfrac{(100 + x^2)(200) - (200x)(2x)}{(100 + x^2)^2}$

$= \dfrac{20,000 + 200x^2 - 400x^2}{(100 + x^2)^2}$

$= \dfrac{200(100 - x^2)}{(100 + x^2)^2}$

b. $\dfrac{dx}{dt} = 2$

$\dfrac{dP}{dt} = \left[\dfrac{200(100 - (4+2t)^2)}{(100 + (4+2t)^2)^2}\right]2$

$= \dfrac{400\left[100 - (4+2t)^2\right]}{\left[100 + (4+2t)^2\right]^2}$

c. When $t = 8$,

$\dfrac{dP}{dt} = \dfrac{400(100 - (4+16)^2)}{(100 + (4+16)^2)^2} = -.48$.

Profits are falling at $480/week

55. $L = 10 + .4x + .0001x^2,\ x = 752 + 23t + .5t^2$

a. $\dfrac{dL}{dx} = .4 + .0002x$

b. $\dfrac{dx}{dt} = 23 + t \Rightarrow \dfrac{dx}{dt}\Big|_{t=2} = 25$

The population is increasing at the rate of 25 thousand persons per year.

c. When $t = 2$, $x = 800$ and $\dfrac{dL}{dx} = .56$.

$(.56)(25) = 14$ ppm per year

57. $\dfrac{d}{dx} f(g(x)) = f'(g(x)) \cdot g'(x)$

$= 3x^2 \cdot f'(x^3 + 1)$

So $g(x) = x^3 + 1$.

59. $\dfrac{d}{dx} f(g(x))\Big|_{x=1} = f'(g(1)) \cdot g'(1)$

$= f'(5) \cdot 6 = 4 \cdot 6 = 24$

61. a. From Fig 1(a), $x = 40$ when $t = 1.5$ and $x = 30$ when $t = 3.5$.

$W(40) = 10\left(\dfrac{12 + 8(40)}{3 + 40}\right) \approx 77.209$

million dollars

$W(30) = 10\left(\dfrac{12 + 8(30)}{3 + 30}\right) \approx 76.364$ million

dollars.

b. The slope of $x(t)$ at $t = 1.5$ is given by

slope$= \dfrac{y_2 - y_1}{x_2 - x_1} = \dfrac{50 - 10}{2 - 0} = 20$.

So, $\dfrac{dx}{dt}\Big|_{t=1.5} = 20$.

A month and a half after the company went public, the share price of $40 was increasing at a rate of $20 per month. The slope of $x(t)$ at $t = 3.5$ is zero, since the graph is horizontal.

So, $\dfrac{dx}{dt}\Big|_{t=3.5} = 0$.

Three and a half months after the company went public, the share price was holding steady at $30.

63. a. $\dfrac{dx}{dt}\Big|_{t=2.5}$ = slope of $x(t)$ at $t = 2.5$.

slope$= \dfrac{30-50}{3-2} = -20 = \dfrac{dx}{dt}\Big|_{t=2.5}$

The share price is dropping at a rate of $20 per month.

$\dfrac{dx}{dt}\Big|_{t=4}$ = slope of $x(t)$ at $t = 4$. Since $x(t)$

is horizontal at $t = 4$, $\dfrac{dx}{dt}\Big|_{t=4} = 0$.

The share price is steady at $30 per share.

b. $W = \dfrac{120+80x}{3+x}$

$\dfrac{dW}{dx} = \dfrac{(3+x)(80)-(120+80x)(1)}{(3+x)^2}$

$= \dfrac{120}{(3+x)^2}$

From Fig 1(a), $x = 40$ when $t = 2.5$ and $x = 30$ when $t = 4$.

$\dfrac{dW}{dt}\Big|_{t=2.5} = \dfrac{dW}{dx}\dfrac{dx}{dt}\Big|_{t=2.5}$

$= \dfrac{120}{(3+40)^2}\cdot(-20) \approx -1.3$

At $t = 2.5$, the value of the company is decreasing at the rate of 1.3 million dollars per month.

$\dfrac{dW}{dt}\Big|_{t=4} = \dfrac{dW}{dx}\dfrac{dx}{dt}\Big|_{t=4} = \dfrac{120}{(3+30)^2}\cdot 0 = 0$

At $t = 4$, the value of +.

65. The derivative of the composite function $f(g(x))$ is the derivative of the outer function evaluated at the inner function and then multiplied by the derivative of the inner function.

3.3 Implicit Differentiation and Related Rates

1. $x^2 - y^2 = 1$

$2x - 2y\dfrac{dy}{dx} = 0 \Rightarrow \dfrac{dy}{dx} = \dfrac{x}{y}$

3. $y^5 - 3x^2 = x$

$5y^4\dfrac{dy}{dx} - 6x = 1 \Rightarrow \dfrac{dy}{dx} = \dfrac{1+6x}{5y^4}$

5. $y^4 - x^4 = y^2 - x^2$

$4y^3\dfrac{dy}{dx} - 4x^3 = 2y\dfrac{dy}{dx} - 2x$

$\dfrac{dy}{dx}(4y^3 - 2y) = 4x^3 - 2x$

$\dfrac{dy}{dx} = \dfrac{2x^3 - x}{2y^3 - y}$

7. $2x^3 + y = 2y^3 + x$

$6x^2 + \dfrac{dy}{dx} = 6y^2\dfrac{dy}{dx} + 1$

$\dfrac{dy}{dx}(1-6y^2) = 1-6x^2$

$\dfrac{dy}{dx} = \dfrac{1-6x^2}{1-6y^2}$

9. $xy = 5$

$1(y) + x\dfrac{dy}{dx} = 0$

$\dfrac{dy}{dx} = -\dfrac{y}{x}$

11. $x(y+2)^5 = 8$

$(1)(y+2)^5 + 5(y+2)^4(x)\dfrac{dy}{dx} = 0$

$\dfrac{dy}{dx} = -\dfrac{(y+2)^5}{5x(y+2)^4}$

$= -\dfrac{y+2}{5x}$

13. $x^3y^2 - 4x^2 = 1$

$3x^2y^2 + 2yx^3\dfrac{dy}{dx} - 8x = 0$

$\dfrac{dy}{dx} = \dfrac{8x - 3x^2y^2}{2yx^3}$

$= \dfrac{8-3xy^2}{2x^2y}$

15. $x^3 + y^3 = x^3y^3$

$3x^2 + 3y^2\dfrac{dy}{dx} = 3x^2y^3 + 3y^2x^3\dfrac{dy}{dx}$

$\dfrac{dy}{dx}(3y^2 - 3y^2x^3) = 3x^2y^3 - 3x^2$

$\dfrac{dy}{dx} = \dfrac{3x^2y^3 - 3x^2}{3y^2 - 3y^2x^3}$

$\dfrac{dy}{dx} = \dfrac{x^2(y^3-1)}{y^2(1-x^3)}$

17. $x^2y + y^2x = 3$

$$2xy + x^2\frac{dy}{dx} + 2yx\frac{dy}{dx} + y^2 = 0$$

$$x\frac{dy}{dx}(2y + x) = -2xy - y^2$$

$$\frac{dy}{dx} = \frac{-2xy - y^2}{x(2y + x)} = \frac{-2xy - y^2}{2xy + x^2} = -\frac{2xy + y^2}{2xy + x^2}$$

19. $4y^3 - x^2 = -5$

$$12y^2\frac{dy}{dx} - 2x = 0$$

$$\text{slope} = \frac{dy}{dx} = \frac{2x}{12y^2} = \frac{x}{6y^2}$$

When $x = 3, y = 1$, slope $= \frac{dy}{dx} = \frac{3}{6(1)^2} = \frac{1}{2}$.

21. $xy^3 = 2$

$$(1)y^3 + 3y^2x\frac{dy}{dx} = 0$$

$$\text{slope} = \frac{dy}{dx} = \frac{-y^3}{3y^2x} = \frac{-y}{3x}$$

When $x = -\frac{1}{4}, y = -2$,

$$\text{slope} = \frac{dy}{dx} = \frac{-(-2)}{3\left(-\frac{1}{4}\right)} = -\frac{8}{3}.$$

23. $xy + y^3 = 14$

$$(1)y + x\frac{dy}{dx} + 3y^2\frac{dy}{dx} = 0$$

$$\text{slope} = \frac{dy}{dx} = \frac{-y}{x + 3y^2}$$

When $x = 3, y = 2$,

$$\text{slope} = \frac{dy}{dx} = \frac{-2}{3 + 3(2)^2} = -\frac{2}{15}.$$

25. $x^2y^4 = 1$

$$2xy^4 + 4y^3x^2\frac{dy}{dx} = 0$$

$$\text{slope} = \frac{dy}{dx} = \frac{-2xy^4}{4y^3x^2} = -\frac{y}{2x}$$

When $x = 4, y = \frac{1}{2}$, slope $= -\frac{1/2}{2(4)} = -\frac{1}{16}$.

Let $(x_1, y_1) = \left(4, \frac{1}{2}\right)$.

$$y - \frac{1}{2} = -\frac{1}{16}(x - 4) \Rightarrow y = -\frac{1}{16}x + \frac{3}{4}$$

When $x = 4, y = -\frac{1}{2}$, slope $= -\frac{-1/2}{2(4)} = \frac{1}{16}$.

Let $(x_1, y_1) = \left(4, -\frac{1}{2}\right)$.

$$y + \frac{1}{2} = \frac{1}{16}(x - 4) \Rightarrow y = \frac{1}{16}x + \frac{1}{4}$$

27. a. $x^4 + 2x^2y^2 + y^4 = 4x^2 - 4y^2$

$$4x^3 + 4xy^2 + 4x^2y\frac{dy}{dx} + 4y^3\frac{dy}{dx}$$
$$= 8x - 8y\frac{dy}{dx}$$

$$\frac{dy}{dx}(4x^2y + 4y^3 + 8y) = 8x - 4x^3 - 4xy^2$$

$$\frac{dy}{dx} = \frac{4(2x - x^3 - xy^2)}{4(x^2y + y^3 + 2y)} = \frac{2x - x^3 - xy^2}{x^2y + y^3 + 2y}$$

b. slope $= \dfrac{dy}{dx} = \dfrac{2x - x^3 - xy^2}{x^2y + y^3 + 2y}$

When

$$x = \frac{\sqrt{6}}{2}, \ y = \frac{\sqrt{2}}{2},$$

$$\text{slope} = \frac{\sqrt{6} - \frac{\sqrt{6^3}}{8} - \frac{\sqrt{6}}{2}\cdot\frac{2}{4}}{\frac{6}{4}\cdot\frac{\sqrt{2}}{2} + \frac{\sqrt{2^3}}{8} + \sqrt{2}} = 0.$$

29. a. $30x^{1/3}y^{2/3} = 1080$

$$10x^{-2/3}y^{2/3} + 20x^{1/3}y^{-1/3}\frac{dy}{dx} = 0$$

$$\frac{dy}{dx} = \frac{-10y^{2/3}x^{-2/3}}{20x^{1/3}y^{-1/3}} = -\frac{y}{2x}$$

b. marginal rate of substitution of x for y is

$$\left|\frac{dy}{dx}\right|_{\substack{x=16 \\ y=54}} = \left|-\frac{27}{16}\right| = \frac{27}{16}.$$

31. $x^4 + y^4 = 1$

$$4x^3\frac{dx}{dt} + 4y^3\frac{dy}{dt} = 0$$

$$\frac{dy}{dt} = -\frac{4x^3}{4y^3}\cdot\frac{dx}{dt}$$

$$\frac{dy}{dt} = -\frac{x^3}{y^3}\frac{dx}{dt}$$

33. $3xy - 3x^2 = 4$

$$3\frac{dx}{dt}y + 3x\frac{dy}{dt} - 6x\frac{dx}{dt} = 0 \Rightarrow \frac{dy}{dt} = \frac{6x - 3y}{3x}\frac{dx}{dt} = \frac{2x - y}{x}\frac{dx}{dt}$$

35. $x^2 + 2xy = y^3$

$$2x\frac{dx}{dt} + 2y\frac{dx}{dt} + 2x\frac{dy}{dt} = 3y^2\frac{dy}{dt} \Rightarrow \frac{dy}{dt} = \frac{2x + 2y}{3y^2 - 2x}\frac{dx}{dt}$$

37. $x^2 - 4y^2 = 9$

$$2x\frac{dx}{dt} - 8y\frac{dy}{dt} = 0 \Rightarrow \frac{dy}{dt} = \frac{2x}{8y}\frac{dx}{dt} = \frac{x}{4y}\frac{dx}{dt}$$

When $x = 5$, $y = -2$, $\frac{dx}{dt} = 3$ per sec, $\frac{dy}{dt} = \frac{5}{-8}(3) = -\frac{15}{8}$.

The y-coordinate is decreasing at $\frac{15}{8}$ units per second.

39. $2p^3 + x^2 = 4500$

$$6p^2\frac{dp}{dt} + 2x\frac{dx}{dt} = 0 \Rightarrow \frac{dx}{dt} = -\frac{6p^2}{2x}\frac{dp}{dt} = -\frac{3p^2}{x}\frac{dp}{dt}$$

When $p = 10$, $x = 50$, $\frac{dp}{dt} = -.5$ per week, $\frac{dx}{dt} = -\frac{3(100)}{50}(-.5) = 3$.

The sales are rising at 3 thousand units per week.

41. $A = 6\sqrt{x^2 - 400}$, $x \geq 20$, $A = 6(x^2 - 400)^{1/2}$

$$\frac{dA}{dx} = 3(x^2 - 400)^{-1/2}(2x) = \frac{6x}{\sqrt{x^2 - 400}}$$

$$\frac{dA}{dt} = \frac{dA}{dx} \cdot \frac{dx}{dt} = \frac{6x}{\sqrt{x^2 - 400}} \cdot \frac{dx}{dt}$$

When $x = 25$, $\frac{dx}{dt} = 2$, $\frac{dA}{dt} = \frac{6(25)}{\sqrt{25^2 - 400}}(2) = 20$.

The revenue is increasing at the rate of 20 thousand dollars per month.

43. a. $x^2 + y^2 = (10)^2 \Rightarrow x^2 + y^2 = 100$

b. $2x\frac{dx}{dt} + 2y\frac{dy}{dt} = 0 \Rightarrow \frac{dy}{dt} = \frac{-x}{y}\frac{dx}{dt}$

When $x = 8$, $y = 6$, $\frac{dx}{dt} = 3$, $\frac{dy}{dt} = -\frac{8}{6}(3) = -4$.

The top end is sliding down at 4 feet per second.

Chapter 3 Review Exercises

1. $\frac{d}{dx}\left[(4x-1)(3x+1)^4\right] = (4x-1)(4)(3x+1)^3(3) + (3x+1)^4(4) = 4(3x+1)^3\left[(3x+1) + 3(4x-1)\right]$

$$= 4(3x+1)^3(15x-2)$$

2. $\frac{d}{dx}\left[2(5-x)^3(6x-1)\right] = 2\left[(5-x)^3(6) + (6x-1)3(5-x)^2(-1)\right] = 2\left[-3(5-x)^2(6x-1) + 6(5-x)^3\right]$

$$= 6\left[2(5-x)^3 - (5-x)^2(6x-1)\right] = 6(5-x)^2\left[10 - 2x - 6x + 1\right] = 6(5-x)^2(11-8x)$$

3. $\dfrac{d}{dx}\left[x(x^5-1)^3\right]=(x)(3)(x^5-1)^2(5x^4)+(x^5-1)^3(1)=(x^5-1)^2(16x^5-1)$

4. $\dfrac{d}{dx}\left[(2x+1)^{5/2}(4x-1)^{3/2}\right]=(2x+1)^{5/2}\left(\dfrac{3}{2}\right)(4x-1)^{1/2}(4)+(4x-1)^{3/2}\left(\dfrac{5}{2}\right)(2x+1)^{3/2}(2)$

$$=5(2x+1)^{3/2}(4x-1)^{3/2}+6(4x-1)^{1/2}(2x+1)^{5/2}$$
$$=(2x+1)^{3/2}(4x-1)^{1/2}\left[5(4x-1)+6(2x+1)\right]$$
$$=(2x+1)^{3/2}(4x-1)^{1/2}(32x+1)$$

5. $\dfrac{d}{dx}\left[5\left(\sqrt{x}-1\right)^4\left(\sqrt{x}-2\right)^2\right]=5\left[\left(\sqrt{x}-1\right)^4(2)\left(\sqrt{x}-2\right)\left(\dfrac{1}{2}x^{-1/2}\right)+\left(\sqrt{x}-2\right)^2(4)\left(\sqrt{x}-1\right)^3\left(\dfrac{1}{2}x^{-1/2}\right)\right]$

$$=5\left[2x^{-1/2}\left(\sqrt{x}-1\right)^3\left(\sqrt{x}-2\right)^2+x^{-1/2}\left(\sqrt{x}-2\right)\left(\sqrt{x}-1\right)^4\right]$$
$$=5x^{-1/2}\left(\sqrt{x}-1\right)^3\left(\sqrt{x}-2\right)\left[2\left(\sqrt{x}-2\right)+\left(\sqrt{x}-1\right)\right]$$
$$=\dfrac{5\left(\sqrt{x}-1\right)^3\left(\sqrt{x}-2\right)\left(3\sqrt{x}-5\right)}{\sqrt{x}}$$

6. $\dfrac{d}{dx}\left[\dfrac{\sqrt{x}}{\sqrt{x}+4}\right]=\dfrac{\left(\sqrt{x}+4\right)\left(\frac{1}{2\sqrt{x}}\right)-\sqrt{x}\left(\frac{1}{2\sqrt{x}}\right)}{\left(\sqrt{x}+4\right)^2}=\dfrac{\frac{1}{2\sqrt{x}}\left(\sqrt{x}+4-\sqrt{x}\right)}{\left(\sqrt{x}+4\right)^2}=\dfrac{2}{\sqrt{x}\left(\sqrt{x}+4\right)^2}$

7. $\dfrac{d}{dx}\left[3(x^2-1)^3(x^2+1)^5\right]=3\left[(x^2-1)^3(5)(x^2+1)^4(2x)+(x^2+1)^5(3)(x^2-1)^2(2x)\right]$

$$=3(x^2-1)^2(2x)(x^2+1)^4\left[3(x^2+1)+5(x^2-1)\right]$$
$$=12x(x^2-1)^2(x^2+1)^4(4x^2-1)$$

8. $\dfrac{d}{dx}\left[\dfrac{1}{(x^2+5x+1)^6}\right]=\dfrac{(x^2+5x+1)^6(0)-(1)(6)(x^2+5x+1)^5(2x+5)}{(x^2+5x+1)^{12}}=\dfrac{-12x-30}{(x^2+5x+1)^7}$

9. $\dfrac{d}{dx}\left[\dfrac{x^2-6x}{x-2}\right]=\dfrac{(x-2)(2x-6)-(x^2-6x)(1)}{(x-2)^2}=\dfrac{2x^2-10x+12-x^2+6x}{(x-2)^2}=\dfrac{x^2-4x+12}{(x-2)^2}$

10. $\dfrac{d}{dx}\left[\dfrac{2x}{2-3x}\right]=\dfrac{(2-3x)(2)-(2x)(-3)}{(2-3x)^2}=\dfrac{4-6x+6x}{(2-3x)^2}=\dfrac{4}{(2-3x)^2}$

11. $\dfrac{d}{dx}\left[\left(\dfrac{3-x^2}{x^3}\right)^2\right]=2\left(\dfrac{3-x^2}{x^3}\right)\left(\dfrac{(x^3)(-2x)-(3-x^2)(3x^2)}{x^6}\right)=2\left(\dfrac{3-x^2}{x^3}\right)\left(\dfrac{-2x^4-9x^2+3x^4}{x^6}\right)$

$$=2\left(\dfrac{3-x^2}{x^3}\right)\left(\dfrac{x^4-9x^2}{x^6}\right)=2\left(\dfrac{3-x^2}{x^3}\right)\left(\dfrac{x^2-9}{x^4}\right)=\dfrac{2(3-x^2)(x^2-9)}{x^7}$$

12. $\dfrac{d}{dx}\left[\dfrac{x^3+x}{x^2-x}\right]=\dfrac{(x^2-x)(3x^2+1)-(x^3+x)(2x-1)}{(x^2-x)^2}=\dfrac{3x^4-3x^3+x^2-x-2x^4-2x^2+x^3+x}{(x^2-x)^2}$

$$=\dfrac{x^4-2x^3-x^2}{x^4-2x^3+x^2}=\dfrac{x^2-2x-1}{x^2-2x+1}=\dfrac{x^2-2x-1}{(x-1)^2}$$

13. $f(x) = (3x+1)^4(3-x)^5$

$f'(x) = (3x+1)^4(5)(3-x)^4(-1) + (3-x)^5(4)(3x+1)^3(3) = 12(3x+1)^3(3-x)^5 - 5(3-x)^4(3x+1)^4$
$\quad = (3x+1)^3(3-x)^4[12(3-x) - 5(3x+1)] = (3x+1)^3(3-x)^4(-27x+31)$

Let $f'(x) = 0$ and solve for x.

$(3x+1)^3(3-x)^4(31-27x) = 0 \Rightarrow x = -\dfrac{1}{3}$ or $x = 3$ or $x = \dfrac{31}{27}$

14. $f(x) = \dfrac{x^2+1}{x^2+5}$

$f'(x) = \dfrac{(x^2+5)(2x) - (x^2+1)(2x)}{(x^2+5)^2} = \dfrac{2x^3 + 10x - 2x^3 - 2x}{(x^2+5)^2} = \dfrac{8x}{(x^2+5)^2}$

Let $f'(x) = 0$ and solve for x.

$\dfrac{8x}{(x^2+5)^2} = 0 \Rightarrow x = 0$

15. $y = (x^3-1)(x^2+1)^4$

slope $= y'$

$y' = (x^3-1)(4)(x^2+1)^3(2x) + (x^2+1)^4(3x^2) = 3x^2(x^2+1)^4 + 8x(x^2+1)^3(x^3-1)$

When $x = -1$, slope $= 3(-1)^2(1+1)^4 + 8(-1)(1+1)^3(-1-1) = 48 + 128 = 176$

When $x = -1$, $y = (-1-1)(1+1)^4 = -32$.

Let $(x_1, y_1) = (-1, -32)$. Then $y + 32 = 176(x+1)$ or $y = 176x + 144$.

16. $y = \dfrac{x-3}{\sqrt{4+x^2}}$

slope $= y'$

$y' = \dfrac{(4+x^2)^{1/2}(1) - (x-3)(\frac{1}{2})(4+x^2)^{-1/2}(2x)}{4+x^2} = \dfrac{(4+x^2)^{1/2}}{4+x^2} - \dfrac{(x^2-3x)}{(4+x^2)^{3/2}} = \dfrac{4+x^2-x^2+3x}{(4+x^2)^{3/2}} = \dfrac{3x+4}{(4+x^2)^{3/2}}$

When $x = 0$, $y = -\dfrac{3}{2}$. When $x = 0$, slope $= \dfrac{4}{8} = \dfrac{1}{2}$. Let $(x_1, y_1) = \left(0, -\dfrac{3}{2}\right)$. Then

$y + \dfrac{3}{2} = \dfrac{1}{2}(x-0)$ or $y = \dfrac{1}{2}x - \dfrac{3}{2}$.

17. The objective is

$A = 2y(2) + (x-4)(2) = 4y + 2x - 8$. The constraint is $(x-4)(y-2) = 800$ or $y = \dfrac{800}{x-4} + 2$.

Thus, $A(x) = 4\left(\dfrac{800}{x-4} + 2\right) + 2x - 8 = \dfrac{3200}{x-4} + 2x \Rightarrow A'(x) = -\dfrac{3200}{(x-4)^2} + 2 \Rightarrow A'(44) = 0$

The minimum value of $A(x)$ for $x > 0$ occurs at $x = 44$. Thus, the optimal values of x and y are $x = 44$ m,

$y = \dfrac{800}{44-4} + 2 = 22$ m.

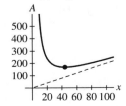

18. The objective is
$A = 2(x-4)(2) + 2(y)(2) = 4x - 16 + 4y$ and
the constraint is $(x-4)(y-4) = 800$ or

$$y = \frac{800}{x-4} + 4.$$

$$A(x) = 4x + 4\left(\frac{800}{x-4} + 4\right) - 16 = 4x + \frac{3200}{x-4}$$

$$A'(x) = 4 - \frac{3200}{(x-4)^2} \Rightarrow A'\left(4 + 20\sqrt{2}\right) = 0$$

The minimum value of $A(x)$ occurs at
$x = 4 + 20\sqrt{2}$. The optimal dimensions are
$x = y = 4 + 20\sqrt{2} \approx 32.3$ meters.

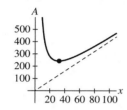

19. We are given that $C(x) = 40x + 30$ and $\dfrac{dx}{dt} = 3$.

By the chain rule, $\dfrac{dC}{dt} = \dfrac{dC}{dx}\dfrac{dx}{dt} = 40 \cdot 3 = 120$

Costs are rising at \$120 per day.

20. $\dfrac{dy}{dt} = \dfrac{dy}{dP} \cdot \dfrac{dP}{dt}$

21. $f'(x) = \dfrac{1}{\left(x^2+1\right)}; g(x) = x^3; g'(x) = 3x^2$

$$\frac{d}{dx}f\left(g(x)\right) = f'\left(g(x)\right)g'(x)$$

$$= f'\left(x^3\right)g'(x)$$

$$= \frac{1}{\left(x^3\right)^2 + 1}\left(3x^2\right) = \frac{3x^2}{x^6+1}$$

22. $f'(x) = \dfrac{1}{\left(x^2+1\right)}; g(x) = \dfrac{1}{x}; g'(x) = -\dfrac{1}{x^2}$

$$\frac{d}{dx}f\left(g(x)\right) = f'\left(g(x)\right)g'(x)$$

$$= f'\left(\frac{1}{x}\right)g'(x) = \frac{1}{\left(\frac{1}{x}\right)^2+1}\left(-\frac{1}{x^2}\right)$$

$$= \frac{1}{\frac{1+x^2}{x^2}}\left(-\frac{1}{x^2}\right) = -\frac{1}{x^2+1}$$

23. $f'(x) = \dfrac{1}{\left(x^2+1\right)}; g(x) = x^2+1; g'(x) = 2x$

$$\frac{d}{dx}f\left(g(x)\right) = f'\left(g(x)\right)g'(x)$$

$$= f'\left(x^2+1\right)g'(x)$$

$$= \frac{1}{\left(x^2+1\right)^2+1}\left(2x\right) = \frac{2x}{\left(x^2+1\right)^2+1}$$

24. $f'(x) = x\sqrt{1-x^2}; g(x) = x^2; g'(x) = 2x$

$$\frac{d}{dx}f\left(g(x)\right) = f'\left(g(x)\right)g'(x)$$

$$= f'\left(x^2\right)g'(x)$$

$$= x^2\sqrt{1-\left(x^2\right)^2}\left(2x\right) = 2x^3\sqrt{1-x^4}$$

25. $f'(x) = x\sqrt{1-x^2}; g(x) = \sqrt{x}; g'(x) = \dfrac{1}{2\sqrt{x}}$

$$\frac{d}{dx}f\left(g(x)\right) = f'\left(g(x)\right)g'(x)$$

$$= f'\left(\sqrt{x}\right)g'(x)$$

$$= \sqrt{x}\sqrt{1-\left(\sqrt{x}\right)^2}\,\frac{1}{2\sqrt{x}}$$

$$= \frac{1}{2}\sqrt{1-x}$$

26. $f'(x) = x\sqrt{1-x^2}; g(x) = x^{3/2}; g'(x) = \dfrac{3}{2}\sqrt{x}$

$$\frac{d}{dx}f\left(g(x)\right) = f'\left(g(x)\right)g'(x)$$

$$= f'\left(x^{3/2}\right)g'(x)$$

$$= x^{3/2}\sqrt{1-\left(x^{3/2}\right)^2}\left(\frac{3}{2}\sqrt{x}\right)$$

$$= \frac{3}{2}x^2\sqrt{1-x^3}$$

27. $\dfrac{dy}{du} = \dfrac{u}{u^2+1}, u = x^{3/2}, \dfrac{du}{dx} = \dfrac{3}{2}x^{1/2},$

$$\frac{dy}{dx} = \frac{dy}{du}\frac{du}{dx} = \left(\frac{u}{u^2+1}\right)\frac{3}{2}x^{1/2}$$

Substitute $x^{3/2}$ for u.

$$\frac{dy}{dx} = \frac{x^{3/2}}{x^3+1}\left(\frac{3}{2}x^{1/2}\right) = \frac{3x^2}{2(x^3+1)}$$

28. $\dfrac{dy}{du} = \dfrac{u}{u^2+1},\ u = x^2+1,\ \dfrac{du}{dx} = 2x,$

$\dfrac{dy}{dx} = \left(\dfrac{u}{u^2+1}\right)(2x)$

Substitute (x^2+1) for u.

$\dfrac{dy}{dx} = \dfrac{(x^2+1)}{(x^2+1)^2+1}(2x) = \dfrac{2x(x^2+1)}{x^4+2x+2}$

29. $\dfrac{dy}{du} = \dfrac{u}{u^2+1},\ u = \dfrac{5}{x},\ \dfrac{du}{dx} = -\dfrac{5}{x^2}$

$\dfrac{dy}{dx} = \dfrac{\frac{5}{x}}{\frac{25}{x^2}+1}\left(-\dfrac{5}{x^2}\right) = -\dfrac{25}{x(25+x^2)}$

30. $\dfrac{dy}{du} = \dfrac{u}{\sqrt{1+u^4}},\ u = x^2,\ \dfrac{du}{dx} = 2x$

$\dfrac{dy}{dx} = \dfrac{x^2}{\sqrt{1+x^8}}(2x) = \dfrac{2x^3}{\sqrt{1+x^8}}$

31. $\dfrac{dy}{du} = \dfrac{u}{\sqrt{1+u^4}},\ u = \sqrt{x},\ \dfrac{du}{dx} = \dfrac{1}{2}x^{-1/2}$

$\dfrac{dy}{dx} = \dfrac{x^{1/2}}{\sqrt{1+x^2}}\left(\dfrac{1}{2}x^{-1/2}\right) = \dfrac{1}{2\sqrt{1+x^2}}$

32. $\dfrac{dy}{du} = \dfrac{u}{\sqrt{1+u^4}},\ u = \dfrac{2}{x},\ \dfrac{du}{dx} = \dfrac{-2}{x^2}$

$\dfrac{dy}{dx} = \dfrac{\frac{2}{x}}{\sqrt{1+\frac{16}{x^4}}}\left(-\dfrac{2}{x^2}\right) = -\dfrac{4}{x^3\sqrt{1+\frac{16}{x^4}}}$

Refer to these graphs for exercises 33–38.

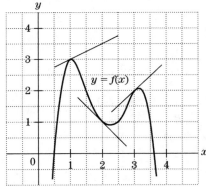

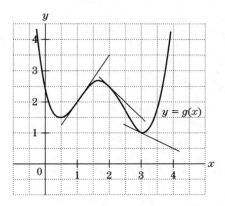

33. $f(1) = 3,\ f'(1) = \dfrac{1}{2},\ g(1) = 2,\ g'(1) = \dfrac{3}{2}$

$h(x) = 2f(x) - 3g(x)$

$h(1) = 2f(1) - 3g(1) = 2(3) - 3(2) = 0$

$h'(1) = 2f'(1) - 3g'(1) = 2\left(\dfrac{1}{2}\right) - 3\left(\dfrac{3}{2}\right) = -\dfrac{7}{2}$

34. $h(x) = f(x)\cdot g(x)$

$h'(x) = f(x)g'(x) + g(x)f'(x)$

$h(1) = 3(2) = 6;\ h'(1) = 3\left(\dfrac{3}{2}\right) + (2)\left(\dfrac{1}{2}\right) = \dfrac{11}{2}$

35. $h(x) = \dfrac{f(x)}{g(x)},\ h'(x) = \dfrac{g(x)f'(x) - g'(x)f(x)}{[g(x)]^2},$

$h(1) = \dfrac{3}{2},\ h'(x) = \dfrac{2\left(\frac{1}{2}\right) - 3\left(\frac{3}{2}\right)}{2^2} = -\dfrac{7}{8}$

36. $h(x) = [f(x)]^2,\ h'(x) = 2f(x)f'(x),$

$h(1) = 3^2 = 9,\ h'(1) = 2(3)\left(\dfrac{1}{2}\right) = 3$

37. $h(x) = f(g(x)),\ h'(x) = f'(g(x))g'(x)$

$g(1) = 2, f(2) = 1,\ f'(2) = -1,\ g'(1) = \dfrac{3}{2}$

$h(1) = 1,\ h'(1) = f'(2)g'(1) = -1\left(\dfrac{3}{2}\right) = -\dfrac{3}{2}$

38. $h(x) = g(f(x)),\ h'(x) = g'(f(x))f'(x)$

$f(1) = 3,\ f'(1) = \dfrac{1}{2},\ g(3) = 1,\ g'(3) = -\dfrac{1}{2}$

$h(1) = 1,\ h'(1) = g'(3)f'(1) = \left(-\dfrac{1}{2}\right)\left(\dfrac{1}{2}\right) = -\dfrac{1}{4}$

39. a. $\dfrac{dR}{dA},\dfrac{dA}{dt},\dfrac{dR}{dx},\dfrac{dx}{dA}$

b. $\dfrac{dR}{dt} = \dfrac{dR}{dx}\dfrac{dx}{dA}\dfrac{dA}{dt}$

40. a. $\dfrac{dP}{dt}, \dfrac{dA}{dP}, \dfrac{dS}{dP}, \dfrac{dA}{dS}$

b. $\dfrac{dA}{dt} = \dfrac{dA}{dS} \cdot \dfrac{dS}{dP} \cdot \dfrac{dP}{dt}$

41. $x^{2/3} + y^{2/3} = 8$

a. $\dfrac{2}{3}x^{-1/3} + \dfrac{2}{3}y^{-1/3}\dfrac{dy}{dx} = 0$

$\dfrac{dy}{dx} = -\dfrac{2}{3}x^{-1/3} \cdot \dfrac{3}{2}y^{1/3} = -\dfrac{y^{1/3}}{x^{1/3}}$

b. slope $= \dfrac{dy}{dx}$
When $x = 8$, $y = -8$, slope $= 1$.

42. $x^3 + y^3 = 9xy$

a. $3x^2 + 3y^2\dfrac{dy}{dx} = 9x\dfrac{dy}{dx} + 9y$

$\dfrac{dy}{dx} = \dfrac{9y - 3x^2}{3y^2 - 9x} = \dfrac{3y - x^2}{y^2 - 3x}$ or $\dfrac{x^2 - 3y}{3x - y^2}$

b. slope $= \dfrac{dy}{dx}$
When $x = 2$, $y = 4$, slope $= \dfrac{3(4) - 4}{16 - 6} = \dfrac{4}{5}$.

43. $x^2y^2 = 9$

$2xy^2 + 2yx^2\dfrac{dy}{dx} = 0 \Rightarrow \dfrac{dy}{dx} = -\dfrac{2xy^2}{2yx^2} = -\dfrac{y}{x}$

When $x = 1$, $y = 3$, $\dfrac{dy}{dx} = -3$.

44. $xy^4 = 48$

$y^4 + 4y^3x\dfrac{dy}{dx} = 0 \Rightarrow \dfrac{dy}{dx} = -\dfrac{y^4}{4y^3x} = -\dfrac{y}{4x}$

When $x = 3$, $y = 2$, $\dfrac{dy}{dx} = \dfrac{-2}{4(3)} = -\dfrac{1}{6}$.

45. $x^2 - xy^3 = 20$

$2x - \left(y^3 + 3y^2x\dfrac{dy}{dx}\right) = 0 \Rightarrow \dfrac{dy}{dx} = \dfrac{2x - y^3}{3y^2x}$

When $x = 5$, $y = 1$, $\dfrac{dy}{dx} = \dfrac{10 - 9}{3(1)(5)} = \dfrac{3}{5}$.

46. $xy^2 - x^3 = 10$

$\left(y^2 + 2yx\dfrac{dy}{dx}\right) - 3x^2 = 0 \Rightarrow \dfrac{dy}{dx} = \dfrac{3x^2 - y^2}{2yx}$

When $x = 2$, $y = 3$, $\dfrac{dy}{dx} = \dfrac{3(4) - 9}{2(3)(2)} = \dfrac{1}{4}$.

47. $y^2 - 5x^3 = 4$

a. $2y\dfrac{dy}{dx} - 15x^2 = 0 \Rightarrow \dfrac{dy}{dx} = \dfrac{15x^2}{2y}$

b. When $x = 4$ and $y = 18$,
$\dfrac{dy}{dx} = \dfrac{15(16)}{2(18)} = \dfrac{20}{3}$ (thousand dollars per thousand-unit increase in production).

c. $\dfrac{dy}{dt} = \dfrac{dy}{dx}\dfrac{dx}{dt} = \dfrac{15x^2}{2y}\dfrac{dx}{dt}$

d. When $x = 4$, $y = 18$, and $\dfrac{dx}{dt} = .3$,
$\dfrac{dy}{dt} = \dfrac{15(16)}{2(18)}(.3) = 2$ (thousand dollars per week).

48. $y^3 - 8000x^2 = 0$

a. $3y^2\dfrac{dy}{dx} - 16,000x = 0 \Rightarrow \dfrac{dy}{dx} = \dfrac{16,000x}{3y^2}$

b. When $x = 27$, $y = 180$,
$\dfrac{dy}{dx} = \dfrac{16,000(27)}{3(32,400)} = \dfrac{40}{9} \approx 4.44$ (thousand books per person).

c. $\dfrac{dy}{dt} = \dfrac{dy}{dx}\dfrac{dx}{dt} = \dfrac{16,000x}{3y^2}\dfrac{dx}{dt}$

d. When $x = 27$, $y = 180$, and $\dfrac{dx}{dt} = 1.8$,
$\dfrac{dy}{dt} = \dfrac{16,000(27)}{3(32,400)}(1.8) = 8$ (thousand books per year).

49. $6p + 5x + xp = 50$

$$6 + 5\frac{dx}{dp} + p\frac{dx}{dp} + x = 0 \Rightarrow \frac{dx}{dp} = \frac{-x-6}{5+p}$$

$$\frac{dx}{dt} = \frac{dx}{dp}\frac{dp}{dt} = \left(\frac{-x-6}{5+p}\right)\frac{dp}{dt}$$

When $x = 4$, $p = 3$, and $\frac{dp}{dt} = -2$,

$$\frac{dx}{dt} = \left(\frac{-4-6}{5+3}\right)(-2) = 2.5 .$$

The quantity x is increasing at the rate of 2.5 units per unit time.

50. $V = .005\pi r^2$, so

$$\frac{dV}{dt} = .005\pi(2)r\frac{dr}{dt} = .01\pi r\frac{dr}{dt} .$$

Now, $\frac{dV}{dt} = 20$, so $20 = .01\pi r\frac{dr}{dt} \Rightarrow$

$$\frac{dr}{dt} = \frac{2000}{\pi r}.$$

When $r = 50$, $\frac{dr}{dt} = \frac{2000}{50\pi} = \frac{40}{\pi}$ m/hr, or approximately 12.73 meters per hour.

51. $S = .1W^{2/3}$, so

$$\frac{dS}{dt} = \frac{2}{3}(.1)W^{-1/3}\frac{dW}{dt} = \frac{.2}{3}W^{-1/3}\frac{dW}{dt}.$$

When $W = 350$ and $\frac{dW}{dt} = 200$,

$$\frac{dS}{dt} = \frac{.2}{3\sqrt[3]{350}}(200) = \frac{40}{3\sqrt[3]{350}} \approx 1.89 \text{ m}^2/\text{yr}.$$

52. $xy - 6x + 20y = 0$,

$$\frac{dx}{dt}y + \frac{dy}{dt}x - 6\frac{dx}{dt} + 20\frac{dy}{dt} = 0$$

Currently, $x = 10$, $y = 2$, $\frac{dx}{dt} = 1.5$. Thus,

$$1.5(2) + \frac{dy}{dt}(10) - 6(1.5) + 20\frac{dy}{dt} = 0, \text{ so}$$

$$30\frac{dy}{dt} = 6 \Rightarrow \frac{dy}{dt} = .2 \text{ or } 200$$

dishwashers/month.

Chapter 4 The Exponential and Natural Logarithm Functions

4.1 Exponential Functions

1. $4^x = \left(2^2\right)^x = 2^{2x}$

$\left(\sqrt{3}\right)^x = \left(3^{1/2}\right)^x = 3^{(1/2)x}$

$\left(\dfrac{1}{9}\right)^x = \left(9^{-1}\right)^x = \left(\left(3^2\right)^{-1}\right)^x = 3^{-2x}$

3. $8^{2x/3} = \left(2^3\right)^{2x/3} = 2^{3(2x/3)} = 2^{2x}$

$9^{3x/2} = \left(3^2\right)^{3x/2} = 3^{2(3x/2)} = 3^{3x}$

$16^{-3x/4} = \left(2^4\right)^{-3x/4} = 2^{4(-3x/4)} = 2^{-3x}$

5. $\left(\dfrac{1}{4}\right)^{2x} = \left(4^{-1}\right)^{2x} = \left(\left(2^2\right)^{-1}\right)^{2x} = 2^{-4x}$

$\left(\dfrac{1}{8}\right)^{-3x} = \left(8^{-1}\right)^{-3x} = \left(\left(2^3\right)^{-1}\right)^{-3x} = 2^{9x}$

$\left(\dfrac{1}{81}\right)^{x/2} = \left(81^{-1}\right)^{x/2} = \left(\left(3^4\right)^{-1}\right)^{x/2} = 3^{-2x}$

7. $6^x \cdot 3^{-x} = (2 \cdot 3)^x \cdot 3^{-x} = 2^x \cdot \left(3^x \cdot 3^{-x}\right)$
$= 2^x \cdot \left(3^{x-x}\right) = 2^x$

$\dfrac{15^x}{5^x} = \dfrac{(3 \cdot 5)^x}{5^x} = \dfrac{3^x \cdot 5^x}{5^x} = 3^x$

$\dfrac{12^x}{2^{2x}} = \dfrac{(3 \cdot 4)^x}{2^{2x}} = \dfrac{3^x \cdot 4^x}{2^{2x}} = \dfrac{3^x \cdot (2^2)^x}{2^{2x}} = 3^x$

9. $\dfrac{3^{4x}}{3^{2x}} = 3^{4x} \cdot 3^{-2x} = 3^{4x-2x} = 3^{2x}$

$\dfrac{2^{5x+1}}{2 \cdot 2^{-x}} = \dfrac{2^{5x+1}}{2^{1-x}} = 2^{5x+1} \cdot 2^{-(1-x)}$
$= 2^{5x+1-(1-x)} = 2^{6x}$

$\dfrac{9^{-x}}{27^{-x/3}} = \dfrac{\left(3^2\right)^{-x}}{\left(3^3\right)^{-x/3}} = \dfrac{3^{-2x}}{3^{-x}} = 3^{-2x} \cdot 3^x$
$= 3^{-2x+x} = 3^{-x}$

11. $2^{3x} \cdot 2^{-5x/2} = 2^{3x-(5x/2)} = 2^{(6x/2)-(5x/2)}$
$= 2^{x/2}$

$3^{2x} \cdot \left(\dfrac{1}{3}\right)^{2x/3} = 3^{2x} \cdot \left(3^{-1}\right)^{2x/3} = 3^{2x} \cdot 3^{-2x/3}$
$= 3^{(6x/3)-(2x/3)} = 3^{4x/3} = 3^{(4/3)x}$

13. $\left(2^{-3x} \cdot 2^{-2x}\right)^{2/5} = \left(2^{-3x-2x}\right)^{2/5}$
$= \left(2^{-5x}\right)^{2/5} = 2^{-2x}$

$\left(9^{1/2} \cdot 9^4\right)^{x/9} = \left(\left(3^2\right)^{1/2} \cdot \left(3^2\right)^4\right)^{x/9}$
$= \left(3^1 \cdot 3^8\right)^{x/9} = \left(3^{1+8}\right)^{x/9}$
$= \left(3^9\right)^{x/9} = 3^x$

15. $f(x) = 3^{-2x} = \left(3^{-2}\right)^x = \left(\dfrac{1}{9}\right)^x \Rightarrow b = \dfrac{1}{9}$

17. $5^{2x} = 5^2 \Rightarrow 2x = 2 \Rightarrow x = 1$

19. $(2.5)^{2x+1} = (2.5)^5 \Rightarrow 2x+1 = 5 \Rightarrow x = \dfrac{5-1}{2} = 2$

21. $10^{1-x} = 100 \Rightarrow 10^{1-x} = 10^2 \Rightarrow 1-x = 2 \Rightarrow$
$x = -1$

23. $3(2.7)^{5x} = 8.1 \Rightarrow 2.7^{5x} = 2.7 \Rightarrow 5x = 1 \Rightarrow$
$x = \dfrac{1}{5}$

25. $\left(2^{x+1} \cdot 2^{-3}\right)^2 = 2 \Rightarrow \left(2^{x+1-3}\right)^2 = 2 \Rightarrow$
$\left(2^{x-2}\right)^2 = 2 \Rightarrow 2^{2x-4} = 2 \Rightarrow 2x - 4 = 1 \Rightarrow$
$x = \dfrac{4+1}{2} = \dfrac{5}{2}$

27. $2^{3x} = 4 \cdot 2^{5x} \Rightarrow 2^{3x} = 2^2 \cdot 2^{5x} \Rightarrow$
$2^{3x} = 2^{2+5x} \Rightarrow 3x = 2 + 5x \Rightarrow 2x = -2 \Rightarrow$
$x = -1$

29. $(1+x)2^{-x} - 5 \cdot 2^{-x} = 0 \Rightarrow 2^{-x}(1+x-5) = 0 \Rightarrow$
$2^{-x}(x-4) = 0$

Since $2^{-x} \neq 0$ for every x, then $x = 4$ is the only solution.

31. $2^x - \dfrac{8}{2^{2x}} = 0 \Rightarrow 2^x - \dfrac{2^3}{2^{2x}} = 0 \Rightarrow$
$2^x - 2^{3-2x} = 0 \Rightarrow 2^x = 2^{3-2x} \Rightarrow$
$x = 3 - 2x \Rightarrow x = 1$

33. $2^{2x} - 6 \cdot 2^x + 8 = 0 \Rightarrow \left(2^x\right)^2 - 6 \cdot 2^x + 8 = 0$

Let $X = 2^x \Rightarrow X^2 - 6X + 8 = 0 \Rightarrow$
$(X-2)(X-4) = 0 \Rightarrow X = 2, X = 4 \Rightarrow$
$2^x = 2$ or $2^x = 4 \Rightarrow x = 1$ or $x = 2$

35. $3^{2x} - 12 \cdot 3^x + 27 = 0 \Rightarrow (3^x)^2 - 12 \cdot 3^x + 27 = 0$

Let $X = 3^x \Rightarrow X^2 - 12X + 27 = 0 \Rightarrow$
$(X - 3)(X - 9) = 0 \Rightarrow X = 3, X = 9 \Rightarrow$
$3^x = 3$ or $3^x = 9 \Rightarrow x = 1$ or $x = 2$

37. $2^{3+h} = 2^3 \cdot 2^h$ The missing factor is 2^h.

39. $2^{x+h} - 2^x = 2^x \cdot 2^h - 2^x = 2^x(2^h - 1)$

The missing factor is $2^h - 1$.

41. $3^{x/2} + 3^{-x/2} = 3^{x-(x/2)} + 3^{-x/2}$
$= 3^x \cdot 3^{-x/2} + 3^{-x/2}$
$= 3^{-x/2}(3^x + 1)$

The missing factor is $3^x + 1$.

43.

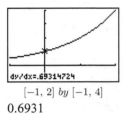

dy/dx=.69314724

$[-1, 2]$ by $[-1, 4]$

0.6931

45. By trial and error, $b = 2.7$.

4.2 The Exponential Function e^x

1. If $h = .1$, then $\dfrac{3^h - 1}{h} \approx 1.16$.

If $h = .01$, then $\dfrac{3^h - 1}{h} \approx 1.10$.

If $h = .001$, then $\dfrac{3^h - 1}{h} \approx 1.10$.

Therefore, $\dfrac{d}{dx}(3^x)\Big|_{x=0} = \lim_{x \to 0} \dfrac{3^h - 1}{h} \approx 1.1$.

3. $m = \dfrac{d}{dx}\left(2^x\right)\Big|_{x=0} \approx .693$ from formula (2) in the text.

a. $\dfrac{d}{dx}(2^x)\Big|_{x=1} = m \cdot 2^1 = 2m$
$\approx 2(.693) = 1.386$

b. $\dfrac{d}{dx}(2^x)\Big|_{x=-2} = m \cdot 2^{-2} = \dfrac{1}{4}m$
$\approx \dfrac{1}{4}(0.693) = 0.1733$

5. a. $\dfrac{d}{dx}(e^x)\Big|_{x=1} = e^1 \approx 2.71828$

b. $\dfrac{d}{dx}(e^x)\Big|_{x=-1} = e^{-1} = \dfrac{1}{e} \approx .367879$

7. $\left(e^2\right)^x = e^{2x}$

$\left(\dfrac{1}{e}\right)^x = \left(e^{-1}\right)^x = e^{-x}$

9. $\left(\dfrac{1}{e^3}\right)^{2x} = \left(e^{-3}\right)^{2x} = e^{-6x}$

$e^{1-x} \cdot e^{3x-1} = e^{2x}$

11. $\left(e^{4x} \cdot e^{6x}\right)^{3/5} = \left(e^{10x}\right)^{3/5} = e^{6x}$

$\dfrac{1}{e^{-2x}} = \left(e^{-2x}\right)^{-1} = e^{2x}$

13. $e^{5x} = e^{20} \Rightarrow 5x = 20 \Rightarrow x = 4$

15. $e^{x^2 - 2x} = e^8 \Rightarrow x^2 - 2x = 8 \Rightarrow$
$x^2 - 2x - 8 = 0 \Rightarrow (x - 4)(x + 2) = 0 \Rightarrow$
$x = 4$ or $x = -2$

17. $e^x(x^2 - 1) = 0 \Rightarrow e^x = 0$ (no solution) or
$x^2 - 1 = 0 \Rightarrow x = \pm 1$

19. The tangent passes through the point
$\left(-1, e^{-1}\right) = \left(-1, \dfrac{1}{e}\right)$, or $(-1, 0.37)$. The slope
of the tangent is given by
$\dfrac{dy}{dx}\Big|_{x=-1} = e^x\Big|_{x=-1} = \dfrac{1}{e} \approx 0.37$. Thus, the
equation of the tangent is $y - \dfrac{1}{e} = \dfrac{1}{e}(x + 1)$ or
$y = 0.37x + 0.74$.

21. $\dfrac{d}{dx}e^x = e^x$ and $\dfrac{d^2}{dx^2}e^x = e^x$

e^x is always increasing and $e^x > 0$, so by the
first derivative test, there are no relative
extreme points. Because $\dfrac{d^2}{dx^2}e^x = e^x > 0$ for
all values of x, then e^x is concave up.

23. The slope of the graph of e^x at (a, b) is
$e^a = b$.

25. $\dfrac{d}{dx}\left(3e^x - 7x\right) = \dfrac{d}{dx}\left(3e^x\right) - \dfrac{d}{dx}\left(7x\right) = 3e^x - 7$

27. $\dfrac{d}{dx}\left(xe^x\right) = x\dfrac{d}{dx}\left(e^x\right) + e^x\dfrac{d}{dx}x$

$\qquad = xe^x + e^x = (x+1)e^x$

29. $\dfrac{d}{dx}\left[\left(8e^x\right)\left(1+2e^x\right)^2\right]$

$\qquad = \left(8e^x\right)\dfrac{d}{dx}\left[\left(1+2e^x\right)^2\right] + \left(1+2e^x\right)^2\dfrac{d}{dx}\left(8e^x\right)$

$\qquad = \left(8e^x\right)(2)\left(1+2e^x\right)\left(2e^x\right) + \left(1+2e^x\right)^2\left(8e^x\right)$

$\qquad = 32e^{2x}\left(1+2e^x\right) + 8e^x\left(1+2e^x\right)^2$

$\qquad = 32e^{2x} + 64e^{3x} + 8e^x\left(1+4e^x+4e^{2x}\right)$

$\qquad = 8e^x + 64e^{2x} + 96e^{3x}$

$\qquad = 8e^x\left(1+8e^x+12e^{2x}\right)$

$\qquad = 8e^x\left(1+6e^x\right)\left(1+2e^x\right)$

31. $\dfrac{d}{dx}\left(\dfrac{e^x}{x+1}\right) = \dfrac{(x+1)\left(e^x\right) - e^x(1)}{(x+1)^2}$

$\qquad = \dfrac{xe^x}{(x+1)^2}$

33. $\dfrac{d}{dx}\left(\dfrac{e^x-1}{e^x+1}\right) = \dfrac{\left(e^x+1\right)\left(e^x\right) - \left(e^x-1\right)\left(e^x\right)}{\left(e^x+1\right)^2}$

$\qquad = \dfrac{e^{2x} + e^x - e^{2x} + e^x}{\left(e^x+1\right)^2}$

$\qquad = \dfrac{2e^x}{\left(e^x+1\right)^2}$

35. $y' = 1 - e^x; \; y'' = -e^x$

$1 - e^x = 0 \Rightarrow 1 = e^x \Rightarrow x = 0$

$-e^0 = -1 < 0,$ so there is a maximum point at

$\left(0, \, 0-e^0\right) = (0, \, -1).$

37. $y = (1+x^2)e^x$

$y' = (1+x^2)e^x + e^x(2x) = e^x(x^2 + 2x + 1)$

$y' = 0 \Rightarrow e^x(x^2 + 2x + 1) = 0 \Rightarrow$

$x^2 + 2x + 1 = 0 \Rightarrow (x+1)(x+1) = 0 \Rightarrow x = -1$

The tangent line is horizontal at $\left(-1, 2e^{-1}\right)$

or $\left(-1, \dfrac{2}{e}\right).$

39. $\dfrac{dy}{dx} = xe^x + e^x$

$\left.\dfrac{dy}{dx}\right|_{x=0} = e^0 + 0 \cdot e^0 = 1$

The slope of the tangent line is 1.

41. $\dfrac{dy}{dx} = \dfrac{(1+2e^x)e^x - e^x(2e^x)}{(1+2e^x)^2}$

$\qquad = \dfrac{e^x}{(1+2e^x)^2}$

$\left.\dfrac{dy}{dx}\right|_{x=0} = \dfrac{e^0}{(1+2e^0)^2} = \dfrac{1}{9}$

$(x_1, y_1) = \left(0, \dfrac{1}{3}\right), \; m = \dfrac{1}{9}$

$y - \dfrac{1}{3} = \dfrac{1}{9}(x-0) \Rightarrow y = \dfrac{1}{9}x + \dfrac{1}{3}$

43. $f(x) = e^x(1+x)^2$

$f'(x) = e^x(2)(1+x) + e^x(1+x)^2$

$\qquad = e^x(1+x)(3+x)$

$\qquad = e^x(x^2 + 4x + 3)$

$f''(x) = e^x(2x+4) + (x^2+4x+3)(e^x)$

$\qquad = e^x(x^2 + 6x + 7)$

45. a. $\dfrac{d}{dx}\left(5e^x\right) = 5\left(e^x\right) + e^x(0) = 5e^x$

   b. $\dfrac{d}{dx}\left(e^x\right)^{10} = 10\left(e^x\right)^9 e^x = 10e^{10x}$

   c. $\dfrac{d}{dx}\left(e^{2+x}\right) = \dfrac{d}{dx}\left(e^2 \cdot e^x\right)$

$\qquad\qquad = e^2\left(e^x\right) + e^x(0) = e^{2+x}$

47.

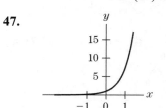

49.

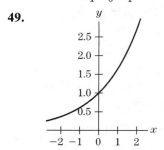

51.

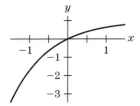

53. $\dfrac{dy}{dx} = e^x = 1$ at $x = 0$.

$y = 1$ at $x = 0$

$y - 1 = 1(x - 0) \Rightarrow y = x + 1$

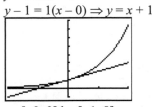

$[-2, 2]$ by $[-1, 8]$

The graph confirms the answer.

55.

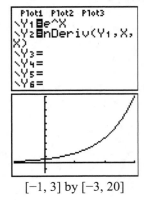

$[-1, 3]$ by $[-3, 20]$

4.3 Differentiation of Exponential Functions

1. $\dfrac{d}{dx}\left(e^{2x+3}\right) = e^{2x+3}\dfrac{d}{dx}(2x+3)$

$= e^{2x+3}(2) = 2e^{2x+3}$

3. $\dfrac{d}{dx}\left(e^{4x^2-x}\right) = e^{4x^2-x}\dfrac{d}{dx}\left(4x^2 - x\right)$

$= (8x-1)e^{4x^2-x}$

5. $\dfrac{d}{dx}\left(e^{e^x}\right) = e^{e^x}\dfrac{d}{dx}\left(e^x\right) = e^{e^x}e^x$

7. $\dfrac{d}{dx}e^{\sqrt{x}} = e^{\sqrt{x}}\dfrac{d}{dx}\sqrt{x} = e^{\sqrt{x}}\left(\dfrac{1}{2}x^{-1/2}\right)$

$= \dfrac{e^{\sqrt{x}}}{2\sqrt{x}}$

9. $\dfrac{d}{dx}\left(-7e^{x/7}\right) = -7\dfrac{d}{dx}\left(e^{x/7}\right) = -7e^{x/7}\dfrac{d}{dx}\left(\dfrac{x}{7}\right)$

$= -7e^{x/7}\left(\dfrac{1}{7}\right) = -e^{x/7}$

11. $\dfrac{d}{dt}\left(4e^{0.05t} - 23e^{0.01t}\right)$

$= \dfrac{d}{dt}\left(4e^{0.05t}\right) - \dfrac{d}{dt}\left(23e^{0.01t}\right)$

$= 4\dfrac{d}{dt}\left(e^{0.05t}\right) - 23\dfrac{d}{dt}\left(e^{0.01t}\right)$

$= 4e^{0.05t}\dfrac{d}{dt}(0.05t) - 23e^{0.01t}\dfrac{d}{dt}(0.01t)$

$= 4e^{0.05t}(0.05) - 23e^{0.01t}(0.01)$

$= 0.2e^{0.05t} - 0.23e^{0.01t}$

13. $f(t) = \left(t^2 + 2e^t\right)e^{t-1}$

To differentiate, use the product rule.

$\dfrac{d}{dt}\left[\left(t^2 + 2e^t\right)e^{t-1}\right]$

$= e^{t-1}\dfrac{d}{dt}\left(t^2 + 2e^t\right) + \left(t^2 + 2e^t\right)\dfrac{d}{dt}\left(e^{t-1}\right)$

$= e^{t-1}\left(2t + 2e^t\right) + \left(t^2 + 2e^t\right)\left(e^{t-1}\right)$

$= e^{t-1}\left(2t + 2e^t + t^2 + 2e^t\right)$

$= e^{t-1}\left(t^2 + 2t + 4e^t\right)$

15. $f(x) = \left(x + \dfrac{1}{x}\right)e^{2x}$

To differentiate, use the product rule.

$\dfrac{d}{dx}\left[\left(x + \dfrac{1}{x}\right)e^{2x}\right]$

$= e^{2x}\dfrac{d}{dx}\left(x + \dfrac{1}{x}\right) + \left(x + \dfrac{1}{x}\right)\dfrac{d}{dx}\left(e^{2x}\right)$

$= e^{2x}\left(1 - \dfrac{1}{x^2}\right) + \left(x + \dfrac{1}{x}\right)\left(2e^{2x}\right)$

$= e^{2x}\left[1 - \dfrac{1}{x^2} + 2\left(x + \dfrac{1}{x}\right)\right]$

$= e^{2x}\left(1 + 2x + \dfrac{2}{x} - \dfrac{1}{x^2}\right)$

17. $f(x) = \dfrac{e^x + e^{-x}}{e^x - e^{-x}}$

To differentiate, use the quotient rule.

$\dfrac{d}{dx}\left(\dfrac{e^x + e^{-x}}{e^x - e^{-x}}\right)$

$= \dfrac{\left(e^x - e^{-x}\right)\dfrac{d}{dx}\left(e^x + e^{-x}\right) - \left(e^x + e^{-x}\right)\dfrac{d}{dx}\left(e^x - e^{-x}\right)}{\left(e^x - e^{-x}\right)^2}$

$= \dfrac{\left(e^x - e^{-x}\right)\left(e^x - e^{-x}\right) - \left(e^x + e^{-x}\right)\left(e^x + e^{-x}\right)}{\left(e^x - e^{-x}\right)^2}$

$= \dfrac{\left(e^{2x} - 2e^x e^{-x} + e^{-2x}\right) - \left(e^{2x} + 2e^x e^{-x} + e^{-2x}\right)}{\left(e^x - e^{-x}\right)^2}$

$= \dfrac{-2 - 2}{\left(e^x - e^{-x}\right)^2} = -\dfrac{4}{\left(e^x - e^{-x}\right)^2}$

19. $\dfrac{d}{dx}\sqrt{e^x + 1} = \dfrac{d}{dx}\left(e^x + 1\right)^{1/2}$

$= \dfrac{1}{2}\left(e^x + 1\right)^{-1/2}\dfrac{d}{dx}\left(e^x + 1\right)$

$= \dfrac{1}{2}\left(e^x + 1\right)^{-1/2}\left(e^x\right)$

$= \dfrac{e^x}{2\sqrt{e^x + 1}}$

21. $f(x) = \left(e^{3x}\right)^5 = e^{15x}$

$\dfrac{d}{dx}\left(e^{3x}\right)^5 = \dfrac{d}{dx}\left(e^{15x}\right) = e^{15x}\dfrac{d}{dx}(15x) = 15e^{15x}$

23. $f(x) = \dfrac{1}{\sqrt{e^x}} = \left(e^x\right)^{-1/2} = e^{-x/2}$

$\dfrac{d}{dx}\left(\dfrac{1}{\sqrt{e^x}}\right) = \dfrac{d}{dx}\left(e^{-x/2}\right) = e^{-x/2}\dfrac{d}{dx}\left(-\dfrac{x}{2}\right)$

$= -\dfrac{e^{-x/2}}{2} = -\dfrac{1}{2\sqrt{e^x}}$

25. $f(x) = \dfrac{e^x + 5e^{2x}}{e^x} = 1 + 5e^x$

$\dfrac{d}{dx}\dfrac{e^x + 5e^{2x}}{e^x} = \dfrac{d}{dx}\left(1 + 5e^x\right) = 5e^x$

27. $\dfrac{d}{dx}\left[(1+x)e^{-3x}\right] = (1+x)e^{-3x}(-3) + e^{-3x}(1)$

$= (-3 - 3x)e^{-3x} + e^{-3x}$

$= (-2 - 3x)e^{-3x}$

$(-2 - 3x)e^{-3x} = 0 \Rightarrow -2 - 3x = 0 \Rightarrow x = -\dfrac{2}{3}$

$\dfrac{d}{dx}\left[(-2 - 3x)e^{-3x}\right]$

$= (-2 - 3x)e^{-3x}(-3) + e^{-3x}(-3)$

$= (6 + 9x)e^{-3x} - 3e^{-3x}$

$= (3 + 9x)e^{-3x} = 3(1 + 3x)e^{-3x}$

At $x = -\dfrac{2}{3}$, $3(1 + 3x)e^{-3x} < 0$, so there is a

maximum at $x = -\dfrac{2}{3}$.

29. $\dfrac{d}{dx}\left(\dfrac{3 - 4x}{e^{2x}}\right) = \dfrac{d}{dx}\left[(3 - 4x)e^{-2x}\right]$

$= (3 - 4x)e^{-2x}(-2) + e^{-2x}(-4)$

$= (-6 + 8x)e^{-2x} - 4e^{-2x}$

$= (8x - 10)e^{-2x}$

$(8x - 10)e^{-2x} = 0 \Rightarrow 8x - 10 = 0 \Rightarrow x = \dfrac{5}{4}$

$\dfrac{d}{dx}\left[(8x - 10)e^{-2x}\right]$

$= (8x - 10)e^{-2x}(-2) + e^{-2x}(8)$

$= (-16x + 20)e^{-2x} + 8e^{-2x}$

$= (-16x + 28)e^{-2x}$

At $x = \dfrac{5}{4}$, $(-16x + 28)e^{-2x} > 0$, so there is a

minimum at $x = \dfrac{5}{4}$.

31. $\dfrac{d}{dx}\left[(5x - 2)e^{1-2x}\right]$

$= (5x - 2)e^{1-2x}(-2) + e^{1-2x}(5)$

$= (-10x + 4)e^{1-2x} + 5e^{1-2x}$

$= (9 - 10x)e^{1-2x}$

$(9 - 10x)e^{1-2x} = 0 \Rightarrow 9 - 10x = 0 \Rightarrow x = \dfrac{9}{10}$

$\dfrac{d}{dx}\left[(9 - 10x)e^{1-2x}\right]$

$= (9 - 10x)e^{1-2x}(-2) + e^{1-2x}(-10)$

$= (-18 + 20x)e^{1-2x} - 10e^{1-2x}$

$= (20x - 28)e^{1-2x}$

At $x = \dfrac{9}{10}$, $(20x - 28)e^{1-2x} < 0$, so there is a

maximum at $x = \dfrac{9}{10}$.

33. a. $f(t) = 3e^{0.06t} + 2e^{0.02t}$

$f(0) = 3e^{0.06(0)} + 2e^{0.02(0)} = 5$

The initial dollar amount invested is $5000.

b. $f(5) = 3e^{0.06(5)} + 2e^{0.02(5)} \approx 6.25992$

After five years, the value of the portfolio is $6259.92.

c. $f'(t) = 3e^{0.06t}(0.06) + 2e^{0.02t}(0.02)$

$= 0.18e^{0.06t} + 0.04e^{0.02t}$

$f'(5) = 0.18e^{0.06(5)} + 0.04e^{0.02(5)}$

≈ 0.28718

After five years, the investment is appreciating at about $287.18 per year.

35. a. $f(t) = 31.87e^{0.096t}$

If $t = 0$ represents 1997, then $t = 18$ represents 2015.

$f(18) = 31.87e^{0.096(18)}$

$\approx \$179.408$ million

b. $f'(t) = 31.87e^{0.096t}(0.096)$

$= 3.05952e^{0.096t}$

$f'(18) = 3.05952e^{0.096(18)} \approx 17.223$

The painting is appreciating at about $17.223 million per year in 2015.

c. $f(23) = 31.87e^{0.096(23)}$

$\approx \$289.937$ million

$f'(23) = 3.05952e^{0.096(23)} \approx 27.834$

The painting is appreciating at about $27.834 million per year in 2020.

37. a. $f(8) \approx 45$ m/sec

b. $f'(0) = 10$ m/sec^2

c. $f(t) = 30$ when $t \approx 4$ sec

d. $f'(t) = 5$ when $t \approx 4$ sec

39. $f'(t) = (-1)(0.05 + e^{-0.4t})^{-2}(e^{-0.4t})(-0.4)$

$= \dfrac{0.4e^{-0.4t}}{(0.05 + e^{-0.4t})^2}$

$f'(7) = \dfrac{.4e^{-2.8}}{(.05 + e^{-2.8})^2} \approx 2$

The rate of growth is about 2 inches per week.

41. $y = e^{-2e^{-0.01x}}$

$\dfrac{dy}{dx} = e^{-2e^{-0.01x}}\left[-2e^{-0.01x}(-.01)\right]$

$= .02e^{-2e^{-0.01x}}e^{-0.01x}$

43. a.

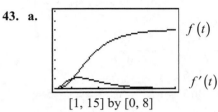

[1, 15] by [0, 8]

The tumor's volume appears to stabilize around 6 ml.

b. $f(5) \approx 3.2$ ml

c. $f(t) = 5$ when $t \approx 7.7$ weeks

d. $f'(5) \approx 0.97$ ml/week

e. $f'(t)$ is maximum when $t \approx 3.74$ weeks

f.   The maximum growth rate appears to be approximately 1.13 ml/week.

4.4   The Natural Logarithm Function

1. $\ln(\sqrt{e}) = \ln e^{1/2} = \dfrac{1}{2}$

3. $e^x = 5 \Rightarrow \ln(e^x) = \ln(5) \Rightarrow x = \ln 5$

5. $\ln x = -1 \Rightarrow e^{\ln x} = e^{-1} \Rightarrow x = \dfrac{1}{e}$

7. $\ln e^{-3} = -3$

9. $e^{e^{\ln 1}} = e^1 = e$

11. $\ln(\ln e) = \ln(1) = 0$

13. $e^{2\ln x} = \left(e^{\ln x}\right)^2 = (x)^2 = x^2$

15. $e^{-2\ln 7} = \left(e^{\ln 7}\right)^{-2} = (7)^{-2} = \dfrac{1}{49}$

17. $e^{\ln x + \ln 2} = e^{\ln x}e^{\ln 2} = (x)(2) = 2x$

19. $e^{2x} = 5 \Rightarrow \ln(e^{2x}) = \ln 5 \Rightarrow 2x = \ln 5 \Rightarrow$

$x = \dfrac{1}{2}\ln 5$

21. $\ln(4 - x) = \dfrac{1}{2} \Rightarrow e^{\ln(4-x)} = e^{1/2} \Rightarrow$

$4 - x = e^{1/2} \Rightarrow x = 4 - e^{1/2}$

23. $\ln x^2 = 9 \Rightarrow e^{\ln x^2} = e^9 \Rightarrow x^2 = e^9 \Rightarrow x = \pm e^{9/2}$

25. $6e^{-.00012x} = 3 \Rightarrow e^{-.00012x} = \dfrac{1}{2} \Rightarrow$

$\ln(e^{-.00012x}) = \ln\left(\dfrac{1}{2}\right) \Rightarrow -.00012x = \ln\dfrac{1}{2} \Rightarrow$

$x = \dfrac{\ln\dfrac{1}{2}}{-.00012} = -\dfrac{\ln.5}{.00012}$

27. $\ln 3x = \ln 5 \Rightarrow e^{\ln 3x} = e^{\ln 5} \Rightarrow 3x = 5 \Rightarrow x = \dfrac{5}{3}$

29. $\ln(\ln 3x) = 0 \Rightarrow e^{\ln(\ln 3x)} = e^0 \Rightarrow \ln 3x = 1 \Rightarrow$

$e^{\ln 3x} = e^1 \Rightarrow 3x = e \Rightarrow x = \dfrac{e}{3}$

31. $2e^{x/3} - 9 = 0 \Rightarrow e^{x/3} = \dfrac{9}{2} \Rightarrow$

$\ln(e^{x/3}) = \ln\left(\dfrac{9}{2}\right) \Rightarrow x = 3\ln\left(\dfrac{9}{2}\right)$

33. $5\ln 2x = 8 \Rightarrow \ln 2x = \dfrac{8}{5} \Rightarrow e^{\ln 2x} = e^{8/5} \Rightarrow$

$2x = e^{8/5} \Rightarrow x = \dfrac{1}{2}e^{8/5}$

35. $\left(e^2\right)^x \cdot e^{\ln 1} = 4 \Rightarrow e^{2x} \cdot 1 = 4 \Rightarrow$

$\ln(e^{2x}) = \ln 4 \Rightarrow 2x = \ln 4 \Rightarrow x = \dfrac{1}{2}\ln 4$

37. $4e^x \cdot e^{-2x} = 6 \Rightarrow e^x \cdot e^{-2x} = \dfrac{3}{2} \Rightarrow e^{x-2x} = \dfrac{3}{2} \Rightarrow$

$\ln(e^{-x}) = \ln\left(\dfrac{3}{2}\right) \Rightarrow -x = \ln\left(\dfrac{3}{2}\right) \Rightarrow x = -\ln\dfrac{3}{2}$

39. $f(x) = -5x + e^x; \; f'(x) = -5 + e^x; \; f''(x) = e^x$

$f'(x) = 0 \Rightarrow -5 + e^x = 0 \Rightarrow \ln(e^x) = \ln 5 \Rightarrow$

$x = \ln 5$. Thus $f'(\ln 5) = 0$ and the

y-coordinate is $y = f(\ln 5) = -5\ln 5 + e^{\ln 5}$

$= -5\ln 5 + 5 = 5 - 5\ln 5$. Using the second

derivative test, $f''(\ln 5) = e^{\ln 5} = 5 > 0$, so

$(\ln 5, 5 - 5\ln 5)$ is the minimum.

41. a. $f(x) = -5x + e^x; \; f'(x) = -5 + e^x;$

$f''(x) = e^x$

Solve $f'(x) = 3$ to find the x-coordinate
of the point(s) on the graph where the
tangent line has slope 3.

$-5 + e^x = 3 \Rightarrow e^x = 8 \Rightarrow x = \ln 8 \approx 2.08$

b.   Solve $f'(x) = -7$ to find the x-coordinate
of the point(s) on the graph where the
tangent line has slope 3.

$-5 + e^x = -7 \Rightarrow e^x = -2$

This equation has no solution, so there are
no points on the graph where the tangent
has slope -7.

43. a. $y = e^{-x}$

$\dfrac{dy}{dx} = -e^{-x} \Rightarrow -e^{-x} = -2 \Rightarrow e^{-x} = 2 \Rightarrow$

$\ln(e^{-x}) = \ln 2 \Rightarrow -x = \ln 2 \Rightarrow$

$x = -\ln 2 \Rightarrow x = \ln\dfrac{1}{2}$

$y = e^{-(-\ln 2)} = 2$

The tangent line has slope -2 at $\left(\ln\dfrac{1}{2},\, 2\right)$.

b.

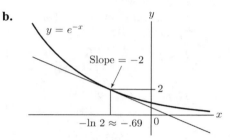

45. $f(x) = e^{-x} + 3x; \; f'(x) = -e^{-x} + 3;$

$f''(x) = e^{-x}$

$f'(x) = 0 \Rightarrow -e^{-x} + 3 = 0 \Rightarrow \ln(e^{-x}) = \ln 3 \Rightarrow$

$-x = \ln 3; \; x = -\ln 3$. Thus $f'(-\ln 3) = 0$ and

the y-coordinate is $f(-\ln 3) = e^{\ln 3} - 3\ln$

$= 3 - 3\ln 3$. Using the second derivative test,

$f''(-\ln 3) = e^{\ln 3} = 3 > 0$, so $(-\ln 3, 3 - 3\ln 3)$

is a relative minimum

47. $e^{0.05t} - 4e^{-0.06t} = 0 \Rightarrow e^{0.05t} = 4e^{-0.06t} \Rightarrow$

$\ln\left(e^{0.05t}\right) = \ln\left(4e^{-0.06t}\right) \Rightarrow$

$.05t = \ln 4 - .06t \Rightarrow .11t = \ln 4 \Rightarrow$

$t = \dfrac{\ln 4}{.11} \approx 12.6$

49. $f(t) = 5(e^{-.01t} - e^{-.51t})$, for $t \geq 0$.

$f'(t) = 5(-.01e^{-.01t} + .51e^{-.51t})$

$f'(t) = 0 \Rightarrow 5(-.01e^{-.01t} + .51e^{-.51t}) = 0 \Rightarrow$

$-e^{-.01t} + 51e^{-.51t} = 0 \Rightarrow 51e^{-.51t} = e^{-.01t} \Rightarrow$

$51e^{-.51t} \cdot e^{.51t} = e^{-.01t} \cdot e^{.51t} \Rightarrow 51(1) = e^{.5t} \Rightarrow$

$\ln 51 = \ln(e^{.5t}) \Rightarrow \ln 51 = .5t \Rightarrow t = 2 \ln 51$.

Thus the maximum must occur at
$t = 2 \ln 51 \approx 7.86$.

51. Using $b^x = e^{kx}$ where $k = \ln b$, we have
$k = \ln 2$.

53. Graph of $y = x$:

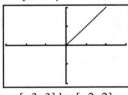

[−3, 3] by [−2, 2]

Graph of $y = e^{\ln x}$:

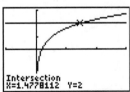

[−3, 3] by [−2, 2]

The graph of $y = e^{\ln x}$ is the same as the graph
of $y = x$ for $x > 0$.

55.

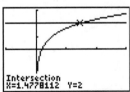

Intersection
X=1.4778112 Y=2

[−1, 3] by [−3, 3]

Using the graph, the x-coordinate of their
point of intersection is approximately
1.4778.

$\ln 5x = 2 \Rightarrow e^{\ln 5x} = e^2 \Rightarrow 5x = e^2 \Rightarrow$

$x = \frac{1}{5}e^2 \approx 1.4778$

4.5 The Derivative of ln x

1. $\dfrac{d}{dx}(3 \ln x + \ln 2) = \dfrac{d}{dx}(3 \ln x) + \dfrac{d}{dx} \ln 2 = \dfrac{3}{x}$

3. $\dfrac{d}{dx}\left(\dfrac{x^2 \ln x}{2}\right) = \dfrac{1}{2}\dfrac{d}{dx}(x^2 \ln x)$

Use the product rule.

$\dfrac{d}{dx}\left(\dfrac{x^2 \ln x}{2}\right) = \dfrac{1}{2}\dfrac{d}{dx}(x^2 \ln x)$

$= \dfrac{1}{2}\left[\ln x \dfrac{d}{dx}x^2 + x^2 \dfrac{d}{dx}\ln x\right]$

$= \dfrac{1}{2}\left(2x \ln x + \dfrac{x^2}{x}\right) = x \ln x + \dfrac{x}{2}$

$= x\left(\ln x + \dfrac{1}{2}\right)$

5. $\dfrac{d}{dx}(e^x \ln x)$

Use the product rule.

$\dfrac{d}{dx}(e^x \ln x) = \ln x \dfrac{d}{dx}e^x + e^x \dfrac{d}{dx}\ln x$

$= e^x \ln x + e^x\left(\dfrac{1}{x}\right)$

$= e^x\left(\ln x + \dfrac{1}{x}\right)$

7. $\dfrac{d}{dx}\dfrac{\ln x}{\sqrt{x}} = \dfrac{d}{dx}\dfrac{\ln x}{x^{1/2}}$

Use the quotient rule.

$\dfrac{d}{dx}\dfrac{\ln x}{\sqrt{x}} = \dfrac{d}{dx}\dfrac{\ln x}{x^{1/2}} = \dfrac{x^{1/2}\dfrac{d}{dx}\ln x - \ln x \dfrac{d}{dx}(x^{1/2})}{(x^{1/2})^2}$

$= \dfrac{x^{1/2}\left(\dfrac{1}{x}\right) - \dfrac{1}{2}\ln x(x^{-1/2})}{x}$

$= \dfrac{x^{-1/2}}{x} - \dfrac{\ln x(x^{-1/2})}{2x} = \dfrac{1}{x\sqrt{x}} - \dfrac{\ln x}{2x\sqrt{x}}$

$= \dfrac{1}{x\sqrt{x}}\left(1 - \dfrac{\ln x}{2}\right)$

9. $\dfrac{d}{dx}\ln x^2 = \dfrac{d}{dx}(2 \ln x) = 2\dfrac{d}{dx}\ln x = \dfrac{2}{x}$

11. $\dfrac{d}{dx}\ln \dfrac{1}{x} = \dfrac{d}{dx}\ln x^{-1} = \dfrac{d}{dx}(-\ln x) = -\dfrac{d}{dx}\ln x$

$= -\dfrac{1}{x}$

13. $\dfrac{d}{dx}\ln\left(3x^4 - x^2\right) = \dfrac{1}{3x^4 - x^2}\dfrac{d}{dx}\left(3x^4 - x^2\right)$

$= \dfrac{1}{3x^4 - x^2}\left(12x^3 - 2x\right)$

$= \dfrac{12x^3 - 2x}{3x^4 - x^2} = \dfrac{12x^2 - 2}{3x^3 - x}$

15. $\dfrac{d}{dx}\left(\dfrac{1}{\ln x}\right) = \dfrac{d}{dx}\left(\ln x\right)^{-1} = -\left(\ln x\right)^{-2}\dfrac{d}{dx}\left(\ln x\right)$

$= -\left(\ln x\right)^{-2}\left(\dfrac{1}{x}\right) = -\dfrac{1}{x\left(\ln x\right)^2}$

17. $\dfrac{d}{dx}\left(\dfrac{\ln x}{\ln 2x}\right)$

Use the quotient rule

$\dfrac{d}{dx}\left(\dfrac{\ln x}{\ln 2x}\right) = \dfrac{\ln 2x\dfrac{d}{dx}\ln x - \ln x\dfrac{d}{dx}\ln 2x}{\left(\ln 2x\right)^2}$

$= \dfrac{\ln 2x\left(\dfrac{1}{x}\right) - \ln x\left(\dfrac{2}{2x}\right)}{\left(\ln 2x\right)^2}$

$= \dfrac{\ln 2x - \ln x}{x\left(\ln 2x\right)^2}$

19. $\dfrac{d}{dx}\left[\left(x^3 + 1\right)\ln\left(x^3 + 1\right)\right]$

Use the product rule.

$\dfrac{d}{dx}\left[\left(x^3 + 1\right)\ln\left(x^3 + 1\right)\right]$

$= \ln\left(x^3 + 1\right)\dfrac{d}{dx}\left(x^3 + 1\right) + \left(x^3 + 1\right)\dfrac{d}{dx}\ln\left(x^3 + 1\right)$

$= 3x^2\ln\left(x^3 + 1\right) + \left(x^3 + 1\right)\dfrac{1}{x^3 + 1}\left(3x^2\right)$

$= 3x^2\ln\left(x^3 + 1\right) + 3x^2$

$= 3x^2\left[\ln\left(x^3 + 1\right) + 1\right]$

21. $\dfrac{d}{dt}\left[t^2\ln t\right] = t^2\left(\dfrac{1}{t}\right) + 2t\ln t = t + 2t\ln t$

$\dfrac{d^2}{dt^2}\left[t^2\ln t\right] = \dfrac{d}{dt}\left[t + 2t\ln t\right]$

$= 1 + 2t\left(\dfrac{1}{t}\right) + 2\ln t$

$= 3 + 2\ln t$

23. $f'(x) = \dfrac{\sqrt{x}\cdot\dfrac{1}{x} - (\ln x)\left(\dfrac{1}{2}x^{-1/2}\right)}{\left(\sqrt{x}\right)^2}$

$= \dfrac{x^{-1/2} - \dfrac{1}{2}x^{-1/2}\ln x}{x} = \dfrac{2 - \ln x}{2x^{3/2}}$

$f'(x) = 0$ when $2 - \ln x = 0$ or

$x = e^2 \approx 7.389$.

From the graph, we see that this value gives a maximum.

$f(e^2) = \dfrac{\ln e^2}{\sqrt{e^2}} = \dfrac{2}{e}$

The coordinates of the maximum point are

$\left(e^2, \dfrac{2}{e}\right)$.

25. $\dfrac{dy}{dx} = \dfrac{1}{x^2 + e}(2x) = \dfrac{2x}{x^2 + e}, \left.\dfrac{dy}{dx}\right|_{x=0} = \dfrac{0}{e} = 0$

At $x = 0$, $y = \ln(0^2 + e) = 1$. Thus the tangent line at $x = 0$ is $y - 1 = 0(x - 0)$ or $y = 1$.

27. **(a)** $f(t) = \ln(\ln t)$

The domain of the inner function, $\ln t$, is $(0, \infty)$, so the domain of the outer function is $(1, \infty)$ or $t > 1$.

(b) $f(t) = \ln(\ln(\ln t))$

From (a), we have the domain of the inner function $\ln(\ln t)$ is $(1, \infty)$, so the domain of the outer function is (e, ∞) or $t > e$.

29. $y = x^2\ln x$

$\dfrac{dy}{dx} = x^2\left(\dfrac{1}{x}\right) + 2x\ln x = x + 2x\ln x$

$\dfrac{dy}{dx} = 0 \Rightarrow x + 2x\ln x = x(1 + 2\ln x) = 0 \Rightarrow$

$x = 0$ or

$1 + 2\ln x = 0 \Rightarrow \ln x = -\dfrac{1}{2} \Rightarrow x = e^{-1/2}$

$\dfrac{d^2y}{dx^2} = 1 + 2x\left(\dfrac{1}{x}\right) + 2\ln x = 3 + 2\ln x$

$\left.\dfrac{d^2y}{dx^2}\right|_{x=e^{-\frac{1}{2}}} = 3 + 2\ln\left(e^{-1/2}\right) = 3 + 2\left(-\dfrac{1}{2}\right) = 2$

Since $\left.\dfrac{d^2y}{dx^2}\right|_{x=e^{-1/2}} > 0$, there is a relative

minimum at $x = e^{-1/2}$.

The relative minimum occurs at $\left(e^{-1/2}, -\dfrac{1}{2}e^{-1}\right)$

or $\left(\dfrac{1}{\sqrt{e}}, -\dfrac{1}{2e}\right)$.

31. a. $y = x + \ln x$

$$y' = \frac{dy}{dx} = 1 + \frac{1}{x}$$

The first derivative is positive for all values of $x > 0$, so the function is increasing.

$y = \ln 2x$

$$y' = \frac{1}{2x}\frac{dy}{dx}(2x) = \frac{1}{2x}(2) = \frac{1}{x}$$

The first derivative is positive for all values of $x > 0$, so the function is increasing.

b. $x + \ln x = \ln 2x \Rightarrow x = \ln 2x - \ln x \Rightarrow$

$$x = \ln\frac{2x}{x} \Rightarrow x = \ln 2$$

The graphs intersect at $x = \ln 2$. This is the point $(\ln 2, \ln(2\ln 2))$ or $(0.6931, 0.3266)$.

33. $y = x^2 - \ln x;\ y' = 2x - \frac{1}{x}$

Solve $y' = 0$ to find the x-coordinate of the minimum point.

$$2x - \frac{1}{x} = 0 \Rightarrow 2x = \frac{1}{x} \Rightarrow x^2 = \frac{1}{2} \Rightarrow x = \pm\frac{1}{\sqrt{2}}$$

Disregard the negative solution. The minimum occurs when $x = \frac{1}{\sqrt{2}} \approx 0.7071$.

$$y = \left(\frac{1}{\sqrt{2}}\right)^2 - \ln\frac{1}{\sqrt{2}} = \frac{1}{2} + \frac{1}{2}\ln 2$$

The minimum point is $\left(\frac{1}{\sqrt{2}}, \frac{1}{2} + \frac{1}{2}\ln 2\right)$.

35. The revenue function is $R(x) = x \cdot \frac{45}{\ln x} = \frac{45x}{\ln x}$.

The marginal revenue function is

$$R'(x) = \frac{45\ln x - 45x\left(\frac{1}{x}\right)}{(\ln x)^2} = \frac{45(\ln x - 1)}{(\ln x)^2}.$$

When $x = 20$, $R'(20) = \frac{45(\ln 20 - 1)}{(\ln 20)^2} \Rightarrow$

$R'(20) \approx 10$.

37.

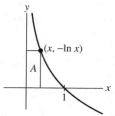

From the graph we see that area $= A = x(-\ln x) = -x\ln x$. To maximize the area, take the first derivative.

$$A' = -x\left(\frac{1}{x}\right) + (\ln x)(-1) = -1 - \ln x$$

Now set $A' = 0$:

$-1 - \ln x = 0 \Rightarrow \ln x = -1 \Rightarrow$

$e^{\ln x} = e^{-1} \Rightarrow x = e^{-1} \approx .36788$. Thus, area is maximized when $x \approx .36788$, and the maximum area is

$$-\left(e^{-1}\right)\ln\left(e^{-1}\right) = \frac{1}{e} \approx .36788.$$

4.6 Properties of the Natural Logarithm Function

1. $\ln 5 + \ln x = \ln(5x)$

3. $\frac{1}{2}\ln 9 = \ln 9^{1/2} = \ln 3$

5. $\ln 4 + \ln 6 - \ln 12 = \ln\left(\frac{4 \cdot 6}{12}\right) = \ln 2$

7. $e^{2\ln x} = e^{\ln x^2} = x^2$

9. $5\ln x - \frac{1}{2}\ln y + 3\ln z = \ln x^5 - \ln y^{1/2} + \ln z^3$

$$= \ln\left(\frac{x^5 z^3}{y^{1/2}}\right)$$

11. $\ln x - \ln x^2 + \ln x^4 = \ln\left(\frac{x \cdot x^4}{x^2}\right) = \ln x^3 = 3\ln x$

13. $2\ln 5 = \ln 5^2 = \ln 25,\ 3\ln 3 = \ln 3^3 = \ln 27$

The natural log function increases as x increases so $3\ln 3 > 2\ln 5$.

15. a. $\ln 4 = \ln 2^2 = 2\ln 2 = 2(.69) = 1.38$

b. $\ln 6 = \ln(2 \cdot 3) = \ln 2 + \ln 3 = .69 + 1.1 = 1.79$

c. $\ln 54 = \ln\left(2 \cdot 3^3\right) = \ln 2 + \ln 3^3$

$$= \ln 2 + 3\ln 3 = .69 + 3(1.1) = 3.99$$

17. a. $\ln \dfrac{1}{6} = \ln 6^{-1} = -\ln 6 = -\ln(2 \cdot 3) = -\left(\ln 2 + \ln 3\right) = -(.69 + 1.1) = -1.79$

b. $\ln \dfrac{2}{9} = \ln 2 - \ln 9 = \ln 2 - \ln 3^2 = \ln 2 - 2\ln 3 = .69 - 2(1.1) = -1.51$

c. $\ln\left(\dfrac{1}{\sqrt{2}}\right) = \ln 2^{-\frac{1}{2}} = -\dfrac{1}{2}\ln 2 = -\dfrac{1}{2}(.69) = -.345$

19.   (d) $4\ln 2x = \ln(2x)^4 = \ln 16x^4$

21.   (d) none of these

23. $\ln x - \ln x^2 + \ln 3 = 0 \Rightarrow \ln\left(\dfrac{3x}{x^2}\right) = 0 \Rightarrow \ln\left(\dfrac{3}{x}\right) = 0 \Rightarrow e^0 = \dfrac{3}{x} \Rightarrow 1 = \dfrac{3}{x} \Rightarrow x = 3$

25. $\ln x^4 - 2\ln x = 1 \Rightarrow 4\ln x - 2\ln x = 1 \Rightarrow 2\ln x = 1 \Rightarrow \ln x = \dfrac{1}{2} \Rightarrow x = e^{1/2} = \sqrt{e}$

27. $(\ln x)^2 - 1 = 0 \Rightarrow (\ln x)^2 = 1 \Rightarrow \ln x = \pm 1 \Rightarrow x = e^1 \text{ or } x = e^{-1}$

29. $\ln \sqrt{x} = \sqrt{\ln x} \Rightarrow \ln\left(x^{1/2}\right) = \sqrt{\ln x} \Rightarrow \dfrac{1}{2}\ln x = \sqrt{\ln x}$

Let $u = \ln x$. Then $\dfrac{1}{2}\ln x = \sqrt{\ln x} \Rightarrow \dfrac{1}{2}u = \sqrt{u} \Rightarrow \dfrac{1}{4}u^2 = u \Rightarrow \dfrac{1}{4}u^2 - u = 0 \Rightarrow$

$u\left(\dfrac{1}{4}u - 1\right) = 0 \Rightarrow u = 0 \text{ or } u = 4$

If $u = 0$, then $0 = \ln x \Rightarrow x = 1$. If $u = 4$, then $4 = \ln x \Rightarrow x = e^4$.

31. $\ln(x+1) - \ln(x-2) = 1 \Rightarrow \ln\left(\dfrac{x+1}{x-2}\right) = 1 \Rightarrow e = \dfrac{x+1}{x-2} \Rightarrow e(x-2) = x+1 \Rightarrow ex - x = 2e + 1 \Rightarrow$

$x(e-1) = 2e + 1 \Rightarrow x = \dfrac{2e+1}{e-1}$

33. $\dfrac{d}{dx}\ln\left[(x+5)(2x-1)(4-x)\right] = \dfrac{d}{dx}\left[\ln(x+5) + \ln(2x-1) + \ln(4-x)\right] = \dfrac{1}{x+5}(1) + \dfrac{1}{2x-1}(2) + \dfrac{1}{4-x}(-1)$

$= \dfrac{1}{x+5} + \dfrac{2}{2x-1} - \dfrac{1}{4-x}$

35. $\dfrac{d}{dx}\ln\left[(1+x)^2(2+x)^3(3+x)^4\right] = \dfrac{d}{dx}\left[\ln(1+x)^2 + \ln(2+x)^3 + \ln(3+x)^4\right]$

$= \dfrac{d}{dx}\left[2\ln(1+x) + 3\ln(2+x) + 4\ln(3+x)\right] = \dfrac{2}{1+x} + \dfrac{3}{2+x} + \dfrac{4}{3+x}$

37. $\dfrac{d}{dx}\ln\left[\sqrt{xe^{x^2+1}}\right] = \dfrac{d}{dx}\left[\ln\left(xe^{x^2+1}\right)^{1/2}\right] = \dfrac{d}{dx}\left[\dfrac{1}{2}\ln\left(xe^{x^2+1}\right)\right]$

$= \dfrac{1}{2} \cdot \dfrac{1}{xe^{x^2+1}} \cdot \left(2x^2 e^{x^2+1} + e^{x^2+1}\right) = x + \dfrac{1}{2x}$

39. $\dfrac{d}{dx}\ln\left[\dfrac{(x+1)^4}{e^{x-1}}\right] = \dfrac{d}{dx}\left[\ln(x+1)^4 - \ln\left(e^{x-1}\right)\right] = \dfrac{d}{dx}\left[4\ln(x+1) - (x-1)\right] = \dfrac{4}{x+1} - 1$

41. $\dfrac{d}{dx}\left[\ln(3x+1)\ln(5x+1)\right] = \ln(3x+1)\left(\dfrac{5}{5x+1}\right) + \ln(5x+1)\left(\dfrac{3}{3x+1}\right) = \dfrac{5\ln(3x+1)}{5x+1} + \dfrac{3\ln(5x+1)}{3x+1}$

43. $f(x) = (x+1)^4(4x-1)^2 \Rightarrow \ln f(x) = \ln\left[(x+1)^4(4x-1)^2\right] = 4\ln(x+1) + 2\ln(4x-1)$

Differentiating both sides, we have

$\dfrac{f'(x)}{f(x)} = \dfrac{4}{x+1} + \dfrac{8}{4x-1} \Rightarrow f'(x) = (x+1)^4(4x-1)^2\left(\dfrac{4}{x+1} + \dfrac{8}{4x-1}\right).$

45. $f(x) = \dfrac{(x+1)(2x+1)(3x+1)}{\sqrt{4x+1}} \Rightarrow$

$\ln f(x) = \ln\left(\dfrac{(x+1)(2x+1)(3x+1)}{\sqrt{4x+1}}\right) = \ln(x+1) + \ln(2x+1) + \ln(3x+1) - \dfrac{1}{2}\ln(4x+1)$

Differentiating both sides, we have

$\dfrac{f'(x)}{f(x)} = \dfrac{1}{x+1} + \dfrac{2}{2x+1} + \dfrac{3}{3x+1} - \dfrac{2}{4x+1} \Rightarrow f'(x) = \dfrac{(x+1)(2x+1)(3x+1)}{\sqrt{4x+1}}\cdot\left(\dfrac{1}{x+1} + \dfrac{2}{2x+1} + \dfrac{3}{3x+1} - \dfrac{2}{4x+1}\right).$

47. $f(x) = 2^x \Rightarrow \ln f(x) = \ln 2^x = x\ln 2$

Differentiating both sides, we have

$\dfrac{f'(x)}{f(x)} = \ln 2 \Rightarrow f'(x) = 2^x \ln 2.$

49. $f(x) = x^x \Rightarrow \ln f(x) = \ln x^x = x\ln x$

Differentiating both sides, we have

$\dfrac{f'(x)}{f(x)} = x\left(\dfrac{1}{x}\right) + \ln x = 1 + \ln x \Rightarrow$

$f'(x) = x^x\left[1 + \ln x\right]$

51. $\ln y - k\ln x = \ln c;$

$\ln y = \ln c + k\ln x = \ln c + \ln x^k = \ln(cx^k) \Rightarrow$

$e^{\ln y} = e^{\ln(cx^k)} \Rightarrow y = cx^k$

53. $y = he^{kx}$

$\left.\begin{array}{l} 6 = he^k \\ 48 = he^{4k} \end{array}\right\} \begin{array}{l} \ln\left(\dfrac{6}{h}\right) = k \\ \ln\left(\dfrac{48}{h}\right) = 4k \end{array} \right\} 4\ln\left(\dfrac{6}{h}\right) = \ln\left(\dfrac{48}{h}\right) \Rightarrow$

$\ln\left(\dfrac{6}{h}\right)^4 = \ln\left(\dfrac{48}{h}\right) \Rightarrow \dfrac{6^4}{h^4} = \dfrac{48}{h} \Rightarrow h^3 = \dfrac{6^4}{48} \Rightarrow$

$h^3 = 27 \Rightarrow h = 3$

$6 = 3e^k \Rightarrow e^k = 2 \Rightarrow k = \ln 2$

Chapter 4 Review Exercises

1. $27^{4/3} = (3^3)^{4/3} = 3^4 = 81$

2. $4^{1.5} = (2^2)^{3/2} = 2^3 = 8$

3. $5^{-2} = \dfrac{1}{5^2} = \dfrac{1}{25}$

4. $16^{-.25} = 16^{-1/4} = \dfrac{1}{(2^4)^{1/4}} = \dfrac{1}{2}$

5. $(2^{5/7})^{14/5} = 2^{14/7} = 2^2 = 4$

6. $8^{1/2}\cdot 2^{1/2} = (2^3)^{1/2}\cdot 2^{1/2} = 2^{3/2}\cdot 2^{1/2}$
$= 2^{4/2} = 4$

7. $\dfrac{9^{5/2}}{9^{3/2}} = \dfrac{(3^2)^{5/2}}{(3^2)^{3/2}} = \dfrac{3^5}{3^3} = 3^2 = 9$

8. $4^{.2}\cdot 4^{.3} = 4^{.5} = 4^{1/2} = 2$

9. $(e^{x^2})^3 = e^{3x^2}$

10. $e^{5x}\cdot e^{2x} = e^{7x}$

11. $\dfrac{e^{3x}}{e^x} = e^{3x-x} = e^{2x}$

12. $2^x\cdot 3^x = (2\cdot 3)^x = 6^x$

13. $(e^{8x} + 7e^{-2x})e^{3x} = e^{11x} + 7e^x$

14. $\dfrac{e^{5x/2} - e^{3x}}{\sqrt{e^x}} = (e^{5x/2} - e^{3x})e^{(-1/2)x}$
$= e^{4x/2} - e^{5x/2} = e^{2x} - e^{5x/2}$

15. $e^{-3x} = e^{-12} \Rightarrow \ln e^{-3x} = \ln e^{-12} \Rightarrow$
$-3x = -12 \Rightarrow x = 4$

16. $e^{x^2 - x} = e^2 \Rightarrow \ln e^{x^2 - x} = \ln e^2 \Rightarrow x^2 - x = 2 \Rightarrow$
$x^2 - x - 2 = 0 \Rightarrow (x - 2)(x + 1) = 0 \Rightarrow x = 2$ or
$x = -1$

17. $(e^x \cdot e^2)^3 = e^{-9} \Rightarrow e^{3x+6} = e^{-9} \Rightarrow$
$\ln e^{3x+6} = \ln e^{-9} \Rightarrow 3x + 6 = -9 \Rightarrow x = -5$

18. $e^{-5x} \cdot e^4 = e \Rightarrow e^{-5x+4} = e \Rightarrow$
$\ln e^{-5x+4} = \ln e \Rightarrow -5x + 4 = 1 \Rightarrow x = \dfrac{3}{5}$

19. $\dfrac{d}{dx}[10e^{7x}] = 10e^{7x}(7) = 70e^{7x}$

20. $\dfrac{d}{dx}e^{\sqrt{x}} = \dfrac{d}{dx}e^{x^{1/2}} = e^{x^{1/2}} \cdot \dfrac{1}{2}x^{-1/2} = \dfrac{e^{\sqrt{x}}}{2\sqrt{x}}$

21. $\dfrac{d}{dx}[xe^{x^2}] = xe^{x^2}(2x) + e^{x^2}(1) = e^{x^2}(2x^2 + 1)$

22. $\dfrac{d}{dx}\left[\dfrac{e^x + 1}{x - 1}\right] = \dfrac{(x - 1)e^x - (e^x + 1)(1)}{(x - 1)^2}$

$= \dfrac{-2e^x + xe^x - 1}{(x - 1)^2} = \dfrac{(x - 2)e^x - 1}{(x - 1)^2}$

23. $\dfrac{d}{dx}[e^{e^x}] = e^{e^x}(e^x) = e^{x + e^x}$

24. $\dfrac{d}{dx}\left[\left(\sqrt{x} + 1\right)e^{-2x}\right]$

$= \left(\sqrt{x} + 1\right)e^{-2x}(-2) + e^{-2x}\left(\dfrac{1}{2}x^{-1/2}\right)$

$= e^{-2x}\left(\dfrac{1}{2\sqrt{x}} - 2\sqrt{x} - 2\right)$

25. $\dfrac{d}{dx}\left[\dfrac{x^2 - x + 5}{e^{3x} + 3}\right]$

$= \dfrac{(e^{3x} + 3)(2x - 1) - (x^2 - x + 5)3e^{3x}}{(e^{3x} + 3)^2}$

$= \dfrac{(2x - 1)(e^{3x} + 3) - 3e^{3x}(x^2 - x + 5)}{(e^{3x} + 3)^2}$

26. $\dfrac{d}{dx}x^e = ex^{e-1}$

27. $f(x) = e^{x^2} - 4x^2;\ f'(x) = 2xe^{x^2} - 8x$

Solve $f'(x) = 0$ to find the first coordinates of the relative extreme point.s

$2xe^{x^2} - 8x = 0 \Rightarrow 2x\left(e^{x^2} - 4\right) \Rightarrow$

$2x = 0$ or $e^{x^2} - 4 = 0 \Rightarrow$

$x = 0$ or $x^2 = \ln 4 \Rightarrow x = \pm\sqrt{\ln 4}$

The relative extreme points occur at $x = 0$ and
$x = \pm\sqrt{\ln 4} \approx \pm 1.1774$.

28. $f(x) = e^{x^2} - 4x^2;\ f'(x) = 2xe^{x^2} - 8x$

$f''(x) = 2e^{x^2}\dfrac{d}{dx}x + 2x\dfrac{d}{dx}e^{x^2} - \dfrac{d}{dx}(8x)$

$= 2e^{x^2} + (2x)^2 e^{x^2} - 8$

$= 2e^{x^2}\left(1 + 2x^2\right) - 8$

Evaluate $f''(0)$ to determine the concavity at $x = 0$.

$f''(0) = 2e^{0^2}\left(1 + 2(0)^2\right) - 8 = 2(1) - 8 = -6$

Since $f''(0) < 0$, the graph is concave down at $x = 0$ and there is a relative maximum there.

29. $4e^{0.03t} - 2e^{0.06t} = 0 \Rightarrow$

$2e^{0.03t}\left(2 - e^{0.03t}\right) = 0 \Rightarrow$

$2e^{0.03t} = 0$ (not possible) or $2 - e^{0.03t} = 0 \Rightarrow$

$2 = e^{0.03t} \Rightarrow \ln 2 = .03t \Rightarrow t = \dfrac{\ln 2}{.03} \approx 23.1049$

30. $e^t - 8e^{0.02t} = 0 \Rightarrow e^t = 8e^{0.02t} \Rightarrow$

$t = \ln 8 + .02t \Rightarrow .98t = \ln 8 \Rightarrow$

$t = \dfrac{\ln 8}{.98} \approx 2.1219$

31. $4 \cdot 2^x = e^x \Rightarrow 2^2 \cdot 2^x = e^x \Rightarrow 2^{2+x} = e^x \Rightarrow$

$(2 + x)\ln 2 = x \Rightarrow 2\ln 2 + x\ln 2 = x \Rightarrow$

$2\ln 2 = x - x\ln 2 \Rightarrow 2\ln 2 = x(1 - \ln 2) \Rightarrow$

$x = \dfrac{2\ln 2}{1 - \ln 2} = \dfrac{\ln 2^2}{1 - \ln 2} = \dfrac{\ln 4}{1 - \ln 2} \approx 4.52$

32. $3^x = 2e^x \Rightarrow x\ln 3 = \ln 2 + x \Rightarrow$

$x\ln 3 - x = \ln 2 \Rightarrow x(\ln 3 - 1) = \ln 2 \Rightarrow$

$x = \dfrac{\ln 2}{\ln 3 - 1} \approx 7.0290$

33. Solve $y' = 4$.

$y = e^x;\ y' = e^x$

$e^x = 4 \Rightarrow x = \ln 4$

The tangent line has slope 4 when $x = \ln 4$.

34. When the tangent line is horizontal, its slope is 0. Solve $y' = 0$.

$$y = e^x + e^{-2x}; \ y' = e^x - 2e^{-2x}$$

$$e^x - 2e^{-2x} = 0 \Rightarrow e^x = 2e^{-2x} \Rightarrow$$

$$x = \ln 2 - 2x \Rightarrow 3x = \ln 2 \Rightarrow x = \frac{\ln 2}{3}$$

35. $f(x) = \ln(x^2 + 1)$

First, determine the relative extreme and inflection points using the first derivative test.

$$f'(x) = \frac{2x}{x^2 + 1}$$

$$f'(x) = \frac{2x}{x^2 + 1} = 0 \Rightarrow x = 0$$

The only relative extreme occurs at $x = 0$. Now use the second derivative test to determine if the graph is increasing or decreasing.

$$f''(x) = \frac{2(x^2 + 1) - 2x(2x)}{(x^2 + 1)^2} = \frac{-2x^2 + 2}{(x^2 + 1)^2}$$

$$f''(0) = \frac{-2(0)^2 + 2}{(0^2 + 1)^2} = 2$$

Since $f''(0) > 0$, the graph is concave up and there is a relative minimum at $x = 0$. Thus, the graph is decreasing for $x < 0$ and increasing for $x > 0$.

36. $f(x) = x \ln x, \ x > 0$

First, determine the relative extreme and inflection points using the first derivative test.

$$f'(x) = \ln x + x\left(\frac{1}{x}\right) = \ln x + 1$$

$$f'(x) = \ln x + 1 = 0 \Rightarrow \ln x = -1 \Rightarrow$$

$$x = e^{-1} = \frac{1}{e}$$

The only relative extreme occurs at $x = \frac{1}{e}$.

Now use the second derivative test to determine if the graph is increasing or decreasing.

$$f''(x) = \frac{1}{x}$$

$$f''\left(\frac{1}{e}\right) = e$$

$f''(0) > 0$, so the graph is concave up and there is a relative minimum at $x = \frac{1}{e}$. Thus, the graph is decreasing for $0 < x < \frac{1}{e}$ and increasing for $x > \frac{1}{e}$.

37. $\dfrac{dy}{dx} = \dfrac{(1 + e^x)e^x - e^x(e^x)}{(1 + e^x)^2} = \dfrac{e^x}{(1 + e^x)^2}$

$$\left.\frac{dy}{dx}\right|_{x=0} = \frac{e^0}{(1 + e^0)^2} = \frac{1}{4}$$

The tangent line at $(0, .5)$ is

$$y - .5 = \frac{1}{4}(x - 0) \text{ or } y = \frac{1}{4}x + \frac{1}{2}.$$

38. $y = \dfrac{e^x - e^{-x}}{e^x + e^{-x}}$

$\dfrac{dy}{dx}$

$$= \frac{(e^x + e^{-x})(e^x + e^{-x}) - (e^x - e^{-x})(e^x - e^{-x})}{\left(e^x + e^{-x}\right)^2}$$

$$= \frac{4}{\left(e^x + e^{-x}\right)^2}$$

$$\left.\frac{dy}{dx}\right|_{x=1} = \frac{4}{\left(e^1 + e^{-1}\right)^2}; \left.\frac{dy}{dx}\right|_{x=-1}$$

$$= \frac{4}{\left(e^{-1} + e^{-(-1)}\right)^2} = \frac{4}{\left(e^1 + e^{-1}\right)^2}$$

The tangent lines at $x = 1$ and $x = -1$ have the same slope, so they are parallel.

39. $e^{(\ln 5)/2} = e^{\ln \sqrt{5}} = \sqrt{5}$

40. $e^{\ln(x^2)} = x^2$

41. $\dfrac{\ln x^2}{\ln x^3} = \dfrac{2 \ln x}{3 \ln x} = \dfrac{2}{3}$

42. $e^{2 \ln 2} = e^{\ln 2^2} = 2^2 = 4$

43. $e^{-5 \ln 1} = e^{-5(0)} = 1$

44. $[e^{\ln x}]^2 = x^2$

45. $t^{\ln t} = e \Rightarrow \ln t^{\ln t} = \ln e \Rightarrow \ln t(\ln t) = 1 \Rightarrow$
$(\ln t)^2 = 1$
Taking the square root of both sides,

$$|\ln t| = 1 \Rightarrow t = e \text{ or } t = \frac{1}{e}.$$

46. $\ln(\ln 3t) = 0 \Rightarrow e^{\ln(\ln 3t)} = e^0 \Rightarrow \ln 3t = 1 \Rightarrow e^{\ln 3t} = e \Rightarrow 3t = e \Rightarrow t = \dfrac{e}{3}$

47. $3e^{2t} = 15 \Rightarrow e^{2t} = 5 \Rightarrow \ln e^{2t} = \ln 5 \Rightarrow 2t = \ln 5 \Rightarrow t = \dfrac{1}{2}\ln 5$

48. $3e^{t/2} - 12 = 0 \Rightarrow 3(e^{t/2} - 4) = 0 \Rightarrow e^{t/2} = 4 \Rightarrow \ln e^{t/2} = \ln 4 \Rightarrow t = 2\ln 4 \Rightarrow t = \ln 16$

49. $2\ln t = 5 \Rightarrow \ln t = \dfrac{5}{2} \Rightarrow e^{\ln t} = e^{5/2} \Rightarrow t = e^{5/2}$

50. $2e^{-0.3t} = 1 \Rightarrow e^{-0.3t} = \dfrac{1}{2} \Rightarrow \ln e^{-0.3t} = \ln\dfrac{1}{2} \Rightarrow -.3t = \ln\dfrac{1}{2} \Rightarrow t = -\dfrac{1}{.3}\ln\dfrac{1}{2} = \dfrac{\ln 2}{.3} = \dfrac{10\ln 2}{3}$

51. $\dfrac{d}{dx}\ln(x^6 + 3x^4 + 1) = \dfrac{1}{x^6 + 3x^4 + 1}(6x^5 + 12x^3) = \dfrac{6x^5 + 12x^3}{x^6 + 3x^4 + 1}$

52. $\dfrac{d}{dx}\left[\dfrac{x}{\ln x}\right] = \dfrac{\ln x - x\left(\frac{1}{x}\right)}{(\ln x)^2} = \dfrac{\ln x - 1}{(\ln x)^2}$

53. $\dfrac{d}{dx}\ln(5x - 7) = \dfrac{1}{5x - 7}(5) = \dfrac{5}{5x - 7}$

54. $\dfrac{d}{dx}\ln(9x) = \dfrac{1}{9x}(9) = \dfrac{1}{x}$

55. $\dfrac{d}{dx}[(\ln x)^2] = 2(\ln x)\dfrac{1}{x} = \dfrac{2\ln x}{x}$

56. $\dfrac{d}{dx}\left[(x\ln x)^3\right] = 3(x\ln x)^2\left(x\cdot\dfrac{1}{x} + \ln x\right) = 3(x\ln x)^2(1 + \ln x)$

57. $\dfrac{d}{dx}\ln\left[\dfrac{xe^x}{\sqrt{1+x}}\right] = \dfrac{d}{dx}\left[\ln\left(xe^x\right) - \ln\sqrt{1+x}\right] = \dfrac{1}{xe^x}(xe^x + e^x) - \dfrac{1}{\sqrt{1+x}}\cdot\dfrac{1}{2}(1+x)^{-1/2} = 1 + \dfrac{1}{x} - \dfrac{1}{2(1+x)}$

58. $\dfrac{d}{dx}\ln\left[e^{6x}(x^2 + 3)^5(x^3 + 1)^{-4}\right] = \dfrac{d}{dx}\left[6x + 5\ln(x^2 + 3) - 4\ln(x^3 + 1)\right]$

$$= 6 + \dfrac{5}{x^2 + 3}(2x) - \dfrac{4}{x^3 + 1}(3x^2) = 6 + \dfrac{10x}{x^2 + 3} - \dfrac{12x^2}{x^3 + 1}$$

59. $\dfrac{d}{dx}[x\ln x - x] = x\left(\dfrac{1}{x}\right) + \ln x - 1 = \ln x$

60. $\dfrac{d}{dx}\left[e^{2\ln(x+1)}\right] = \dfrac{d}{dx}\left[e^{\ln(x+1)^2}\right] = \dfrac{d}{dx}\left[(x+1)^2\right] = 2(x+1)$

61. $\dfrac{d}{dx}\ln(\ln\sqrt{x}) = \dfrac{1}{\ln\sqrt{x}}\cdot\dfrac{1}{\sqrt{x}}\cdot\dfrac{1}{2}x^{-1/2} = \dfrac{1}{2x\ln\sqrt{x}} = \dfrac{1}{x\ln x}$

62. $\dfrac{d}{dx}\left[\dfrac{1}{\ln x}\right] = \dfrac{d}{dx}\left[(\ln x)^{-1}\right] = -1(\ln x)^{-2}\left(\dfrac{1}{x}\right) = -\dfrac{1}{x(\ln x)^2}$

63. $\dfrac{d}{dx}[e^x \ln x] = e^x\left(\dfrac{1}{x}\right) + e^x\ln x = \dfrac{e^x}{x} + e^x\ln x$

64. $\dfrac{d}{dx}\ln(x^2+e^x)=\dfrac{1}{x^2+e^x}(2x+e^x)=\dfrac{2x+e^x}{x^2+e^x}$

65. $\dfrac{d}{dx}\ln\sqrt{\dfrac{x^2+1}{2x+3}}=\dfrac{d}{dx}\ln\left(\dfrac{x^2+1}{2x+3}\right)^{1/2}=\dfrac{d}{dx}\dfrac{1}{2}\left[\ln\left(x^2+1\right)-\ln(2x+3)\right]$

$=\dfrac{1}{2}\left[\dfrac{1}{x^2+1}(2x)-\dfrac{1}{2x+3}(2)\right]=\dfrac{x}{x^2+1}-\dfrac{1}{2x+3}$

66. $-2x+1>0$ gives us $x<\dfrac{1}{2}$. $-2x+1<0$ gives us $x>\dfrac{1}{2}$.

For $x<\dfrac{1}{2}$, we have $\dfrac{d}{dx}\ln|-2x+1|=\dfrac{d}{dx}\ln(-2x+1)=\dfrac{1}{-2x+1}(-2)=\dfrac{2}{2x-1}$.

For $x>\dfrac{1}{2}$, we have $\dfrac{d}{dx}\ln|-2x+1|=\dfrac{d}{dx}\ln(-(-2x+1))=\dfrac{d}{dx}\ln(2x-1)=\dfrac{1}{2x-1}(2)=\dfrac{2}{2x-1}$.

So, for $x\neq\dfrac{1}{2}$, $\dfrac{d}{dx}\ln|-2x+1|=\dfrac{2}{2x-1}$.

67. $\dfrac{d}{dx}\ln\left(\dfrac{e^{x^2}}{x}\right)=\dfrac{d}{dx}\left[\ln\left(e^{x^2}\right)-\ln x\right]=\dfrac{d}{dx}\left[x^2-\ln x\right]=2x-\dfrac{1}{x}$

68. $\dfrac{d}{dx}\ln\sqrt[3]{x^3+3x-2}=\dfrac{d}{dx}\left[\dfrac{1}{3}\ln(x^3+3x-2)\right]=\dfrac{1}{3}\dfrac{1}{x^3+3x-2}(3x^2+3)=\dfrac{x^2+1}{x^3+3x-2}$

69. $\dfrac{d}{dx}\ln(2^x)=\dfrac{d}{dx}(x\ln 2)=\ln 2$

70. $\dfrac{d}{dx}\left[\ln(3^{x+1})-\ln 3\right]=\dfrac{d}{dx}\left[(x+1)\ln 3-\ln 3\right]=\ln 3$

71. $x-1>0$ gives $x>1$. $x-1<0$ gives $x<1$.

For $x>1$, we have $\dfrac{d}{dx}\ln|x-1|=\dfrac{d}{dx}\ln(x-1)=\dfrac{1}{x-1}$.

For $x<1$, we have $\dfrac{d}{dx}\ln|x-1|=\dfrac{d}{dx}\ln(-(x-1))=\dfrac{d}{dx}\ln(-x+1)=\dfrac{1}{-x+1}(-1)=\dfrac{1}{x-1}$.

So, for $x\neq1$, $\dfrac{d}{dx}\ln|x-1|=\dfrac{1}{x-1}$.

72. $\dfrac{d}{dx}e^{2\ln(2x+1)}=\dfrac{d}{dx}e^{\ln(2x+1)^2}=\dfrac{d}{dx}(2x+1)^2=2(2x+1)(2)=8x+4$

73. $\dfrac{d}{dx}\ln\left(\dfrac{1}{e^{\sqrt{x}}}\right)=\dfrac{d}{dx}\ln\left(e^{-\sqrt{x}}\right)=\dfrac{d}{dx}(-\sqrt{x})=-\dfrac{1}{2}x^{-1/2}=-\dfrac{1}{2\sqrt{x}}$

74. $\dfrac{d}{dx}\ln(e^x+3e^{-x})=\dfrac{1}{e^x+3e^{-x}}\left(e^x+3e^{-x}(-1)\right)=\dfrac{e^x-3e^{-x}}{e^x+3e^{-x}}$

75. $\ln f(x) = \ln \sqrt[5]{\dfrac{x^5+1}{x^5+5x+1}} = \dfrac{1}{5}\ln(x^5+1) - \dfrac{1}{5}\ln(x^5+5x+1)$

Differentiating both sides, we have $\dfrac{f'(x)}{f(x)} = \dfrac{1}{5}\dfrac{1}{x^5+1}(5x^4) - \dfrac{1}{5}\dfrac{1}{x^5+5x+1}(5x^4+5) \Rightarrow$

$f'(x) = \sqrt[5]{\dfrac{x^5+1}{x^5+5x+1}}\left(\dfrac{x^4}{x^5+1} - \dfrac{x^4+1}{x^5+5x+1}\right).$

76. $\ln f(x) = \ln 2^x = x \ln 2$

Differentiating both sides, we have $\dfrac{f'(x)}{f(x)} = \ln 2 \Rightarrow f'(x) = 2^x \ln 2.$

77. $\ln f(x) = \ln x^{\sqrt{x}} = \sqrt{x}\ln x$

Differentiating both sides, we have

$\dfrac{f'(x)}{f(x)} = \sqrt{x}\left(\dfrac{1}{x}\right) + \left(\dfrac{1}{2}\right)x^{-1/2}\ln x = \dfrac{1}{\sqrt{x}} + \dfrac{\ln x}{2\sqrt{x}} \Rightarrow f'(x) = x^{\sqrt{x}}\left(\dfrac{1}{\sqrt{x}} + \dfrac{\ln x}{2\sqrt{x}}\right) = x^{\sqrt{x}-1/2}\left(1 + \dfrac{1}{2}\ln x\right)$

78. $\ln f(x) = \ln b^x = x \ln b$

Differentiating both sides, we have $\dfrac{f'(x)}{f(x)} = \ln b \Rightarrow f'(x) = b^x \ln b.$

79. $\ln f(x) = \ln\left[(x^2+5)^6(x^3+7)^8(x^4+9)^{10}\right] = 6\ln(x^2+5) + 8\ln(x^3+7) + 10\ln(x^4+9)$

Differentiating both sides, we have

$\dfrac{f'(x)}{f(x)} = \dfrac{6}{x^2+5}(2x) + \dfrac{8}{x^3+7}(3x^2) + \dfrac{10}{x^4+9}(4x^3) \Rightarrow$

$f'(x) = (x^2+5)^6(x^3+7)^8(x^4+9)^{10}\left[\dfrac{12x}{x^2+5} + \dfrac{24x^2}{x^3+7} + \dfrac{40x^3}{x^4+9}\right]$

80. $\ln f(x) = \ln x^{1+x} = (1+x)\ln x = \ln x + x\ln x$

Differentiating both sides, we have $\dfrac{f'(x)}{f(x)} = \dfrac{1}{x} + x\left(\dfrac{1}{x}\right) + \ln x \Rightarrow f'(x) = x^{1+x}\left(\dfrac{1}{x} + 1 + \ln x\right).$

81. $\ln f(x) = \ln 10^x = x \ln 10$

Differentiating both sides, we have $\dfrac{f'(x)}{f(x)} = \ln 10 \Rightarrow f'(x) = 10^x \ln 10.$

82. $\ln f(x) = \ln\left(\sqrt{x^2+5}\,e^{x^2}\right) = \dfrac{1}{2}\ln(x^2+5) + \ln e^{x^2} = \dfrac{1}{2}\ln(x^2+5) + x^2$

Differentiating both sides, we have

$\dfrac{f'(x)}{f(x)} = \dfrac{1}{2}\cdot\dfrac{1}{x^2+5}(2x) + 2x = \dfrac{x}{x^2+5} + 2x \Rightarrow f'(x) = \sqrt{x^2+5}\,e^{x^2}\left(\dfrac{x}{x^2+5} + 2x\right).$

83. $\ln f(x) = \ln\sqrt{\dfrac{xe^x}{x^3+3}} = \dfrac{1}{2}\left[\ln x + \ln e^x - \ln(x^3+3)\right] = \dfrac{1}{2}\left[\ln x + x - \ln(x^3+3)\right]$

Differentiating both sides, we have $\dfrac{f'(x)}{f(x)} = \dfrac{1}{2}\left[\dfrac{1}{x} + 1 - \dfrac{1}{x^3+3}(3x^2)\right] \Rightarrow f'(x) = \dfrac{1}{2}\sqrt{\dfrac{xe^x}{x^3+3}}\left(\dfrac{1}{x} + 1 - \dfrac{3x^2}{x^3+3}\right).$

84. $\ln f(x) = \ln\left[\dfrac{e^x\sqrt{x+1}(x^2+2x+3)^2}{4x^2}\right] = x + \dfrac{1}{2}\ln(x+1) + 2\ln(x^2+2x+3) - 2\ln(2x)$

Differentiating both sides, we have $\dfrac{f'(x)}{f(x)} = 1 + \dfrac{1}{2(x+1)} + 2\dfrac{1}{x^2+2x+3}(2x+2) - 2\dfrac{1}{2x}(2) \Rightarrow$

$f'(x) = \dfrac{e^x\sqrt{x+1}(x^2+2x+3)^2}{4x^2}\left[1 + \dfrac{1}{2x+2} + \dfrac{4x+4}{x^2+2x+3} - \dfrac{2}{x}\right].$

85. $\ln f(x) = \ln\left[e^{x+1}(x^2+1)x\right]$

$\qquad = x + 1 + \ln(x^2+1) + \ln x$

Differentiating both sides, we have

$\dfrac{f'(x)}{f(x)} = 1 + \dfrac{1}{x^2+1}(2x) + \dfrac{1}{x} \Rightarrow f'(x) = e^{x+1}(x^2+1)x\left(1 + \dfrac{2x}{x^2+1} + \dfrac{1}{x}\right) = e^{x+1}(x^3+3x^2+x+1)$

86. $\ln f(x) = \ln(e^x x^2 2^x) = x + 2\ln x + x\ln 2$

Differentiating both sides, we have $\dfrac{f'(x)}{f(x)} = 1 + 2\left(\dfrac{1}{x}\right) + \ln 2 \Rightarrow f'(x) = e^x x^2 2^x\left(1 + \ln 2 + \dfrac{2}{x}\right).$

87. a. $f(2) \approx 800$ g/cm^2

Actual answer: 814.160 g/cm^2

b. $f(x) = 200$ when $x \approx 14$ km

c. $f'(8) \approx -50$ g/cm^2 per km

Actual answer: $f'(x) = 1035e^{-0.12x}(-.12) = -124.2e^{-0.12x}$

$\qquad\qquad\qquad f'(8) = -124.2e^{-0.12(8)} = -47.6$

At an altitude of 8 km, the atmospheric pressure is dropping at the rate of 9.266 g/cm^2.

d. $f'(x) = -100$ when $x \approx 2$ km

88. a. $f(18) \approx 180$ billion dollars

Actual answer: \$181.969 billion

b. $f'(12) \approx 10$ billion dollars per year

Actual answer:

$f'(t) = 27e^{0.106t}(.106) = 2.862e^{0.106t}$

$f'(12) = 2.862e^{0.106(12)} = 10.212$

Expenditures were rising at \$10.212 billion per year.

c. $f(t) = 120$ when $t \approx 14$, so in 2004

d. $f'(t) = 20$ when $t \approx 18$, so expenditures were rising at the rate of \$20 billion per year in 2008

Chapter 5 Applications of the Exponential and Natural Logarithm Functions

5.1 Exponential Growth and Decay

For exercises 1–9, refer to Theorem 1 and Example 1 in your text.

1. $y' = y \Rightarrow k = 1$

   Then, $y = Ce^t$.

3. $y' = 1.7y \Rightarrow k = 1.7$

   Then, $y = Ce^{1.7t}$.

5. $y' - \dfrac{y}{2} = 0 \Rightarrow y' = \dfrac{y}{2} \Rightarrow k = \dfrac{1}{2}$

   Then, $y = Ce^{0.5t}$.

7. $2y' - \dfrac{y}{2} = 0 \Rightarrow 2y' = \dfrac{y}{2} \Rightarrow y' = \dfrac{y}{4} \Rightarrow k = \dfrac{1}{4}$

   Then, $y = Ce^{0.25t}$.

9. $\dfrac{y}{3} = 4y' \Rightarrow \dfrac{y}{12} = y' \Rightarrow k = \dfrac{1}{12}$

   Then, $y = Ce^{t/12}$.

For exercises 11–17, refer to Theorem 2 and Example 3 in your text.

11. $y' = 3y,\ y(0) = 1$

    Then, $k = 3$ and $y = P(t) = P_0 e^{kt} \Rightarrow y = e^{3t}$.

13. $y' = 2y,\ y(0) = 2$

    Then, $k = 2$ and $y = P(t) = P_0 e^{kt} \Rightarrow y = 2e^{2t}$.

15. $y' - .6y = 0 \Rightarrow y' = .6y,\ y(0) = 5$

    Then, $k = .6$ and $y = P(t) = P_0 e^{kt} \Rightarrow$

    $y = 5e^{0.6t}$.

17. $6y' = y \Rightarrow y' = \dfrac{y}{6},\ y(0) = 12$

    Then, $k = \frac{1}{6}$ and $y = P(t) = P_0 e^{kt} \Rightarrow$

    $y = 12e^{t/6}$.

19. a. $P' = .01P(t),\ P(0) = 2$

    $k = .01$, so $P(t) = P_0 e^{kt} \Rightarrow$

    $P(t) = 2e^{0.01t}$.

    b. The initial population in 2015 was 2 million.

    c. The year 2019 corresponds to $t = 4$.

    $P(4) = 2e^{0.01(4)} \approx 2.0816$ million.

21. a. $P(t) = P_0 e^{kt}$

    $P(2) = 4P_0 \Rightarrow 4P_0 = P_0 e^{2k} \Rightarrow 4 = e^{2k} \Rightarrow$

    $\ln 4 = 2k \Rightarrow \ln\left(2^2\right) = 2k \Rightarrow 2\ln 2 = 2k \Rightarrow$

    $k = \ln 2$

    So, $P(t) = P_0 e^{t\ln 2}$.

    b. $P(0.5) = 20000e^{0.5\ln 2} \approx 28{,}284$ bacteria

23. $P'(t) = 0.03P(t),\ P(0) = 4$

    a. $P(t) = 5$

    $P'(t) = 0.03P(t) = 0.03(5) = 0.15$

    When the population reaches 5 million people, it is growing at the rate of 0.15 million people per year.

    b. $P'(t) = 0.03P(t)$

    $400000 = 0.4\,\text{million} = 0.03P(t)$

    $P(t) = \dfrac{0.4}{0.03} \approx 13.33$ million

    The population is about 13.33 million when it is growing at the rate of 400,000 people per year.

    c. $k = 0.03$, so $P(t) = P_0 e^{0.03t} = 4e^{0.03t}$.

25. a. $P(0) = 5000e^{(0.2)(0)} = 5000$, so 5000 cells were present initially.

    b. $k = 0.2$, so a differential equation is $P'(t) = .2P(t)$.

    c. $10{,}000 = 5000e^{0.2t} \Rightarrow 2 = e^{0.2t} \Rightarrow$

    $\ln 2 = 0.2t \Rightarrow t = \dfrac{\ln 2}{0.2} \approx 3.5$ hours

    d. $20{,}000 = 5000e^{0.2t} \Rightarrow 4 = e^{0.2t} \Rightarrow$

    $\ln 4 = 0.2t \Rightarrow t = \dfrac{\ln 4}{0.2} \approx 6.9$ hours

27. Let $P(t)$ be the population after t days,

    $P(t) = P_0 e^{kt}$. It is given that $P(40) = 2P(0)$, so

    $P_0 e^{40k} = 2P_0 e^{0(k)} = 2P_0 \Rightarrow e^{40k} = 2 \Rightarrow$

    $\ln e^{40k} = \ln 2 \Rightarrow k = \dfrac{\ln 2}{40} \approx .017$

29. Let $P(t)$ be the population after t years.

$P(t) = P_0 e^{.05t} \Rightarrow 3P_0 = P_0 e^{.05t} \Rightarrow 3 = e^{.05t} \Rightarrow$

$\ln 3 = .05t \Rightarrow t = \dfrac{\ln 3}{.05} \approx 22$ years

31. Let $P(t)$ be the cell population (in millions) after t hours, $P(t) = P_0 e^{kt}$. It is given that $P(0) = 1$ and that $P(10) = 9$. Thus,

$9 = 1 \cdot e^{k(10)} \Rightarrow \ln 9 = \ln e^{10k} \Rightarrow k = \dfrac{\ln 9}{10} \approx .22$.

$P(t) = e^{.22t}$ so $P(15) = e^{.22(15)} = e^{3.3}$

≈ 27 (million) cells.

33. Let $P(t)$ be the population (in millions) t years after the beginning of 1990.

$P(0) = 20.2$, so $P(t) = 20.2 e^{kt}$.

Solve $P(5) = 23$ for k.

$20.2 e^{k(5)} = 23 \Rightarrow 5k = \ln\left(\dfrac{23}{20.2}\right) \Rightarrow$

$k = \dfrac{1}{5}\ln\left(\dfrac{23}{20.2}\right) \Rightarrow P(t) = 20.2 e^{(1/5)\ln(23/20.2)t}$

$P(20) = 20.2 e^{(1/5)\ln(23/20.2)(20)} \approx 34.0$ million

35. a. $P(t) = 8e^{-0.021t}$

b. $P(0) = 8$ g

c. $.021$

d. $P(10) = 8e^{(-0.021)(10)} \approx 6.5$ grams

e. $P'(t) = (-.021)(1) = -.021$
The sample is disintegrating at a rate of .021 gram per year

f. $-.105 = -.021 P(t) \Rightarrow$

$P(t) = \dfrac{-.105}{-.021} = 5$ grams remaining

g. 4 grams will remain after 33 years.
2 grams will remain after 66 years.
1 gram will remain after 99 years.

37. a. $f'(t) = -.6 f(t)$

b. $f(5) = 300 e^{(-0.6)(5)} \approx 14.9$ mg

c. $150 = 300 e^{-0.6t} \Rightarrow .5 = e^{-0.6t} \Rightarrow$

$\ln .5 = -.6t \Rightarrow t = \dfrac{\ln .5}{-.6} \approx 1.2$ hours

39. $.5 = e^{-0.023t} \Rightarrow \ln .5 = -.023t \Rightarrow$

$t = \dfrac{\ln 5}{-.023} \approx 30.1$ years

41. $.5 = e^{8k} \Rightarrow \ln .5 = 8k \Rightarrow$

$k = \dfrac{\ln .5}{8} \approx -.0866$

$P(t) = 10 e^{-0.0866t} \Rightarrow 1 = 10 e^{-0.0866t} \Rightarrow$

$.1 = e^{-0.0866t} \Rightarrow \ln .1 = -.0866t \Rightarrow$

$t = \dfrac{\ln .1}{-.0866} \approx 26.6$ days

43. $f'(t) = kf(t)$ so $f(t) = Ce^{kt}$ where $C = 8$

$f(50) = 4 = 8e^{k(50)} \Rightarrow 0.5 = e^{50k} \Rightarrow$

$\ln .5 = 50k \Rightarrow k = \dfrac{\ln .5}{50} \approx -.014$

So, $.5 = e^{50k}$. Thus, $f(t) = 8e^{-0.014t}$.

45. a. From the graph $P(1) = 8$ grams.

b. From the graph, the half-life appears to be about 3.5 hours.

c. $P'(6) = -.2P(6) = (-.2)(3) = -.6$
The sample was decaying at the rate of .6 grams per hour.

d. $-.4 = -.2P(t) \Rightarrow P(t) = \dfrac{-.4}{-.2} = 2$

$P(t) = 2$ when $t = 8$ hours

47. Let the original amount be 1, $P(t) = e^{-0.00012t}$.

Then for some t_0, $P(t_0) = .20 = e^{-0.00012t_0}$.

$\dfrac{\ln .20}{-.00012} = t_0 \Rightarrow t_0 \approx 13,412$ years

That was the estimated age of the cave paintings more than 65 years ago. The estimated age now is about 13,500 years.

49. Let the original amount be 1. Then

$P(t) = 1 \cdot e^{-0.00012t}$.

$P(4500) = e^{-0.00012(4500)} = e^{-0.54} \approx .583$.

Thus about 58.3% remains.

51. $P(t) = e^{-0.00012t}$; $P(t_0) = 0.27 = e^{-0.00012t_0}$;

$\dfrac{\ln .27}{-.00012} = t_0$; $t_0 \approx 10,911$ years ago.

53. a-D, b-G, c-E, d-B, e-H, f-F, g-A, h-C

55. $y'(t) = -.5y(t); \quad y(0) = 10$

    a. $P(0) = 10$ and $P'(0) = -.5(10) = -5$.
Thus, the slope of the tangent is -5 and the y-intercept of the tangent is 10, so the equation of the tangent is $y = -5t + 10$.

    b. $P(t) = 10e^{-.5t}$

    c. $T = \dfrac{1}{\lambda} = \dfrac{1}{.5} = 2$

5.2   Compound Interest

1. a. $5000

    b. $.04 = 4\%$

    c. $A(10) = 5000e^{(0.04)(10)} \approx \$7459.12$

    d. $A'(t) = .04A(t)$

    e. $A'(10) = (.04)(7459.12)$
           $\approx \$298.36$ per year

    f. $280 = .04A(t) \Rightarrow A(t) = \dfrac{280}{.04} = \$7000$

3. a. $r = .035, P = 4000$
       $A(t) = 4000e^{0.035t}$

    b. $A'(t) = .035A(t)$

    c. $A(2) = 4000e^{(0.035)(2)} \approx \$4290.03$

    d. $5000 = 4000e^{0.035t} \Rightarrow 1.25 = e^{0.035t} \Rightarrow$
       $\ln 1.25 = .035t \Rightarrow t = \dfrac{\ln 1.25}{.035} \approx 6.4$ years

    e. $A'(t) = .035A(t) = (.035)(5000)$
              $= \$175$ per year

5. $r = .042; \quad A'(t) = .042A(t)$
   $A'(t) = (.042)(9000) = \$378$ per year

7. $A(t) = 1000e^{0.06t} \Rightarrow 2500 = 1000e^{0.06t} \Rightarrow$
   $2.5 = e^{0.06t} \Rightarrow \ln 2.5 = .06t \Rightarrow$
   $t = \dfrac{\ln 2.5}{.06} \approx 15.3$ years

9. $A(t) = 1200e^{rt} \Rightarrow 12,500 = 1200e^{r(8)} \Rightarrow$
   $\dfrac{125}{12} = e^{8r} \Rightarrow \ln \dfrac{125}{12} = 8r \Rightarrow$
   $r = \dfrac{1}{8}\ln \dfrac{125}{12} \approx 0.293 = 29.3\%$

11. $2 = e^{0.04t} \Rightarrow \ln 2 = 0.04t \Rightarrow$
   $t = \dfrac{\ln 2}{.04} \approx 17.3$ years

13. $3 = e^{r(15)} \Rightarrow \ln 3 = 15r \Rightarrow$
   $r = \dfrac{\ln 3}{15} \approx .073 = 7.3\%$

15. $A(t) = P_0 e^{-rt} = 10000e^{-0.009t}$
   $A(2) = 10000e^{-0.009(2)} \approx 9822$ SFr

17. $A(t) = e^{rt}$ (in millions) where t is the number of years since 2000.
   $3 = e^{r(10)} \Rightarrow \ln 3 = 10r \Rightarrow r = \dfrac{\ln 3}{10} \approx .11$
   $10 = e^{[(\ln 3)/10]t} \Rightarrow \ln 10 = \dfrac{\ln 3}{10}t \Rightarrow$
   $t = \dfrac{10\ln 10}{\ln 3} \approx 21$ years
   It will be worth $10 million in 2021.

19. $A(t) = Pe^{0.08t}$
   $1000 = Pe^{(0.08)(3)} \Rightarrow P = \dfrac{1000}{e^{0.24}} \approx \$786.63$

21. $A(t) = Pe^{0.045t}$
   $10,000 = Pe^{(0.045)(5)} \Rightarrow$
   $P = \dfrac{10,000}{e^{0.225}} \approx \$7985.16$

23. $A(t) = 70,200e^{0.13t}; \; B(t) = 60,000e^{0.14t}$
   Set $A(t) = B(t)$ and solve for t.
   $70,200e^{0.13t} = 60,000e^{0.14t} \Rightarrow$
   $\dfrac{70,200}{60,000} = \dfrac{e^{0.14t}}{e^{0.13t}} \Rightarrow 1.17 = e^{0.01t} \Rightarrow$
   $\ln 1.17 = .01t \Rightarrow t = \dfrac{\ln 1.17}{.01} \approx 15.7$ years

25. a-B, b-D, c-G, d-A, e-F, f-E, g-H, h-C

27. a. From the graph in Fig. 2(a), the balance will be $200.

    b. From the graph in Fig. 2(b), the rate is $8 per year.

    c. $A'(20) = rA(200) \Rightarrow 8 = r(200) \Rightarrow$
       $r = \dfrac{8}{200} = .04 = 4\%$

    d. From the graph in Fig. 2(a), $A(t) = 300$ when $t = 30$ years.

e. From the graph in Fig. 2(b), $A'(t) = 12$ when $t = 30$ years.

f. $A'(t) = rA(t)$, so the graph of $A'(t)$ is a constant multiple of the graph of $A(t)$.

29.

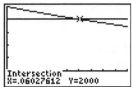

31. Use a graphing calculator to graph
$y_1 = 1200e^{-3x} + 800e^{-4x} + 500e^{-5x}$ and
$y_2 = 2000$ and find the intersection point.

[0, .1] by [−500, 2500]
This occurs when $x \approx .06$, so $r \approx .06$.

5.3 Applications of the Natural Logarithm Function to Economics

1. $f(t) = t^2$; $\dfrac{f'(t)}{f(t)} = \dfrac{2t}{t^2} = \dfrac{2}{t}$

$\dfrac{f'(10)}{f(10)} = \dfrac{2}{10} = \dfrac{1}{5} = 20\%$

$\dfrac{f'(50)}{f(50)} = \dfrac{2}{50} = \dfrac{1}{25} = 4\%$

3. $f(t) = e^{0.3x}$; $\dfrac{f'(t)}{f(t)} = \dfrac{0.3e^{0.3x}}{e^{0.3x}} = 0.3 = 30\%$

Thus for all x, the percentage rate of change is 30%.

5. $f(t) = e^{0.3t^2}$; $\dfrac{f'(t)}{f(t)} = \dfrac{.6te^{.3t^2}}{e^{0.3t^2}} = .6t$

$\dfrac{f'(1)}{f(1)} = .6 = 60\%$; $\dfrac{f'(5)}{f(5)} = 3 = 300\%$

7. $f(p) = \dfrac{1}{p+2}$; $\dfrac{f'(p)}{f(p)} = \dfrac{-\frac{1}{(p+2)^2}}{\frac{1}{p+2}} = -\dfrac{1}{p+2}$

$\dfrac{f'(2)}{f(2)} = -\dfrac{1}{2+2} = -.25 = -25\%$

$\dfrac{f'(8)}{f(8)} = -\dfrac{1}{8+2} = -.1 = -10\%$

9. $S = 50{,}000\sqrt{e^{\sqrt{t}}} = 50{,}000\left(e^{t^{1/2}}\right)^{1/2}$

$\ln S = \ln\left[50{,}000\left(e^{t^{1/2}}\right)^{1/2}\right]$

$\quad = \ln 50{,}000 + \dfrac{1}{2}\ln e^{t^{1/2}} = \ln 50{,}000 + \dfrac{t^{1/2}}{2}$

Differentiating both sides,

$\dfrac{S'}{S} = 0 + \left(\dfrac{1}{2}\right)\left(\dfrac{1}{2}\right)t^{-1/2} = \dfrac{1}{4\sqrt{t}}$

At $t = 4$, the percentage rate of growth is

$\dfrac{1}{4\sqrt{4}} = \dfrac{1}{8} = 12.5\%$.

11. $f(t) = 3.08 + .57t - .1t^2 + .01t^3$
$f'(t) = .57 - .2t + .03t^2$

a. $f(1) = 3.08 + .57(1) - .1(1)^2 + .01(1)^3$
$\quad = 3.56$
In 2011, the wholesale price of one pound of ground beef was $3.56.
$f'(1) = 0.57 - .2 + .03 = .4$
In 2011, the wholesale price of one pound of ground beef was changing at $0.40 per year.

b. $\dfrac{f'(1)}{f(1)} = \dfrac{.4}{3.56} \approx .11 = 11\%$
In 2011, the price was rising at about 11% per year.

c. $f(6) = 3.08 + .57(6) - .1(6)^2 + .01(6)^3$
$\quad = 5.06$
In 2016, the wholesale price of one pound of ground beef was $5.06.
$f'(6) = .57 - .2(6) + .03(6)^2 = .45$
In 2016, the wholesale price of one pound of ground beef was changing at $0.45 per year.
$\dfrac{f'(6)}{f(6)} = \dfrac{.45}{5.06} \approx .089 \approx 9\%$
In 2016, the price was rising at about 9% per year.

For exercises 13–23, the elasticity of demand $E(p)$ at price p for the demand function $q = f(p)$ is

$E(p) = \dfrac{-pf'(p)}{f(p)}$. Demand is elastic if $E(p_0) > 1$ and inelastic if $E(p_0) < 1$.

13. $E(p) = \dfrac{-p(-5)}{700 - 5p} = \dfrac{5p}{700 - 5p} = \dfrac{p}{140 - p}$

$E(80) = \dfrac{80}{140 - 80} = \dfrac{4}{3} > 1$ so demand is elastic.

15. $E(p) = \dfrac{-p(-800p)}{400(116 - p^2)} = \dfrac{2p^2}{116 - p^2}$;

$E(6) = \dfrac{2 \cdot 36}{116 - 36} = \dfrac{72}{80} < 1$ so demand is inelastic.

17. $q' = p^2 e^{-(p+3)}(-1) + e^{-(p+3)} 2p$

$\quad = p e^{-(p+3)}(2 - p)$

$E(p) = \dfrac{-p\left(p e^{-(p+3)}\right)(2 - p)}{p^2 e^{-(p+3)}}$

$\quad = -(2 - p) = p - 2$

$E(4) = 4 - 2 = 2 > 1$ so demand is elastic.

19. a. $q = 3000 - 600 p^{1/2}; q' = -300 p^{-1/2}$

$E(p) = \dfrac{-p(-300 p^{-1/2})}{3000 - 600 p^{1/2}}$

$\quad = \dfrac{300 p^{1/2}}{3000 - 600 p^{1/2}} = \dfrac{p^{1/2}}{10 - 2 p^{1/2}}$

$E(4) = \dfrac{4^{1/2}}{10 - 2 \cdot 4^{1/2}} = \dfrac{2}{6} = \dfrac{1}{3} < 1$

The demand is inelastic at $p = 4$.

b. Since demand is inelastic, to increase revenue, the ticket price should be raised.

21. a. $E(p) = \dfrac{-p\left(-\dfrac{18,000}{p^2}\right)}{\dfrac{18,000}{p} - 1500} = \dfrac{\dfrac{18,000}{p}}{\dfrac{18,000}{p} - 1500}$

$E(6) = \dfrac{3000}{3000 - 1500} = \dfrac{3000}{1500} > 1$

Thus, demand is elastic.

b. If the price is lowered, revenue will increase.

23. a. $q = \dfrac{1000}{p^2}; q' = -\dfrac{2000}{p^3}$

$E(p) = \dfrac{\dfrac{2000}{p^2}}{\dfrac{1000}{p^2}} = 2$

b. Yes, the country succeed in raising its revenue since the demand is always elastic.

25. The relative rate of change of cost $C(x)$, with respect to x, is $\dfrac{C'(x)}{C(x)}$. The relative rate of change of quantity x with respect to x is $\dfrac{x'}{x} = \dfrac{1}{x}$. $E_c(x)$ is the ratio of these two:

$E_c(x) = \dfrac{\dfrac{C'(x)}{C(x)}}{\dfrac{1}{x}} = \dfrac{x \cdot C'(x)}{C(x)}$

27. $C(x) = \dfrac{1}{10} x^2 + 5x + 300$

$C'(x) = \dfrac{2}{10} x + 5 = \dfrac{1}{5} x + 5$

$E_c(x) = \dfrac{x\left(\dfrac{1}{5} x + 5\right)}{\dfrac{1}{10} x^2 + 5x + 300} = \dfrac{\dfrac{1}{5} x^2 + 5x}{\dfrac{1}{10} x^2 + 5x + 300}$

$\quad = \dfrac{2x^2 + 50x}{x^2 + 50x + 3000}$

$E_c(50) = \dfrac{500 + 250}{250 + 250 + 300} = \dfrac{750}{800} = \dfrac{15}{16} < 1$

29. a. $q' = -30,000 e^{-.5p}$

$E(p) = \dfrac{-pq'}{q} = \dfrac{-p(-30,000 e^{-.5p})}{60,000 e^{-.5p}} = \dfrac{p}{2}$

$\dfrac{p}{2} < 1 \Rightarrow p < 2$

Demand is inelastic for $p < 2$

5.4 Further Exponential Models

1. $f(x) = 5(1 - e^{-2x}), x \geq 0$

a. $f'(x) = 5\left[-e^{-2x}(-2)\right] = 10 e^{-2x}$

$f''(x) = -20 e^{-2x}$

Since $f'(x) > 0$ for all $x \geq 0$, $f(x)$ is increasing for all $x \geq 0$. Furthermore, since $f''(x) < 0$ for all $x \geq 0$, $f(x)$ is concave down for all $x \geq 0$.

b. Since $\lim\limits_{x \to \infty} e^{-2x} = 0$, then for very large values of x, $f(x) \approx 5(1 - 0) = 5$.

c.

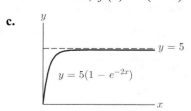

$y = 5$

$y = 5(1 - e^{-2x})$

3. $y' = -2e^{-x}(-1) = 2e^{-x}$

$2 - y = 2 - 2(1 - e^{-x}) = 2 - 2 + 2e^{-x} = 2e^{-x} \Rightarrow$
$2 - y = y'$

5. $y' = f'(x) = -3e^{-10x}(-10) = 30e^{-10x}$

$10(3 - y) = 10\left[3 - \left(3 - 3e^{-10x}\right)\right]$
$\qquad = 30e^{-10x} = y'$

Also, $f(0) = 3(1 - 1) = 0$.

7. The number of people who have heard about the indictment by time t is given by $f(t) = P(1 - e^{-kt})$. It is given that

$f(1) = \dfrac{1}{4}P = P\left(1 - e^{k(1)}\right) \Rightarrow \dfrac{1}{4} = 1 - e^{-k} \Rightarrow$

$e^{-k} = \dfrac{3}{4} \Rightarrow -k = \ln\dfrac{3}{4} \Rightarrow k \approx .29$

Thus, $f(t) = P\left(1 - e^{-.29t}\right)$.

Now, solve $\dfrac{3}{4}P = P\left(1 - e^{-.29t}\right)$ for t.

$\dfrac{3}{4} = 1 - e^{-.29t} \Rightarrow e^{-.29t} = \dfrac{1}{4} \Rightarrow$

$-.29t = \ln.25 \Rightarrow \dfrac{\ln.25}{-.29} = t \approx 4.8$ hours

9. a. $f(7) = \dfrac{10,000}{1 + 50e^{-0.4(7)}} \approx 2475 \approx 2,500$

b. $f'(t) = \dfrac{200,000e^{-0.4t}}{\left(1 + 50e^{-0.4t}\right)^2}$

$f'(14) = \dfrac{200,000e^{-0.4(14)}}{\left(1 + 50e^{-0.4(14)}\right)^2} \approx 526$
$\qquad \approx 500$ people per day

c. $f(t) = 7000 \Rightarrow \dfrac{10,000}{1 + 50e^{-0.4t}} = 7000 \Rightarrow$

$\dfrac{10}{7} = 1 + 50e^{-0.4t} \Rightarrow \dfrac{3}{7} = 50e^{-0.4t} \Rightarrow$

$\dfrac{3}{350} = e^{-0.4t} \Rightarrow \ln\dfrac{3}{350} = -.4t \Rightarrow$

$t = \dfrac{\ln\dfrac{3}{350}}{-.4} \approx 12$

7000 will have heard the news on day 12.

d. $f'(t) = 600 \Rightarrow \dfrac{200,000e^{-0.4t}}{\left(1 + 50e^{-0.4t}\right)^2} = 600$

Using a graphing calculator, we find that $t \approx 6$ and $t \approx 13.5$.

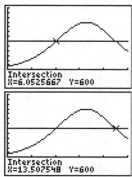

The news will be spreading at the rate of 600 people per day on day 6 and day 14.

e. f' has a maximum at approximately $t = 9.78$, so the news will be spreading at the greatest rate on day 10.

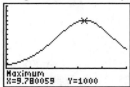

f. Using Equation (9), $y = \dfrac{M}{1 + Be^{-Mkt}}$, we

have $y = \dfrac{M}{1 + Be^{-Mkt}} = \dfrac{10,000}{1 + 50e^{-0.4t}} \Rightarrow$

$M = 10,000$, $B = 50$, and $Mk = .4$ so $k = .00004$.
From Equation (10), we have
$y' = ky(M - y) \Rightarrow$
$\quad f'(t) = .00004f(t)(10,000 - f(t))$.

g. When $f(t) = 5000$,
$f'(t) = 0.00004(5000)(10,000 - 5000)$
$\qquad = 1000$ people per day.

11. a. $f(10) = 50,000\left(1 - e^{-0.3(10)}\right) \approx 47,510$

b. $f'(t) = 15,000e^{-0.3t}$

$f'(0) = 15,000e^{-0.3(0)} = 15,000$

Initially, the news is spreading at 15,000 people/day.

c. $f(t) = 22,500 \Rightarrow$

$50,000\left(1 - e^{-0.3t}\right) = 22,500 \Rightarrow$

$1 - e^{-0.3t} = 0.5 \Rightarrow e^{-0.3t} = 0.5 \Rightarrow$

$-0.3t = \ln 0.5 \Rightarrow t = \dfrac{\ln 0.5}{-0.3} \approx 2.3$

22,500 people will have heard the news on about day 2.

d. $f'(t) = 2500 \Rightarrow 15,000e^{-0.3t} = 2500 \Rightarrow$

$e^{-0.3t} = \dfrac{1}{6} \Rightarrow -0.3t = \ln\left(\dfrac{1}{6}\right) \Rightarrow$

$t = \dfrac{\ln\left(\frac{1}{6}\right)}{-0.3} \approx 6$

The news will be spreading at the rate of 2500 people per day on about day 6.

e. Using Equation (4), we have

$f(t) = P\left(1 - e^{-kt}\right) \Rightarrow P = 50,000$ and

$k = .3$. From Equation (3), we have

$f'(t) = k\left[P - f(t)\right] \Rightarrow$

$f'(t) = .3\left(50,000 - f(t)\right).$

f. When $f(t) = 25,000$,

$f'(t) = .3(50,000 - 25,000)$

$= 7500$ people per day.

13. a. $f(t) = 122\left(e^{-0.2t} - e^{-t}\right)$

$f'(t) = 122\left(-.2e^{-0.2t} + e^{-t}\right)$

$f''(t) = 122\left(.04e^{-0.2t} - e^{-t}\right)$

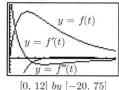

[0, 12] by [−20, 75]

b. $f(7) = 122\left(e^{-0.2(7)} - e^{-7}\right) \approx 30$

There are about 30 units of the drug in the bloodstream after 7 hours.

c. $f'(1) = 122\left(-.2e^{-0.2} + e^{-1}\right) \approx 25$

After 1 hour, the level of the drug is increasing at about 25 units per hour.

d. Use a graphing calculator to find when $f(t) = 20$ for $f(t)$ decreasing.

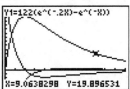

The level of the drug in the bloodstream is 20 units at about 9 hours

e. Use a graphing calculator to find the maximum of $f(t)$. The greatest level is about 65.3 units after 2 hours.

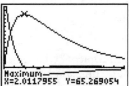

f. Use a graphing calculator to find the minimum of $f'(t)$.

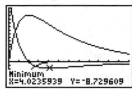

The level of the drug is decreasing the fastest after 4 hours.

Chapter 5 Review Exercises

1. $P(x) = Ce^{-0.2x}$

$P(0) = 29.92 = Ce^{-0.2(0)} = C$

Thus, $P(x) = 29.92e^{-0.2x}$.

2. $P(x) = P_0e^{kt}$ (t in years, P_0 in herring gulls in 1990)

$P(13) = 2P_0 = P_0e^{(k)13} \Rightarrow \ln 2 = \ln e^{13k} \Rightarrow$

$\dfrac{\ln 2}{13} = k \Rightarrow k \approx 0.0533$

$P'(t) = .0533P(t)$

3. $10,000 = P_0e^{(0.12)5} = P_0e^{0.6}$

$P_0 = \dfrac{10,000}{e^{0.6}} \approx \$5488.12$

4. Solve $3000 = 1000e^{0.1t}$ for t.

$$\ln 3 = \ln e^{0.1t} \Rightarrow \frac{\ln 3}{0.1} = t \Rightarrow t \approx 11 \text{ years}$$

5. $\frac{1}{2} = e^{-\lambda(12)}; \frac{\ln .5}{12} = -\lambda \Rightarrow \lambda \approx .058$

6. $.63 = e^{-0.00012t} \Rightarrow \frac{\ln .63}{-.00012} = t \Rightarrow t \approx 3850$ years old

7. a. $A(t) = 17e^{kt} \Rightarrow 19.3 = 17e^{k(7)} \Rightarrow$

$$\frac{19.3}{17} = e^{7k} \Rightarrow \ln\frac{19.3}{17} = 7k \Rightarrow$$

$$k = \frac{1}{7}\ln\frac{19.3}{17} \approx .018$$

$$A(t) = 17e^{0.018t}$$

b. $A(10) = 17e^{(0.018)(10)} \approx 20.4$ million

c. $25 = 17e^{0.018t} \Rightarrow \frac{25}{17} = e^{0.018t} \Rightarrow$

$$\ln\frac{25}{17} = .018t \Rightarrow t = \frac{1}{.018}\ln\frac{25}{17} \approx 21 \text{ years}$$

The population will reach 25 million in the year 2011

8. $A(t) = 100,000e^{kt}$

$$A(2) = 117,000 = 100,000e^{2k} \Rightarrow \frac{\ln\frac{11.7}{10}}{2} = k \Rightarrow$$

$k \approx .0785$ so it earned 7.85%.

9. a. $A(t) = (10,000e^{.2(5)})e^{.06(5)} = (10,000e)e^{.3}$
$$= 10,000e^{1.3} \approx \$36,693$$

b. $A(t) = 10,000e^{.14(10)} = 10,000e^{1.4}$
$$\approx \$40,552$$

The alternative investment is superior by $\$40,552 - \$36,693 = \$3859$.

10. $P_1(t) = 1000e^{k_1 t}$

$$P_1(21) = 2000 = 1000e^{k_1(21)} \Rightarrow \frac{\ln 2}{21} = k_1 \Rightarrow$$

$k_1 \approx .033$

Thus, $P_1(t) = 1000e^{0.033t}$.

$P_2(t) = 710,000e^{k_2 t}$

$P_2(33) = 1,420,000 = 710,000e^{k_2(33)} \Rightarrow$

$\frac{\ln 2}{33} = k_2 \Rightarrow k_2 \approx .021$

Thus $P_2(t) = 710,000e^{0.021t}$.

Equating P_1 and P_2 and solving for t,

$1000e^{0.033t} = 710,000e^{0.021t} \Rightarrow$

$710 = e^{0.012t} \Rightarrow \frac{\ln 710}{.012} = t \Rightarrow t \approx 547$ min.

11. The growth constant is .02.
$y' = (.02)(3) = .06$
When the population reaches 3 million people, the population will be growing at the rate of 60,000 people per year.

$$100,000 = .02y \Rightarrow y = \frac{100,000}{.02} = 5,000,000$$

The population level at the growth rate of 100,000 people per year is 5 million people.

12. $y' = .4y \Rightarrow 200,000 = .4y \Rightarrow$

$$y = \frac{200,000}{.4} = 500,000$$

The size of the colony will be 500,000.
$y' = (.4)(1,000,000) = 400,000$
The colony will be growing at the rate of 400,000 bacteria per hour.

13. a–F, b–D, c–A, d–G, e–H, f–C, g–B, h–E

14. a. From the graph, $f(5) = 25$ grams.

b. From the graph, $f(t) = 10$ when $t = 9$ yr.

c. From the graph, $f(t) = 40$ when $t \approx 3$ yr, so the half-life is about 3 years.

d. From the graph, $f'(1) = -15$ grams/year.

e. From the graph, $f'(t) = -5$ when $t = 6$ yr.

15. $A'(t) = rA(t) \Rightarrow 60 = r(1000) \Rightarrow$

$$r = \frac{60}{1000} = .06 = 6\%$$

16. $A'(t) = rA(t) = (.045)(1230) = \$55.35$ per year

17. $\frac{f'(t)}{f(t)} = \frac{50e^{0.2t^2}(.4t)}{50e^{0.2t^2}} = .4t$

$$\frac{f'(10)}{f(10)} = .4(10) = 4 = 400\%$$

18. $E(p) = \frac{-p(-80p)}{4000 - 40p^2} = \frac{80p^2}{4000 - 40p^2} = \frac{2}{\frac{100}{p^2} - 1}$

$$E(5) = \frac{2}{\frac{100}{25} - 1} = \frac{2}{4 - 1} = \frac{2}{3} < 1 \text{ so demand is}$$

inelastic.

19. Since a price increase of \$0.16 represents a 2% increase in price, the quantity demanded will decrease 1.5(2%) = 3%. Demand is elastic, so revenue will decrease.

20. $f(p) = \dfrac{1}{3p+1}$, $f'(p) = -\dfrac{3}{(3p+1)^2}$

$$\frac{f'(p)}{f(p)} = \frac{-\frac{3}{(3p+1)^2}}{\frac{1}{3p+1}} = -\frac{3}{3p+1}$$

$$\frac{f'(1)}{f(1)} = -\frac{3}{3+1} = -\frac{3}{4} = -75\%$$

21. $q = 1000p^2 e^{-0.02(p+5)}$

$$q' = 1000p^2(e^{-0.02(p+5)})(-.02)$$
$$\qquad\qquad + (e^{-0.02(p+5)})2000p$$
$$= -1000pe^{-0.02(p+5)}(.02p-2)$$

$$E(p) = \frac{1000p^2 e^{-0.02(p+5)}(.02p-2)}{1000p^2 e^{-0.02(p+5)}}$$
$$= .02p - 2$$

$E(200) = .02(200) - 2 = 4 - 2 = 2 > 1$

Thus, demand is elastic, so a decrease in price will increase revenue.

22. $E(p) = \dfrac{-p(ae^{-bp})(-b)}{ae^{-bp}} = pb$. Thus if $p = \dfrac{1}{b}$,

$$E(p) = \frac{1}{b} \cdot b = 1.$$

23. Since for group A, $f'(t) = k(P - f(t))$, it follows that $f(t) = P(1 - e^{-kt})$.

$f(0) = 0 = 100\left(1 - e^{-k(0)}\right)$ and

$f(13) = 66 = 100\left(1 - e^{-k(13)}\right) \Rightarrow$ Thus,

$.66 = 1 - e^{-13k} \Rightarrow e^{-13k} = .34 \Rightarrow$

$\dfrac{\ln 0.34}{-13} = k \Rightarrow k \approx 0.083$

$f(t) = 100\left(1 - e^{-.083t}\right)$.

24. $f(t) = \dfrac{M}{1 + Be^{-Mkt}}$

Since 55 is the maximum height for the weed, $M = 55$.

$$f(9) = 8 = \frac{55}{1 + Be^{-55(9)k}} \Rightarrow$$

$$1 + Be^{-55(9)k} = \frac{55}{8} \Rightarrow B = \frac{47}{8}e^{55(9)k}$$

$$f(25) = 48 = \frac{55}{1 + Be^{-55(25)k}} \Rightarrow$$

$$1 + Be^{-55(25)k} = \frac{55}{48} \Rightarrow B = \frac{7}{48}e^{55(25)k}$$

So, $\dfrac{47}{8}e^{55(9)k} = \dfrac{7}{48}e^{55(25)k} \Rightarrow$

$$\frac{47}{8} = \frac{7}{48}e^{880k} \Rightarrow \frac{\ln\frac{47 \cdot 48}{8 \cdot 7}}{880} = k \Rightarrow$$

$k \approx .0042$ and $-Mk = -.231$

$$B = \frac{47}{8}e^{55(9)k} \Rightarrow B = \frac{47}{8}e^{55(9)(0.0042)} \approx 46.98$$

Thus, $f(t) = \dfrac{55}{1 + 46.98e^{-0.231t}}$.

25. a. From the graph, $f(11) = 400°$

b. From the graph $f'(6) = -100$. The temperature is decreasing at a rate of 100°/sec.

c. From the graph $f(t) = 200$ at $t = 17$ sec.

d. From the graph $f'(t) = -200$ at about $t = 2$ sec.

26. Since the culture grows at a rate proportional to its size, $500 = 10,000k \Rightarrow k = .05$. Then, $P' = .05(15,000) = 750$ bacteria per day.

Chapter 6 The Definite Integral

6.1 Antidifferentiation

1. $f(x) = x \Rightarrow F(x) = \frac{1}{2}x^2 + C$

3. $f(x) = e^{3x} \Rightarrow F(x) = \frac{1}{3}e^{3x} + C$

5. $f(x) = 3 \Rightarrow F(x) = 3x + C$

7. $\int 4x^3 dx = x^4 + C$

9. $\int 7 dx = 7x + C$

11. $\int \frac{x}{c} dx = \int \frac{1}{c} x dx = \frac{1}{2c}x^2 + C$

13. $\int \left(\frac{2}{x} + \frac{x}{2} \right) dx = \int \left(2 \cdot \frac{1}{x} + \frac{1}{2} \cdot x \right) dx$
$= 2 \int \frac{1}{x} dx + \frac{1}{2} \int x dx$
$= 2 \ln|x| + \frac{1}{4}x^2 + C$

15. $\int x\sqrt{x} dx = \int x^{3/2} dx = \frac{2}{5}x^{5/2} + C$

17. $\int \left(x - 2x^2 + \frac{1}{3x} \right) dx = \int \left(x - 2x^2 + \frac{1}{3} \cdot \frac{1}{x} \right) dx$
$= \int x dx - 2 \int x^2 dx + \frac{1}{3} \int \frac{1}{x} dx$
$= \frac{1}{2}x^2 - \frac{2}{3}x^3 + \frac{1}{3}\ln|x| + C$

19. $\int 3e^{-2x} dx = -\frac{3}{2}e^{-2x} + C$

21. $\int e dx = ex + C$

23. $\int -2(e^{2x} + 1) dx = -2 \int e^{2x} dx - 2 \int 1 dx$
$= -2 \left(\frac{1}{2}e^{2x} \right) - 2x + C$
$= -e^{2x} - 2x + C$

25. $\frac{d}{dt} \left[ke^{-2t} \right] = -2ke^{-2t} = 5e^{-2t} \Rightarrow k = -\frac{5}{2}$

27. $\frac{d}{dx} \left[ke^{4x-1} \right] = 4ke^{4x-1} = 2e^{4x-1} \Rightarrow k = \frac{1}{2}$

29. $\frac{d}{dx} \left[k(5x-7)^{-1} \right] = -k(5x-7)^{-2}(5)$
$= -5k(5x-7)^{-2} = (5x-7)^{-2}$
$k = -\frac{1}{5}$

31. $\frac{d}{dx} \left[k\ln|4-x| \right] = \frac{k}{4-x}(-1) = \frac{-k}{4-x} = \frac{1}{4-x} \Rightarrow$
$k = -1$

33. $\frac{d}{dx} \left[k(3x+2)^5 \right] = 5k(3x+2)^4(3)$
$= 15k(3x+2)^4 = (3x+2)^4 \Rightarrow$
$k = \frac{1}{15}$

35. $\frac{d}{dx}[k\ln|2+x|] = \frac{k}{2+x} = \frac{3}{2+x} \Rightarrow k = 3$

37. $f'(t) = t^{3/2} \Rightarrow f(t) = \frac{2}{5}t^{5/2} + C$

39. $f'(t) = 0 \Rightarrow f(t) = C$

41. $f'(x) = 0.5e^{-0.2x} \Rightarrow f(x) = -2.5e^{-0.2x} + C$
$f(0) = 0 \Rightarrow -2.5e^{-0.2 \cdot 0} + C = 0 \Rightarrow C = 2.5$
Thus, $f(x) = -2.5e^{-0.2x} + 2.5$.

43. $f'(x) = x \Rightarrow f(x) = \frac{1}{2}x^2 + C$
$f(0) = 3 \Rightarrow \frac{1}{2} \cdot 0^2 + C = 3 \Rightarrow C = 3$
Thus, $f(x) = \frac{1}{2}x^2 + 3$.

45. $f'(x) = x^{1/2} + 1 \Rightarrow f(x) = \frac{2}{3}x^{3/2} + x + C$
$f(4) = 0 \Rightarrow \frac{2}{3}4^{3/2} + 4 + C = 0 \Rightarrow$
$\frac{2}{3} \cdot 8 + 4 + C = 0 \Rightarrow C = -\frac{28}{3}$
Thus, $f(x) = \frac{2}{3}x^{3/2} + x - \frac{28}{3}$.

47. $f(x) = \int \frac{2}{x} dx = 2\ln|x| + C$
$f(1) = 2 \Rightarrow 2\ln|1| + C = 2 \Rightarrow C = 2$
Thus, $f(x) = 2\ln|x| + 2$.

49. $\dfrac{d}{dx}\left(\dfrac{1}{x}+C\right)=-\dfrac{1}{x^2}\neq\ln x$

$\dfrac{d}{dx}\left(x\ln x-x+C\right)=(\ln x+1)-1=\ln x$

$\dfrac{d}{dx}\left(\dfrac{1}{2}(\ln x)^2+C\right)=\dfrac{\ln x}{x}\neq\ln x$

The answer is (b).

51.

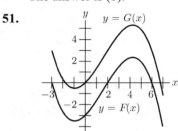

53. $g(x)=f(x)+3$

$g'(x)=f'(x)\Rightarrow g'(5)=f'(5)=\dfrac{1}{4}$

55. a. $\displaystyle\int(96-32t)dt=96t-16t^2+C$

The initial height is 256 feet, so $C=256$.

Thus, $s(t)=-16t^2+96t+256$.

b. Setting $s(t)=0$, $-16t^2+96t+256=0\Rightarrow$

$t^2-6t-16=0\Rightarrow(t-8)(t+2)=0$. The only solution that is sensible is $t=8$ seconds.

c. Since $s'(t)=96-32t$, $s(t)$ has a maximum when

$s'(t)=0\Rightarrow96-32t=0\Rightarrow t=3$.

The ball will reach a maximum height of $s(3)=400$ ft.

57. $P(t)=\displaystyle\int\left(60+2t-\dfrac{1}{4}t^2\right)dt$

$=60t+t^2-\dfrac{1}{12}t^3+C$

$P(0)=0\Rightarrow60\cdot0+0^2-\dfrac{1}{12}\cdot0^3+C=0\Rightarrow$

$C=0$

Thus, $P(t)=60t+t^2-\dfrac{1}{12}t^3$.

59. $f(t)=\displaystyle\int10e^{-0.4t}dt=-\dfrac{100}{4}e^{-0.4t}+C$

$=-25e^{-0.4t}+C$

$f(0)=-5\Rightarrow-25e^{-0.4\cdot0}+C=-5\Rightarrow C=20$

Thus, $f(t)=-25e^{-0.4t}+20$ and the

temperature at time t is $-25e^{-0.4t}+20\,°\text{C}$.

61. $P(x)=\displaystyle\int(1.30+.06x-.0018x^2)\,dx$

$=1.30x+.03x^2-.0006x^3+C$

$P(0)=-95\Rightarrow$

$1.30\cdot0+.03\cdot0^2-.0006\cdot0^3+C=-95\Rightarrow$

$C=-95$

Thus, $P(x)=-.0006x^3+.03x^2+1.30x-95$.

63. $f(t)=\displaystyle\int94e^{0.016t}dt=\dfrac{94}{.016}e^{0.016t}+C$

$=5875e^{0.016t}+C$

Since consumption is reckoned from 1980, we have $f(0)=0=5875(1)+C\Rightarrow C=-5875$.

Thus,

$f(t)=5875e^{0.016t}-5875=5875\left(e^{0.016t}-1\right)$.

65. $\displaystyle\int C'(x)dx=C(x)=1000x+25x^2+C_1$

$C(0)=C_1=$ fixed cost $=10,000\Rightarrow$

$1000\cdot0+25\cdot0^2+C_1=10,000\Rightarrow C_1=10,000$

$C(x)=25x^2+1000x+10,000$

67. $F(x)=\dfrac{1}{2}e^{2x}-e^{-x}+\dfrac{1}{6}x^3$

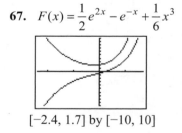

$[-2.4, 1.7]$ by $[-10, 10]$

6.2   The Definite Integral and Net Change of a Function

1. $\int_0^1 \left(2x - \frac{3}{4}\right) dx = \left(x^2 - \frac{3}{4}x\right)\Big|_0^1 = \left(1^2 - \frac{3}{4}(1)\right) - \left(0^2 - \frac{3}{4}(0)\right) = \frac{1}{4}$

3. $\int_1^4 \left(3\sqrt{t} + 4t\right) dt = \int_1^4 \left(3t^{1/2} + 4t\right) dt = \left(2t^{3/2} + 2t^2\right)\Big|_1^4 = \left(2(4)^{3/2} + 2(4)^2\right) - \left(2(1)^{3/2} + 2(1)^2\right) = 48 - 4 = 44$

5. $\int_1^2 \left(-\frac{3}{x^2}\right) dx = \int_1^2 \left(-3x^{-2}\right) dx = \left(3x^{-1}\right)\Big|_1^2 = \left(\frac{3}{x}\right)\Big|_1^2 = \frac{3}{2} - 3 = -\frac{3}{2}$

7. $\int_1^2 \left(\frac{5 - 2x^3}{x^6}\right) dx = \int_1^2 \left(\frac{5}{x^6} - \frac{2x^3}{x^6}\right) dx = \int_1^2 \left(5x^{-6} - 2x^{-3}\right) dx = \left(-x^{-5} + x^{-2}\right)\Big|_1^2$

$\qquad = \left(-\frac{1}{2^5} + \frac{1}{2^2}\right) - \left(-\frac{1}{1^5} + \frac{1}{1^2}\right) = -\frac{1}{32} + \frac{1}{4} = \frac{7}{32}$

9. $\int_{-1}^0 \left(3e^{3t} + t\right) dt = \left(e^{3t} + \frac{t^2}{2}\right)\Big|_{-1}^0 = \left(e^{3(0)} + \frac{0^2}{2}\right) - \left(e^{3(-1)} + \frac{(-1)^2}{2}\right) = 1 - \left(\frac{1}{e^3} + \frac{1}{2}\right) = \frac{1}{2} - \frac{1}{e^3}$

11. $\int_1^2 \frac{2}{x} dx = 2\ln|x|\Big|_1^2 = 2\ln 2 - 2\ln 1 = 2\ln 2 - 0 = \ln 2^2 = \ln 4$

13. $\int_0^1 \frac{e^x + e^{0.5x}}{e^{2x}} dx = \int_0^1 \left(\frac{e^x}{e^{2x}} + \frac{e^{0.5x}}{e^{2x}}\right) dx = \int_0^1 \left(e^{-x} + e^{-1.5x}\right) dx = \left(-e^{-x} - \frac{e^{-1.5x}}{1.5}\right)\Big|_0^1$

$\qquad = \left(-e^{-1} - \frac{e^{-1.5}}{1.5}\right) - \left(-e^0 - \frac{e^0}{1.5}\right) = -\frac{1}{e} - \frac{e^{-1.5}}{1.5} - \left(-1 - \frac{1}{1.5}\right)$

$\qquad = -\frac{1}{e} - \frac{2e^{-1.5}}{3} + \frac{5}{3}$

15. $\int_0^4 f(x) dx = \int_0^1 f(x) dx + \int_1^4 f(x) dx = 3.5 + 5 = 8.5$

17. $\int_1^3 \left(2f(x) - 3g(x)\right) dx = \int_1^3 \left(2f(x)\right) dx - \int_1^3 \left(3g(x)\right) dx = 2\int_1^3 f(x) dx - 3\int_1^3 g(x) dx = 2(3) - 3(-1) = 9$

19. $2\int_1^2 \left(3x + \frac{1}{2}x^2 - x^3\right) dx + 3\int_1^2 \left(x^2 - 2x + 7\right) dx = \int_1^2 \left(6x + x^2 - 2x^3\right) dx + \int_1^2 \left(3x^2 - 6x + 21\right) dx$

$\qquad = \int_1^2 \left(6x + x^2 - 2x^3 + 3x^2 - 6x + 21\right) dx$

$\qquad = \int_1^2 \left(-2x^3 + 4x^2 + 21\right) dx = \left(-\frac{1}{2}x^4 + \frac{4}{3}x^3 + 21x\right)\Big|_1^2$

$\qquad = \left(-\frac{1}{2}(2)^4 + \frac{4}{3}(2)^3 + 21(2)\right) - \left(-\frac{1}{2}(1)^4 + \frac{4}{3}(1)^3 + 21(1)\right)$

$\qquad = \frac{134}{3} - \frac{131}{6} = \frac{137}{6}$

21. $\int_{-1}^{0}\left(x^{3}+x^{2}\right)dx+\int_{0}^{1}\left(x^{3}+x^{2}\right)dx=\int_{-1}^{1}\left(x^{3}+x^{2}\right)dx=\left(\frac{x^{4}}{4}+\frac{x^{3}}{3}\right)\Big|_{-1}^{1}=\left(\frac{1^{4}}{4}+\frac{1^{3}}{3}\right)-\left(\frac{(-1)^{4}}{4}+\frac{(-1)^{3}}{3}\right)$

$$=\frac{7}{12}+\frac{1}{12}=\frac{2}{3}$$

23. $f(3)-f(1)=\int_{1}^{3}(-2x+3)\,dx=\left(-x^{2}+3x\right)\Big|_{1}^{3}=\left(-3^{2}+3(3)\right)-\left(-1^{2}+3(1)\right)=-2$

25. $f(1)-f(-1)=\int_{-1}^{1}\left(-.5t+e^{-2t}\right)dt=\left(-\frac{1}{4}t^{2}-\frac{1}{2}e^{-2t}\right)\Big|_{-1}^{1}=\left(-\frac{1}{4}(1)^{2}-\frac{1}{2}e^{-2(1)}\right)-\left(-\frac{1}{4}(-1)^{2}-\frac{1}{2}e^{-2(-1)}\right)$

$$=-\frac{1}{4}-\frac{e^{-2}}{2}-\left(-\frac{1}{4}-\frac{e^{2}}{2}\right)=\frac{e^{2}-e^{-2}}{2}$$

27. $\int_{0}^{2}f(x)\,dx=\int_{0}^{1}f(x)\,dx+\int_{1}^{2}f(x)\,dx=\int_{0}^{1}1\,dx+\int_{1}^{2}x\,dx=x\Big|_{0}^{1}+\frac{x^{2}}{2}\Big|_{1}^{2}=1+2-\frac{1}{2}=\frac{5}{2}$

29. $\int_{-1}^{1}f(t)\,dt=\int_{-1}^{0}f(t)\,dt+\int_{0}^{1}f(t)\,dt=\int_{-1}^{0}(1+t)\,dt+\int_{0}^{1}(1-t)\,dt$

$$=\left(t+\frac{t^{2}}{2}\right)\Big|_{-1}^{0}+\left(t-\frac{t^{2}}{2}\right)\Big|_{0}^{1}=-\left(-1+\frac{1}{2}\right)+\left(1-\frac{1}{2}\right)=1$$

31. Let $s(t)$ represent the position function. We know that $s'(t)=v(t)=-32t,$ so the change in position is given

by $s(4)-s(2)=\int_{2}^{4}(-32t)\,dt=\left(-16t^{2}\right)\Big|_{2}^{4}=-16(4)^{2}-\left(-16(2)^{2}\right)=-192.$

The rock fell 192 feet during the time interval $2\le t\le4.$

33. a. Let $s(t)$ represent the position function. We know that $s'(t)=v(t)=-32t+75,$ so the change in
position is given by

$$s(3)-s(1)=\int_{1}^{3}(-32t+75)\,dt=\left(-16t^{2}+75t\right)\Big|_{1}^{3}=\left(-16(3)^{2}+75(3)\right)-\left(-16(1)^{2}+75(1)\right)=81-59=22.$$

b. During the time interval $1\le t\le3,$ the ball rose 22 feet. Therefore, at time $t=3,$ the ball is 22 feet higher
than its position at time $t=1.$

c. $s(5)-s(1)=\int_{1}^{5}(-32t+75)\,dt=\left(-16t^{2}+75t\right)\Big|_{1}^{5}=\left(-16(5)^{2}+75(5)\right)-\left(-16(1)^{2}+75(1)\right)=-25-59=-84$

During the time interval $1\le t\le5,$ the ball fell 84 feet. Therefore, at time $t=5,$ the ball is 84 feet lower
than its position at time $t=1.$

35. a. Let $C(x)$ represent the cost function. The cost increase is given by

$$C(3)-C(1)=\int_{1}^{3}C'(x)\,dx=\int_{1}^{3}\left(.1x^{2}-x+12\right)dx=\left(\frac{x^{3}}{30}-\frac{x^{2}}{2}+12x\right)\Big|_{1}^{3}$$

$$=\left(\frac{3^{3}}{30}-\frac{3^{2}}{2}+12(3)\right)-\left(\frac{1^{3}}{30}-\frac{1^{2}}{2}+12(1)\right)=32.4-11.53=20.87$$

The cost will increase $20.87 if the company goes from a production level of 1 to 3 items per day.

b. $C(3)=C(1)+\left(C(3)-C(1)\right)=15+20.87=35.87$
The cost of producing three items is $35.87.

37. Let $T(t)$ represent the value of the investment during a given time interval. Then $T'(t) = R(t)$, and the increase in value is given by

$$T(10) - T(0) = \int_0^{10} T'(t)\,dt = \int_0^{10} R(t)\,dt = \int_0^{10} \left(700e^{0.07t} + 1000\right)dt$$

$$= \left(\frac{700}{.07}e^{0.07t} + 1000t\right)\Bigg|_0^{10} = \left(10,000e^{0.07t} + 1000t\right)\Bigg|_0^{10}$$

$$= \left(10,000e^{0.07(10)} + 1000(10)\right) - \left(10,000e^{0.07(0)} + 1000(0)\right)$$

$$\approx 30137.50 - 10,000 = 20137.50$$

The investment increased by \$20,137.50.

39. a.
$$P(10) - P(0) = \int_0^{10}\left(\frac{7}{300}e^{t/25} - \frac{1}{80}e^{t/16}\right) = \left(\frac{7}{12}e^{t/25} - \frac{1}{5}e^{t/16}\right)\Bigg|_0^{10}$$

$$= \left(\frac{7}{12}e^{10/25} - \frac{1}{5}e^{10/16}\right) - \left(\frac{7}{12}e^{0/25} - \frac{1}{5}e^{0/16}\right) = \left(\frac{7}{12}e^{10/25} - \frac{1}{5}e^{10/16}\right) - \left(\frac{7}{12} - \frac{1}{5}\right)$$

$$= \frac{1}{60}\left(35e^{2/5} - 12e^{5/8} - 23\right) \approx .11325$$

The population increased about .11325 million or 113,250 from 2000 to 2010.

b.
$$P(40) - P(10) = \int_{10}^{40}\left(\frac{7}{300}e^{t/25} - \frac{1}{80}e^{t/16}\right) = \left(\frac{7}{12}e^{t/25} - \frac{1}{5}e^{t/16}\right)\Bigg|_{10}^{40}$$

$$= \left(\frac{7}{12}e^{40/25} - \frac{1}{5}e^{40/16}\right) - \left(\frac{7}{12}e^{10/25} - \frac{1}{5}e^{10/16}\right) = \left(\frac{7}{12}e^{8/5} - \frac{1}{5}e^{5/2}\right) - \left(\frac{7}{12}e^{2/5} - \frac{1}{5}e^{5/8}\right)$$

$$= \frac{7}{12}e^{2/5}\left(35e^{6/5} - 1\right) - \frac{1}{5}\left(e^{5/2} - e^{5/8}\right) \approx -.0438182$$

The population will decrease by about .043812 million or about 43,812 people due to emigration.

41.
$$P(t) = P(0) + \int_0^t -4.1107e^{0.03t}\,dt = 200 + \left(-\frac{4.1107}{.03}e^{0.03t}\right)\Bigg|_0^t = 200 - \frac{4.1107}{.03}e^{0.03t} + \frac{4.1107}{.03}$$

$$\approx 337.023 - 137.023e^{0.03t} \text{ thousand dollars}$$

43. Let $T(t)$ represent the amount of salt in grams during a given time interval. Then $T'(t) = r(t)$, and the amount of salt that was eliminated during the first two minutes is given by

$$T(2) - T(0) = \int_0^2 T'(t)\,dt = \int_0^2 r(t)\,dt = \int_0^2\left(-\left(t + \frac{1}{2}\right)\right)dt = \int_0^2\left(-t - \frac{1}{2}\right)dt$$

$$= \left(-\frac{t^2}{2} - \frac{1}{2}t\right)\Bigg|_0^2 = \left(-\frac{2^2}{2} - \frac{1}{2}(2)\right) - \left(-\frac{0^2}{2} - \frac{1}{2}(0)\right) = -3$$

Three grams of salt were eliminated in the first two minutes.

6.3 The Definite Integral and Area Under a Graph

1. a. $A = lw = 3(2) = 6$

b. $A = \int_1^4 2\, dx = 2x\big|_1^4 = 2(4) - 2(1) = 6$

3. a. $A = \frac{1}{2}bh = \frac{1}{2}(2)(2) = 2$

b. $A = \int_{-2}^0 (-x)\, dx = -\frac{x^2}{2}\Big|_{-2}^0 = 0 - (-2) = 2$

5. a. $A = \frac{1}{2}b_1h_1 + \frac{1}{2}b_2h_2 = \frac{1}{2}(1)(1) + \frac{1}{2}(1)(1) = 1$

b. $A = \int_0^1 (1-x)\, dx + \int_1^2 (x-1)\, dx$

$= \left(x - \frac{x^2}{2}\right)\Big|_0^1 + \left(\frac{x^2}{2} - x\right)\Big|_1^2$

$= \frac{1}{2} + \left[\left(\frac{2^2}{2} - 2\right) - \left(\frac{1^2}{2} - 1\right)\right] = \frac{1}{2} + \frac{1}{2} = 1$

7. $\int_1^2 \frac{1}{x}\, dx$

9. $\int_1^2 \ln x\, dx$

11. $\int_1^3 \left(x + \frac{1}{x}\right) dx$

13. $\int_1^2 \frac{1}{x}\, dx = \ln|x|\big|_1^2 = \ln 2 - \ln 1 = \ln 2$

15. $\int_1^3 \left(x + \frac{1}{x}\right) dx = \left(\frac{x^2}{2} + \ln|x|\right)\Big|_1^3$

$= \left(\frac{3^2}{2} + \ln|3|\right) - \left(\frac{1^2}{2} + \ln|1|\right)$

$= \left(\frac{9}{2} + \ln 3\right) - \frac{1}{2} - = 4 + \ln 3$

17.

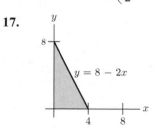

y

8

$y = 8 - 2x$

x

$4 \qquad 8$

19. $\int_2^3 4x\, dx = \left(2x^2\right)\big|_2^3 = 2(9) - 2(4) = 10$

21. $\int_0^1 \left(3x^2 + x + 2e^{x/2}\right) dx$

$= \left(x^3 + \frac{x^2}{2} + 4e^{x/2}\right)\Big|_0^1$

$= \left(\frac{3}{2} + 4e^{1/2}\right) - 4 = -\frac{5}{2} + 4e^{1/2}$

23. $\int_1^4 (x-3)^4\, dx = \frac{(x-3)^5}{5}\Big|_1^4 = \frac{(4-3)^5}{5} - \frac{(1-3)^5}{5}$

$= \frac{1}{5} - \left(-\frac{32}{5}\right) = \frac{33}{5}$

25. $\int_0^b x^3\, dx = 4 \Rightarrow \frac{x^4}{4}\Big|_0^b = 4 \Rightarrow \frac{b^4}{4} = 4 \Rightarrow$

$b^4 = 16 \Rightarrow b = 2$

27. $\Delta x = \frac{2-0}{4} = .5$

The first midpoint is that of $[0, .5]$ which is .25, so the midpoints are .25, .75, 1.25, 1.75.

29. $\Delta x = \frac{4-1}{5} = .6$

The first midpoint is that of $[1, 1.6]$ which is 1.3, so the midpoints are 1.3, 1.9, 2.5, 3.1, 3.7.

31. $\Delta x = .5$

The midpoints are 1.25, 1.75, 2.25, 2.75.

Area

$= .5\left[f(1.25) + f(1.75) + f(2.25) + f(2.75)\right]$

$= .5\left[(1.25)^2 + (1.75)^2 + (2.25)^2 + (2.75)^2\right]$

$= 8.625$

33. $\Delta x = .4$; the left endpoints are 1, 1.4, 1.8, 2.2, 2.6.

Area

$= .4\left[1^3 + (1.4)^3 + (1.8)^3 + (2.2)^3 + (2.6)^3\right]$

$= 15.12$

35. $\Delta x = .2$; the right endpoints are 2.2, 2.4, 2.6, 2.8, 3.

Area $= .2\left[e^{-2.2} + e^{-2.4} + e^{-2.6} + e^{-2.8} + e^{-3}\right]$

$\approx .077278$

37. midpoints: 1, 3, 5, 7; $\Delta x = 2$

$\left[f(1) + f(3) + f(5) + f(7)\right]\Delta x$

$= [4 + 8 + 6 + 2]2 = 40$

39. right endpoints: 5, 6, 7, 8, 9; $\Delta x = 1$

$$\left[f(5) + f(6) + f(7) + f(8) + f(9)\right]\Delta x = \left[6 + 4 + 2 + 1 + 2\right]1 = 15$$

41. $\Delta x = .75$; the left endpoints are 1, 1.75, 2.5, 3.25.

Area $= .75\left[(4-1) + (4-1.75) + (4-2.5) + (4-3.25)\right] = 5.625$

The midpoints are 1.375, 2.125, 2.875, 3.625

Area $= .75\left[(4-1.375) + (4-2.125) + (4-2.875) + (4-3.625)\right] = 4.5$

43. $\Delta x = .4$; the midpoints are $-.8, -.4, 0, .4, .8$.

Area $= 0.4\left[\left(1-(-.8)^2\right)^{1/2} + \left(1-(-.4)^2\right)^{1/2} + \left(1-(0)^2\right)^{1/2} + \left(1-(.4)^2\right)^{1/2} + \left(1-(.8)^2\right)^{1/2}\right] \approx 1.61321$

The error is $1.61321 - 1.57080 = .04241$.

45. $A = 20(106) + 40(101) + 40(100) + 40(113) + 20(113) = 16{,}940 \text{ ft}^2$

47. $1^2 + 2^2 + 3^2 + \cdots + n^2 = \dfrac{n(n+1)(2n+1)}{6}$

$n = 1 : 1^2 = \dfrac{1(1+1)(2(1)+1)}{6} \Rightarrow 1 = \dfrac{6}{6}$

$n = 2 : 1^2 + 2^2 = \dfrac{2(2+1)(2(2)+1)}{6} \Rightarrow 5 = \dfrac{30}{6}$

$n = 3 : 1^2 + 2^2 + 3^2 = \dfrac{3(3+1)(2(3)+1)}{6} \Rightarrow 14 = \dfrac{84}{6}$

$n = 4 : 1^2 + 2^2 + 3^2 + 4^2 = \dfrac{4(4+1)(2(4)+1)}{6} \Rightarrow 30 = \dfrac{180}{6}$

The formula can be proven for all values of n using mathematical induction.

49.

```
fnInt(e^(-X²),X,
-1,1)
        1.493648266
```

The area under the graph is about 1.494

In exercise 51, the figures were created on a TI-84 Plus using the program RIEMANN.8xp downloaded from http://www.calcblog.com. Similar programs are available at www.ticalc.org.

51. There are 20 intervals, so $\Delta x = \dfrac{2}{20} = .1$. Since we are using the midpoints of the subintervals, $x_1 = 1.05$. On

the calculator, set $Y_1 = x\sqrt{1+x^2}$. Use the **sum(** and **seq(** as shown to find the sum.

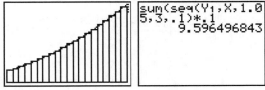

The area is approximately 9.60 square units.

6.4 Areas in the *xy*-Plane

1. $A = \int_1^2 f(x)\,dx + \int_3^4 \left[-f(x)\right]dx$

3.

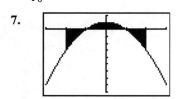

5. $\int_0^7 f(x)\,dx$ is clearly positive since there is more area above the *x*-axis.

7.

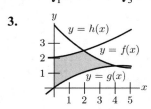

$$A = -\int_{-2}^{-1}\left(1-x^2\right)dx + \int_{-1}^{1}\left(1-x^2\right)dx - \int_{1}^{2}\left(1-x^2\right)dx = -\left(x-\frac{x^3}{3}\right)\Bigg|_{-2}^{-1} + \left(x-\frac{x^3}{3}\right)\Bigg|_{-1}^{1} - \left(x-\frac{x^3}{3}\right)\Bigg|_{1}^{2}$$

$$= -\left[\left(-1+\frac{1}{3}\right)-\left(-2+\frac{8}{3}\right)\right] + \left[\left(1-\frac{1}{3}\right)-\left(-1+\frac{1}{3}\right)\right] - \left[\left(2-\frac{8}{3}\right)-\left(1-\frac{1}{3}\right)\right]$$

$$= -\left(-\frac{4}{3}\right)+\frac{4}{3}-\left(-\frac{4}{3}\right)=4$$

9.

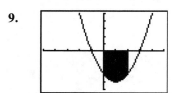

$$A = -\int_0^2\left(x^2-2x-3\right)dx = -\left(\frac{x^3}{3}-x^2-3x\right)\Bigg|_0^2 = -\left(\frac{8}{3}-4-6\right)-0=\frac{22}{3}$$

11.

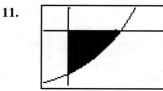

$$A = -\int_0^{\ln 3}\left(e^x-3\right)dx = -\left(\left(e^x-3x\right)\Big|_0^{\ln 3}\right)=-\left[\left(e^{\ln 3}-3\ln 3\right)-1\right]=-\left(3-3\ln 3-1\right)=3\ln 3-2$$

13.

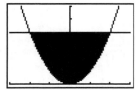

Since $y = 8$ lies above $y = 2x^2$ on $[-2, 2]$, we calculate

$$\int_{-2}^{2} (8 - 2x^2)\, dx = \left(8x - \frac{2}{3}x^3\right)\Bigg|_{-2}^{2}$$

$$= 16 - \frac{16}{3} - \left(-16 + \frac{16}{3}\right)$$

$$= 32 - \frac{32}{3} = \frac{96}{3} - \frac{32}{3} = \frac{64}{3}$$

15.

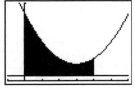

Since $y = x^2 - 6x + 12$ lies above $y = 1$ on $[0, 4]$, we calculate

$$\int_{0}^{4} \left((x^2 - 6x + 12) - 1\right) dx = \left(\frac{1}{3}x^3 - 3x^2 + 11x\right)\Bigg|_{0}^{4}$$

$$= \frac{64}{3} - 48 + 44 - 0 = \frac{52}{3}$$

17.

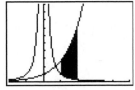

Since $y = e^x$ lies above $y = \dfrac{1}{x^2}$ on $[1, 2]$, we calculate

$$\int_{1}^{2}\left(e^x - \frac{1}{x^2}\right) dx = \int_{1}^{2}\left(e^x - x^{-2}\right) dx$$

$$= \left(e^x + \frac{1}{x}\right)\Bigg|_{1}^{2}$$

$$= \left(e^2 + \frac{1}{2}\right) - (e + 1) = e^2 - e - \frac{1}{2}$$

19.

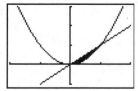

To find the points of intersection, solve

$x^2 = x \Rightarrow x^2 - x = 0 \Rightarrow x(x - 1) = 0 \Rightarrow$

$x = 0$ or $x = 1$

Thus, we want to integrate from $x = 0$ to $x = 1$

with $y = x$ above $y = x^2$, so $\int_{0}^{1}\left(x - x^2\right) dx$.

$$\int_{0}^{1}\left(x - x^2\right) dx = \left(\frac{1}{2}x^2 - \frac{1}{3}x^3\right)\Bigg|_{0}^{1} = \frac{1}{2} - \frac{1}{3} = \frac{1}{6}$$

21.

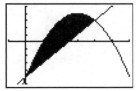

Solve:

$2x - 5 = -x^2 + 6x - 5 \Rightarrow$

$2x - 5 - (-x^2 + 6x - 5) = 0 \Rightarrow x^2 - 4x = 0 \Rightarrow$ Th

$x(x - 4) = 0 \Rightarrow x = 0$ or $x = 4$

us, we should integrate over $[0, 4]$ and since

$y = -x^2 + 6x - 5$ lies above $y = 2x - 5$ on

$[0, 4]$, we have $\int_{0}^{4}\left[-x^2 + 6x - 5 - (2x - 5)\right] dx$.

$$\int_{0}^{4}\left[-x^2 + 6x - 5 - (2x - 5)\right] dx$$

$$= \int_{0}^{4}(-x^2 + 4x)\, dx = \left(-\frac{1}{3}x^3 + 2x^2\right)\Bigg|_{0}^{4}$$

$$= -\frac{64}{3} + 32 = \frac{32}{3}$$

23.

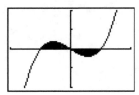

Solve $x(x^2 - 1) = 0 \Rightarrow x(x-1)(x+1) = 0 \Rightarrow$
$x = 0$ or $x = 1$ or $x = -1$
We should integrate over $[-1, 0]$ and $[0, 1]$.
Since $y = x(x^2 - 1)$ lies above the x-axis on
$[-1, 0]$, we have

$$\int_{-1}^{0} \left[x(x^2 - 1) - 0 \right] dx = \int_{-1}^{0} \left(x^3 - x \right) dx$$

$$= \left(\frac{1}{4}x^4 - \frac{1}{2}x^2 \right) \Big|_{-1}^{0}$$

$$= 0 - \left(\frac{1}{4} - \frac{1}{2} \right) = \frac{1}{4}$$

The x-axis lies above $y = x(x^2 - 1)$ on $[0, 1]$, so
we have

$$\int_{0}^{1} \left[0 - x(x^2 - 1) \right] dx = \int_{0}^{1} \left(x - x^3 \right) dx$$

$$= \left(\frac{1}{2}x^2 - \frac{1}{4}x^4 \right) \Big|_{0}^{1}$$

$$= \left(\frac{1}{2} - \frac{1}{4} \right) = \frac{1}{4}$$

Thus, the total area bounded by these curves is
$\frac{1}{4} + \frac{1}{4} = \frac{1}{2}$.

25.

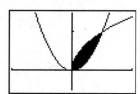

$8x^2 = \sqrt{x} \Rightarrow 64x^4 - x = 0 \Rightarrow$
$x(4x - 1)(16x^2 + 4x + 1) = 0 \Rightarrow x = 0$ or $x = \frac{1}{4}$
(Note that there is no real solution for
$16x^2 + 4x + 1 = 0$.) Thus, we should integrate
over $\left[0, \frac{1}{4} \right]$ and since $y = \sqrt{x}$ lies above

$y = 8x^2$ on $\left[0, \frac{1}{4} \right]$, we have

$$\int_{0}^{1/4} \left(\sqrt{x} - 8x^2 \right) dx = \int_{0}^{1/4} \left(x^{1/2} - 8x^2 \right) dx$$

$$= \left(\frac{2}{3}x^{3/2} - \frac{8}{3}x^3 \right) \Big|_{0}^{1/4}$$

$$= \frac{2}{3}\left(\frac{1}{4} \right)^{3/2} - \frac{8}{3}\left(\frac{1}{4} \right)^3$$

$$= \frac{1}{12} - \frac{1}{24} = \frac{1}{24}$$

27.

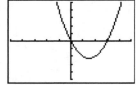

First solve $x^2 - 3x = 0$ to find any x-intercepts.
$x(x - 3) = 0$ so the x-intercepts are at $x = 0, 3$.

a. $y = x^2 - 3x$ lies under the x-axis on the

interval, so calculate $\int_{0}^{3} [0 - (x^2 - 3x)] \, dx$

$$= \int_{0}^{3} (-x^2 + 3x) \, dx = \left(-\frac{1}{3}x^3 + \frac{3}{2}x^2 \right) \Big|_{0}^{3}$$

$$= -\frac{27}{3} + \frac{27}{2} = \frac{27}{6} = \frac{9}{2}$$

b. On $[3, 4]$ $y = x^2 - 3x$ lies above the
x-axis so, using results from part (a), we
wish to calculate

$$\int_{0}^{3} \left[-(x^2 - 3x) \right] dx + \int_{3}^{4} (x^2 - 3x) \, dx$$

$$= \frac{9}{2} + \left(\frac{1}{3}x^3 - \frac{3}{2}x^2 \right) \Big|_{3}^{4}$$

$$= \frac{9}{2} + \left(\frac{64}{3} - 24 - \left(\frac{27}{3} - \frac{27}{2} \right) \right)$$

$$= \frac{9}{2} + \left(\frac{64}{3} - 24 + \frac{9}{2} \right) = \frac{19}{3}$$

c. On $[-2, 0]$, $y = x^2 - 3x$ lies above the
x-axis so, using results from (a) and (b)

$$\int_{-2}^{0} (x^2 - 3x) \, dx + \int_{0}^{3} -(x^2 - 3x) \, dx$$

$$= \left(\frac{1}{3}x^3 - \frac{3}{2}x^2 \right) \Big|_{-2}^{0} + \frac{9}{2}$$

$$= -\left(-\frac{8}{3} - \frac{12}{2} \right) + \frac{9}{2}$$

$$= \frac{26}{3} + \frac{9}{2} = \frac{79}{6}$$

29.

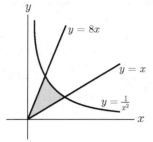

First solve: $\dfrac{1}{x^2} = 8x$. This has a solution at

$x = \dfrac{1}{2}$. Next solve: $\dfrac{1}{x^2} = x$. This has a solution

at $x = 1$. Thus, the area should be calculated by the following sum of integrals:

$$\int_0^{1/2} (8x - x)\, dx + \int_{1/2}^1 \left(\frac{1}{x^2} - x\right) dx$$

$$= \left(4x^2 - \frac{1}{2}x^2\right)\Bigg|_0^{1/2} + \left(-\frac{1}{x} - \frac{1}{2}x^2\right)\Bigg|_{1/2}^1$$

$$= 1 - \frac{1}{8} + \left(-1 - \frac{1}{2} - \left(-2 - \frac{1}{8}\right)\right) = \frac{3}{2}$$

31. a. $\displaystyle\int_0^5 (2t+1)\, dt = \left(t^2 + t\right)\Big|_0^5 = 30 \text{ ft}$

b.

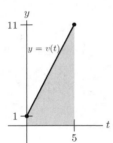

33. a. $\displaystyle\int_2^8 \left(\frac{3}{32}x^2 - x + 200\right) dx$

$$= \left(\frac{1}{32}x^3 - \frac{1}{2}x^2 + 200x\right)\Bigg|_2^8$$

$$= 16 - 32 + 1600 - \left(\frac{1}{4} - 2 + 400\right)$$

$$= \$1185.75$$

b. It is the area under the marginal cost curve from $x = 2$ to $x = 8$.

35. $\displaystyle\int_{44}^{48} M(x)\,dx$ represents the increase in profits resulting from increasing the production level from 44 to 48 units.

37. a. $\displaystyle\int_0^2 \left(12 + \frac{4}{(t+3)^2}\right) dt = \left(12t - \frac{4}{t+3}\right)\Bigg|_0^2$

$$= 24 - \frac{4}{5} - \left(-\frac{4}{3}\right) = \frac{368}{15}$$

$$\approx 24.5$$

b. The area represents the amount the temperature falls during the first 2 hours.

39. $\displaystyle\int_0^{20} 76.2e^{0.03t}\, dt = 2540e^{0.03t}\Big|_0^{20}$

$$\approx 2088 \text{ million cubic meters}$$

41. $\displaystyle\int_5^{10} M_2(x)\, dx - \int_5^{10} M_1(x)\, dx$

$$= \int_5^{10} (M_2(x) - M_1(x))\, dx$$

$$= \int_5^{10} \left((2x^2 - 2.4x + 8) - (2x^2 - 3x + 11)\right) dx$$

$$= \int_5^{10} (.6x - 3)\, dx = \left(\frac{.6x^2}{2} - 3x\right)\Bigg|_5^{10}$$

$$= \left(.3x^2 - 3x\right)\Big|_5^{10}$$

$$= \left(.3(10)^2 - 3(10)\right) - \left(.3(5)^2 - 3(5)\right) = 7.5$$

The net change in profit is \$7.5 thousand or \$7500.

43. A is the difference between the two heights after 10 seconds.

45. a.

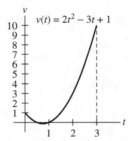

Let $s(t)$ represent the position of the object at time t, measured from its initial position. The required displacement is $s(3) - s(0)$, the net change in position over the interval $0 \le t \le 3$.

$$s(3) - s(0) = \int_0^3 \left(2t^2 - 3t + 1\right) dt$$

$$= \left(\frac{2t^3}{3} - \frac{3t^2}{2} + t\right)\Bigg|_0^3$$

$$= \frac{2(3)^3}{3} - \frac{3(3)^2}{2} + 3 = \frac{15}{2}$$

The object is displaced 7.5 feet higher after three seconds than it was at the start.

b. Note that $v(t) \le 0$ for $\frac{1}{2} \le t \le 1$. So, the object changes direction at time $t = \frac{1}{2}$, so we must compute the net displacements on the three intervals $\left[0, \frac{1}{2}\right], \left[\frac{1}{2}, 1\right]$, and $[1, 3]$. On the interval $\left[0, \frac{1}{2}\right]$ the net displacement is

$$\int_0^{1/2} \left(2t^2 - 3t + 1\right) dt = \left(\frac{2t^3}{3} - \frac{3t^2}{2} + t\right)\Bigg|_0^{1/2}$$

$$= \frac{2\left(\frac{1}{2}\right)^3}{3} - \frac{3\left(\frac{1}{2}\right)^2}{2} + \frac{1}{2}$$

$$= \frac{5}{24}$$

The object moved $\frac{5}{24}$ ft upward during this time interval.

On the interval $\left[\frac{1}{2}, 1\right]$, we have

$$\int_{1/2}^1 \left(2t^2 - 3t + 1\right) dt$$

$$= \left(\frac{2t^3}{3} - \frac{3t^2}{2} + t\right)\Bigg|_{1/2}^1$$

$$= \left(\frac{2(1)^3}{3} - \frac{3(1)^2}{2} + 1\right) - \frac{5}{24} = -\frac{1}{24}$$

The object moved $\frac{1}{24}$ ft downward during this time interval.

On the interval $[1, 3]$, we have

$$\int_1^3 \left(2t^2 - 3t + 1\right) dt$$

$$= \left(\frac{2t^3}{3} - \frac{3t^2}{3} + t\right)\Bigg|_1^3$$

$$= \left(\frac{2(3)^3}{3} - \frac{3(3)^2}{2} + 3\right) - \frac{1}{6} = \frac{22}{3}$$

The object moved $\frac{22}{3}$ ft upward during this time interval. Thus, the total distance traveled was

$$\frac{5}{24} + \frac{1}{24} + \frac{22}{3} = \frac{91}{12} \approx 7.583 \text{ ft.}$$

47.

 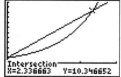

49.

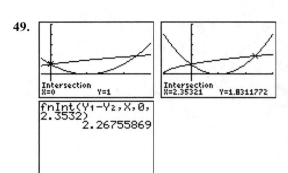

Intersect at $x = 0$ and $x \approx 2.3532$
area ≈ 2.2676

6.5 Applications of the Definite Integral

1. Average $= \dfrac{1}{b-a}\displaystyle\int_a^b f(x)\, dx = \dfrac{1}{3}\int_0^3 x^2 dx$

$$= \frac{x^3}{9}\Bigg|_0^3 = 3$$

3. $\dfrac{1}{4-0}\displaystyle\int_0^4 100e^{-.5x}\, dx = \dfrac{1}{4}\left(-200e^{-0.5x}\right)\Bigg|_0^4$

$$= \frac{1}{4}\left(-200e^{-2} + 200\right)$$

$$= 50 - 50e^{-2} = 50\left(1 - e^{-2}\right)$$

5. $\dfrac{1}{3-\frac{1}{3}}\displaystyle\int_{1/3}^3 \dfrac{1}{x}\, dx = \dfrac{3}{8}\left(\ln x\right)\Bigg|_{1/3}^3 = \dfrac{3}{8}\left(\ln 3 - \ln\dfrac{1}{3}\right)$

$$= \frac{3}{8}\left(\ln 3 + \ln 3\right) = \frac{3}{4}\ln 3$$

7. $\dfrac{1}{12}\displaystyle\int_0^{12}\left(47 + 4t - \dfrac{1}{3}t^2\right) dt$

$$= \frac{1}{12}\left(47t + 2t^2 - \frac{1}{9}t^3\right)\Bigg|_0^{12} = 55°$$

9. $P(t) = P(0)e^{kt}$

Find k: $1 = 2e^{k(1690)} \Rightarrow \dfrac{1}{2} = e^{1690k} \Rightarrow$

$1690k = \ln\dfrac{1}{2} \Rightarrow k = -.00041.$

The average value

$$= \frac{1}{1000}\int_0^{1000} 100e^{-0.00041t}\, dt$$

$$= \frac{1}{10}\cdot\frac{e^{-0.00041t}}{-.00041}\Bigg|_0^{1000} \approx 82 \text{ grams}$$

11. $p(20) = 3 - \dfrac{20}{10} = 1$

$\displaystyle\int_0^{20}\left(3 - \dfrac{x}{10} - 1\right)dx = \int_0^{20}\left(2 - \dfrac{x}{10}\right)dx$

$\qquad = \left(2x - \dfrac{x^2}{20}\right)\Big|_0^{20}$

$\qquad = 40 - 20 = \$20$

13. $p(40) = \dfrac{500}{40+10} - 3 = 7$

$\displaystyle\int_0^{40}\left(\dfrac{500}{x+10} - 3 - 7\right)dx$

$\qquad = \int_0^{40}\left(\dfrac{500}{x+10} - 10\right)dx$

$\qquad = \left[500\ln(x+10) - 10x\right]_0^{40}$

$\qquad = 500\ln 50 - 400 - 500\ln 10 \approx \$404.72$

15. $p(200) = 0.01(200) + 3 = 5$

$\displaystyle\int_0^{200}\left[5 - (.01x+3)\right]dx$

$\qquad = \int_0^{200}(2 - .01x)\,dx$

$\qquad = \left(2x - \dfrac{.01}{2}x^2\right)\Big|_0^{200}$

$\qquad = 400 - .005(40,000) = \$200$

17. $p(10) = \dfrac{10}{2} + 7 = 12$

$\displaystyle\int_0^{10}\left(12 - \left(\dfrac{x}{2}+7\right)\right)dx = \int_0^{10}\left(5 - \dfrac{x}{2}\right)dx$

$\qquad = \left(5x - \dfrac{x^2}{4}\right)\Big|_0^{10}$

$\qquad = 50 - 25 = \$25$

19. First find the point of intersection of the functions:

$12 - \dfrac{x}{50} = \dfrac{x}{20} + 5 \Rightarrow 7 = \dfrac{x}{50} + \dfrac{x}{20} \Rightarrow$

$7 = \dfrac{2x+5x}{100} \Rightarrow 700 = 7x \Rightarrow x = 100$

$p(100) = 12 - \dfrac{100}{50} = 10$

Thus, the functions intersect at $(100, 10)$.

$\text{C.S.} = \displaystyle\int_0^{100}\left(12 - \dfrac{x}{50} - 10\right)dx$

$\qquad = \int_0^{100}\left(2 - \dfrac{x}{50}\right)dx = \left(2x - \dfrac{x^2}{100}\right)\Big|_0^{100}$

$\qquad = 200 - 100 = \$100$

$\text{P.S.} = \displaystyle\int_0^{100}\left(10 - \left(\dfrac{x}{20}+5\right)\right)dx$

$\qquad = \int_0^{100}\left(5 - \dfrac{x}{20}\right)dx = \left(5x - \dfrac{x^2}{40}\right)\Big|_0^{100}$

$\qquad = 500 - 250 = \$250$

For exercises 21–25, the future value of a continuous income stream of k dollars per year for N years at interest rate r compounded continuously is

$\displaystyle\int_0^N Ke^{r(N-t)}\,dt.$

21. $\displaystyle\int_0^3 1000e^{0.05(3-t)}\,dt = -20,000e^{0.05(3-t)}\Big|_0^3$

$\qquad = \$3236.68$

23. $\displaystyle\int_0^4 16,000e^{0.08(4-t)}\,dt = -200,000e^{0.08(4-t)}\Big|_0^4$

$\qquad = \$75,426$

25. Solve $140,000 = \displaystyle\int_0^x 5000e^{.1(x-t)}\,dt$ for x.

$140,000 = -50,000e^{0.1(x-t)}\Big|_0^x \Rightarrow$

$140,000 = -50,000\left(1 - e^{0.1x}\right)$

$2.8 = e^{0.1x} - 1 \Rightarrow e^{0.1x} = 3.8 \Rightarrow$

$0.1x = \ln 3.8 \Rightarrow x = 10\ln 3.8 \approx 13.35$

It will take about 13.35 years until the value of the investment reaches \$140,000.

For exercises 27–35, recall that the volume of the solid of revolution obtained form revolving the region below the graph of $y = g(x)$ from $x = a$ to $x = b$ about the x-axis is $\displaystyle\int_a^b \pi\left[g(x)\right]^2 dx.$

27. $\displaystyle\int_0^2 \pi(x+1)^2\,dx = \pi\int_0^2\left(x^2 + 2x + 1\right)dx$

$\qquad = \pi\left(\dfrac{x^3}{3} + x^2 + x\right)\Big|_0^2$

$\qquad = \pi\left(\dfrac{8}{3} + 4 + 2\right) - 0 = \dfrac{26\pi}{3}$

29. $\displaystyle\int_{-2}^2 \pi\left[\sqrt{4 - x^2}\right]^2 dx$

$\qquad = \int_{-2}^2 \pi(4 - x^2)\,dx = \left(4\pi x - \dfrac{\pi x^3}{3}\right)\Big|_{-2}^2$

$\qquad = \left(8\pi - \dfrac{8\pi}{3}\right) - \left(-8\pi + \dfrac{8\pi}{3}\right) = \dfrac{32\pi}{3}$

31. $\int_1^2 \pi\left(x^2\right)^2 dx = \int_1^2 \pi x^4 dx = \frac{\pi}{5} x^5 \Big|_1^2$

$$= \frac{32\pi}{5} - \frac{\pi}{5} = \frac{31\pi}{5}$$

33. $\int_0^4 \pi\left(\sqrt{x}\right)^2 dx = \int_0^4 \pi x \, dx = \frac{\pi}{2} x^2 \Big|_0^4 = 8\pi$

35. $\int_0^1 \pi(2x+1)^2 dx = \int_0^1 \pi(4x^2 + 4x + 1) \, dx$

$$= \pi\left(\frac{4}{3}x^3 + 2x^2 + x\right)\Big|_0^1$$

$$= \pi\left(\frac{4}{3} + 2 + 1\right) = \frac{13\pi}{3}$$

37. $\left[8.25^3 + 8.75^3 + 9.25^3 + 9.75^3\right](.5)$

$n = 4; \ a = 8 \Rightarrow \dfrac{b-8}{4} = .5 \Rightarrow b = 10; \ f(x) = x^3$

39. $\left[\left(5 + e^5\right) + \left(6 + e^6\right) + \left(7 + e^7\right)\right](1)$

$n = 3; \ a = 4 \Rightarrow \dfrac{b-4}{3} = 1 \Rightarrow b = 7$

$f(x) = x + e^x$

41. The sum is approximately

$\int_0^3 (3-x)^2 dx = \int_0^3 (x^2 - 6x + 9) \, dx$

$$= \left(\frac{1}{3}x^3 - 3x^2 + 9x\right)\Big|_0^3$$

$$= \frac{27}{3} - 27 + 27 = 9$$

43. a. $\dfrac{1}{3}\int_0^3 1000 e^{rt} dt = \dfrac{1000}{3r} e^{rt} \Big|_0^3 = \dfrac{1000}{3r}\left(e^{3r} - 1\right)$

     b. Solve $\dfrac{1000}{3r}\left(e^{3r} - 1\right) = 1070.60$ using a graphing utility.

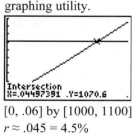

[0, .06] by [1000, 1100]

$r \approx .045 = 4.5\%$

45. a. $\int_0^6 1000 e^{r(6-t)} dt = \dfrac{1000}{-r} e^{r(6-t)} \Big|_0^6$

$$= \frac{1000}{-r}\left(1 - e^{6r}\right)$$

$$= \frac{1000}{r}\left(e^{6r} - 1\right)$$

     b. Solve $\dfrac{1000}{r}\left(e^{6r} - 1\right) = 6997.18$ using a graphing utility.

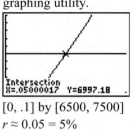

[0, .1] by [6500, 7500]

$r \approx 0.05 = 5\%$

Chapter 6 Review Exercises

1. $\int 3^2 dx = 9x + C$

2. $\int (x^2 - 3x + 2) dx = \dfrac{1}{3}x^3 - \dfrac{3}{2}x^2 + 2x + C$

3. $\int \sqrt{x+1}\, dx = \dfrac{2}{3}(x+1)^{3/2} + C$

4. $\int \dfrac{2}{x+4} dx = 2\ln|x+4| + C$

5. $2\int (x^3 + 3x^2 - 1) dx = \dfrac{1}{2}x^4 + 2x^3 - 2x + C$

6. $\int \sqrt[5]{x+3}\, dx = \dfrac{5}{6}(x+3)^{6/5} + C$

7. $\int e^{-x/2} dx = -2e^{-x/2} + C$

8. $\int \dfrac{5}{\sqrt{x-7}} dx = 10\sqrt{x-7} + C$

9. $\int (3x^4 - 4x^3) \, dx = \dfrac{3}{5}x^5 - x^4 + C$

10. $\int (2x+3)^7 dx = \dfrac{1}{16}(2x+3)^8 + C$

11. $\int \sqrt{4-x} \, dx = -\dfrac{2}{3}(4-x)^{3/2} + C$

12. $\int \left(\dfrac{5}{x} - \dfrac{x}{5}\right) dx = 5\ln|x| - \dfrac{x^2}{10} + C$

13. $\int_{-1}^{1}(x+1)^2\,dx = \frac{1}{3}(x+1)^3\Big|_{-1}^{1} = \frac{8}{3}-0 = \frac{8}{3}$

14. $\int_{0}^{1/8}\sqrt[3]{x}\,dx = \frac{3}{4}x^{4/3}\Big|_{0}^{1/8} = \frac{3}{64}-0 = \frac{3}{64}$

15. $\int_{-1}^{2}\sqrt{2x+4}\,dx = \sqrt{2}\int_{-1}^{2}\sqrt{x+2}\,dx = \frac{2}{3}\sqrt{2}(x+2)^{3/2}\Big|_{-1}^{2} = \frac{2}{3}\sqrt{2}\left(4^{3/2}-1\right) = \frac{14}{3}\sqrt{2}$

16. $2\int_{0}^{1}\left(\frac{2}{x+1}-\frac{1}{x+4}\right)dx = \left[4\ln(x+1)-2\ln(x+4)\right]\Big|_{0}^{1} = 4\ln 2 - 2\ln 5 - (4\ln 1 - 2\ln 4) = 2\ln\frac{16}{5}$

17. $\int_{1}^{2}\frac{4}{x^5}\,dx = -\frac{1}{x^4}\Big|_{1}^{2} = -\frac{1}{16}-(-1) = \frac{15}{16}$

18. $\frac{2}{3}\int_{0}^{8}\sqrt{x+1}\,dx = \frac{4}{9}(x+1)^{3/2}\Big|_{0}^{8} = 12 - \frac{4}{9} = \frac{104}{9}$

19. $\int_{1}^{4}\frac{1}{x^2}\,dx = -\frac{1}{x}\Big|_{1}^{4} = -\frac{1}{4}+1 = \frac{3}{4}$

20. $\int_{3}^{6}e^{2-(x/3)}\,dx = -3e^{2-(x/3)}\Big|_{3}^{6} = -3+3e = 3(e-1)$

21. $\int_{0}^{5}(5+3x)^{-1}\,dx = \frac{1}{3}\ln(5+3x)\Big|_{0}^{5} = \frac{1}{3}\ln 20 - \frac{1}{3}\ln 5 = \frac{1}{3}\ln 4$

22. $\int_{-2}^{2}\frac{3}{2e^{3x}}\,dx = -\frac{1}{2e^{3x}}\Big|_{-2}^{2} = -\frac{1}{2e^6}-\left(-\frac{1}{2e^{-6}}\right) = \frac{1}{2}\left(e^6 - e^{-6}\right)$

23. $\int_{0}^{\ln 2}(e^x - e^{-x})\,dx = (e^x + e^{-x})\Big|_{0}^{\ln 2} = 2+\frac{1}{2}-(1+1) = \frac{1}{2}$

24. $\int_{\ln 2}^{\ln 3}\left(e^x + e^{-x}\right)dx = \left(e^x - e^{-x}\right)\Big|_{\ln 2}^{\ln 3} = 3-\frac{1}{3}-\left(2-\frac{1}{2}\right) = \frac{7}{6}$

25. $\int_{0}^{\ln 3}\frac{e^x + e^{-x}}{e^{2x}}\,dx = \left(-e^{-x}-\frac{1}{3}e^{-3x}\right)\Big|_{0}^{\ln 3} = -\frac{1}{3}-\frac{1}{81}-\left(-1-\frac{1}{3}\right) = \frac{80}{81}$

26. $\int_{0}^{1}\frac{3+e^{2x}}{e^x}\,dx = \left(-3e^{-x}+e^x\right)\Big|_{0}^{1} = -3e^{-1}+e-(-3+1) = 2+e-\frac{3}{e}$

27. $\int_{1}^{2}(3x-2)^{-3}\,dx = \left(-\frac{1}{6}(3x-2)^{-2}\right)\Big|_{1}^{2} = -\frac{1}{6}\cdot\frac{1}{16}+\frac{1}{6} = \frac{15}{96} = \frac{5}{32}$

28. $\int_{1}^{9}\left(1+\sqrt{x}\right)dx = \left(x+\frac{2}{3}x^{3/2}\right)\Big|_{1}^{9} = 9+18-\left(1+\frac{2}{3}\right) = 26-\frac{2}{3} = \frac{76}{3}$

29. $\int_{0}^{1}\left(\sqrt{x}-x^2\right)dx = \left(\frac{2}{3}x^{3/2}-\frac{1}{3}x^3\right)\Big|_{0}^{1} = \frac{2}{3}-\frac{1}{3}-(0-0) = \frac{1}{3}$

30. $y = x^3$ lies above $y = \frac{1}{2}x^3 + 2x$ on $[-2, 0]$ and below on $[0, 2]$. Thus, we calculate

$$\int_{-2}^{0}\left[x^3 - \left(\frac{1}{2}x^3 + 2x\right)\right]dx + \int_{0}^{2}\left[\frac{1}{2}x^3 + 2x - x^3\right]dx = \int_{-2}^{0}\left(\frac{1}{2}x^3 - 2x\right)dx + \int_{0}^{2}\left(-\frac{1}{2}x^3 + 2x\right)dx$$

$$= \left(\frac{1}{8}x^4 - x^2\right)\Big|_{-2}^{0} + \left(-\frac{1}{8}x^4 + x^2\right)\Big|_{0}^{2}$$

$$= 0 - 0 - (2 - 4) + (-2) + 4 - (0 + 0) = 4$$

31. $\int_{0}^{\ln 2}(e^x - e^{-x})dx = (e^x + e^{-x})\Big|_{0}^{\ln 2} = 2 + \frac{1}{2} - (1 + 1) = \frac{1}{2}$

32. Set $\sqrt{x} = x^2 \Rightarrow x = x^4 \Rightarrow x(x^3 - 1) = 0 \Rightarrow x = 0$ or $x = 1$. Thus, the graphs intersect at $x = 0, 1$. On $(0, 1)$, $y = \sqrt{x}$ lies above $y = x^2$, and below on $(1, 1.21)$. Thus, we calculate

$$\int_{0}^{1}\left[\sqrt{x} - x^2\right]dx + \int_{1}^{1.21}\left(x^2 - \sqrt{x}\right)dx = \left(\frac{2}{3}x^{3/2} - \frac{1}{3}x^3\right)\Big|_{0}^{1} + \left(\frac{1}{3}x^3 - \frac{2}{3}x^{3/2}\right)\Big|_{1}^{1.21}$$

$$\approx \frac{2}{3} - \frac{1}{3} - (0 - 0) + .5905 - .8873 - \left(\frac{1}{3} - \frac{2}{3}\right) = \frac{1.109561}{3} \approx 0.370$$

33. $4 - x^2$ and $1 - x^2$ are even, so the area is given by

$$2\int_{0}^{2}\left(4 - x^2\right)dx - 2\int_{0}^{1}\left(1 - x^2\right)dx = 2\left[\left(4x - \frac{1}{3}x^3\right)\Big|_{0}^{2} - \left(x - \frac{1}{3}x^3\right)\Big|_{0}^{1}\right] = 2\left[8 - \frac{8}{3} - \left(1 - \frac{1}{3}\right)\right] = \frac{28}{3}$$

34. $\int_{1/2}^{2}\left[\left(1 - \frac{1}{x}\right) - \left(x^2 - \frac{3}{2}x - \frac{1}{2}\right)\right]dx = \int_{1/2}^{2}\left(-x^2 + \frac{3}{2}x + \frac{3}{2} - \frac{1}{x}\right)dx = \left(-\frac{1}{3}x^3 + \frac{3}{4}x^2 + \frac{3}{2}x - \ln x\right)\Big|_{1/2}^{2}$

$$= -\frac{8}{3} + 3 + 3 - \ln 2 - \left(-\frac{1}{24} + \frac{3}{16} + \frac{3}{4} - \ln\frac{1}{2}\right) = \frac{39}{16} - \ln 4$$

35. $\int_{0}^{1}(e^x - ex)dx = \left(e^x - \frac{e}{2}x^2\right)\Big|_{0}^{1} = e - \frac{e}{2} - (1 - 0) = \frac{e}{2} - 1$

36. $y = 2x^3 - x^2 - 6x$ lies above $y = x^3$ on $[-2, 0]$ and below on $[0, 3]$. Thus, we calculate

$$\int_{-2}^{0}\left[2x^3 - x^2 - 6x - (x^3)\right]dx + \int_{0}^{3}\left[x^3 - (2x^3 - x^2 - 6x)\right]dx$$

$$= \int_{-2}^{0}\left[x^3 - x^2 - 6x\right]dx + \int_{0}^{3}\left[-x^3 + x^2 + 6x)\right]dx = \left(\frac{1}{4}x^4 - \frac{1}{3}x^3 - 3x^2\right)\Big|_{-2}^{0} + \left(-\frac{1}{4}x^4 + \frac{1}{3}x^3 + 3x^2\right)\Big|_{0}^{3}$$

$$= 0 - \left(4 + \frac{8}{3} - 12\right) + \left(-\frac{81}{4}\right) + 9 + 27 - 0 = \frac{253}{12}$$

37.

$x^3 - 3x + 1 = x + 1 \Rightarrow x^3 - 4x = 0 \Rightarrow x(x^2 - 4) = 0 \Rightarrow x = -2$ or $x = 0$ or $x = 2$

Thus, the graphs intersect at $x = 0, \pm 2$. On $[-2, 0]$, $y = x^3 - 3x + 1$ lies above $y = x + 1$, and below on $[0, 2]$.

$$\int_{-2}^{0}\left[(x^3 - 3x + 1) - (x + 1)\right]dx + \int_{0}^{2}\left[(x + 1) - (x^3 - 3x + 1)\right]dx$$

$$= \int_{-2}^{0}(x^3 - 4x)\,dx + \int_{0}^{2}(-x^3 + 4x)\,dx = \left(\frac{1}{4}x^4 - 2x^2\right)\Big|_{-2}^{0} + \left(-\frac{1}{4}x^4 + 2x^2\right)\Big|_{0}^{2} = 0 - (4 - 8) + (-4 + 8) = 8$$

38.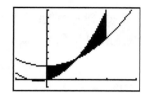

$2x^2 + x = x^2 + 2 \Rightarrow x^2 + x - 2 = 0 \Rightarrow (x+2)(x-1) = 0 \Rightarrow x = -2$ or $x = 1$

Thus, on the interval $[0, 2]$, the graphs intersect at $x = 1$. On $(0, 1)$, $y = x^2 + 2$ lies above $y = 2x^2 + x$ and below on $(1, 2)$.

$\int_0^1 \left[x^2 + 2 - (2x^2 + x) \right] dx + \int_1^2 \left[2x^2 + x - (x^2 + 2) \right] dx$

$= \int_0^1 \left(-x^2 - x + 2 \right) dx + \int_1^2 \left(x^2 + x - 2 \right) dx = \left(-\frac{1}{3}x^3 - \frac{1}{2}x^2 + 2x \right) \Big|_0^1 + \left(\frac{1}{3}x^3 + \frac{1}{2}x^2 - 2x \right) \Big|_1^2$

$= -\frac{1}{3} - \frac{1}{2} + 2 + \frac{8}{3} + 2 - 4 - \left(\frac{1}{3} + \frac{1}{2} - 2 \right) = \frac{6}{3} - \frac{2}{2} + 2 = 3$

39. $\int (x-5)^2 \, dx = \frac{1}{3}(x-5)^3 + C$

$f(8) = 2 = \frac{1}{3}(3)^3 + C = 9 + C \Rightarrow C = -7$

$f(x) = \frac{1}{3}(x-5)^3 - 7$

40. $\int e^{-5x} dx = -\frac{1}{5}e^{-5x} + C$

$f(0) = 1 = -\frac{1}{5} + C \Rightarrow C = \frac{6}{5}$

$f(x) = \frac{6}{5} - \frac{1}{5}e^{-5x}$

41. a. $y' = 4t \Rightarrow y = 2t^2 + C$

b. $y' = 4y \Rightarrow y = Ce^{4t}$

c. $y' = e^{4t} \Rightarrow y = \frac{1}{4}e^{4t} + C$

42. Theorem II of section 6.1 states if $F'(x) = 0$ for all x in an interval I, then there is a constant C such that $F(x) = C$ for all x in I.

$y' = f'(t) = kt(f(t))$.

Using the hint, we have

$\frac{d}{dt} \left[f(t)e^{-kt^2/2} \right]$

$= f(t)(-kt)e^{-kt^2/2} + f'(x)e^{-kt^2/2}$

$= e^{-kt^2/2} \left[-f(t)kt + f'(t) \right]$

$= e^{-kt^2/2} \left[-f(t)kt + kt(f(t)) \right]$

$= e^{-kt^2/2}(0) = 0$

Since only constant functions have a zero derivative, $f(t)e^{-kt^2/2} = C$ for some C. Thus,

$f(t) = Ce^{kt^2/2}$.

43. $C(x) = \int (.04x + 150) \, dx = .02x^2 + 150x + C$

$C(0) = 500 = 0 + 0 + C = C$

Thus, $C(x) = .02x^2 + 150x + 500$ dollars.

44. $\int_{10}^{20} (400 - 3x^2) \, dx = (400x - x^3) \Big|_{10}^{20} = -3000$

Thus, a loss of \$3000 would result.

45. It represents the total quantity of drug (in cubic centimeters) injected during the first 4 minutes.

46. $v(t) = -9.8t + 20$

a. $\int_0^2 (-9.8t + 20) \, dt = (-4.9t^2 + 20t) \Big|_0^2$

$= -19.6 + 40 = 20.4$ m

b.

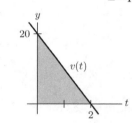

Use the figure below for exercises 47 and 48.

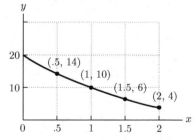

47. $\left[f(0) + f(.5) + f(1) + f(1.5) \right] \Delta x$

$= (20 + 14 + 10 + 6)(.5) = 25$

48. $\left[f(.5) + f(1) + f(1.5) + f(2) \right] \Delta x$

$= (14 + 10 + 6 + 4)(.5) = 17$

49. $\Delta x = 1$; the midpoints are .5, 1.5

$$\text{Area} = \left[\frac{1}{0.5+2} + \frac{1}{1.5+2}\right] \approx 0.68571$$

$$\int_0^2 \frac{1}{x+2}\, dx = \ln(x+2)\Big|_0^2 = \ln 4 - \ln 2 = \ln\frac{4}{2}$$
$$= \ln 2 \approx .69315$$

50. $\Delta x = 0.2$; the midpoints are 0.1, 0.3, 0.5, 0.7, 0.9

Area =
$$\left[e^{2(0.1)} + e^{2(0.3)} + e^{2(0.5)} + e^{2(0.7)} + e^{2(0.9)}\right]\cdot(.2)$$
$$\approx 3.17333$$

$$\int_0^1 e^{2x}\, dx = \frac{1}{2}e^{2x}\Big|_0^1 = \frac{1}{2}e^2 - \frac{1}{2} \approx 3.1945$$

51. $p(400) = \sqrt{25 - 0.04(400)} = 3$

$$\text{C.S.} = \int_0^{400} (\sqrt{25 - 0.04x} - 3)\, dx$$
$$= \left(\frac{2}{-0.12}(25 - 0.04x)^{3/2} - 3x\right)\Big|_0^{400}$$
$$= \frac{2}{-0.12}(27) - 1200 - \frac{2}{-0.12}(125)$$
$$\approx \$433.33$$

52. $\frac{1}{10}\int_0^{10} 3000e^{.04t}\, dt = \int_0^{10} 300e^{.04t}\, dt = 7500e^{.04t}\Big|_0^{10}$
$$= 7500(e^{.4} - e^0) \approx \$3688.69$$

53. $\frac{1}{\frac{1}{2}-\frac{1}{3}}\int_{1/3}^{1/2}\frac{1}{x^3}\, dx = 6\int_{1/3}^{1/2}\frac{1}{x^3}\, dx = 6\left[-\frac{1}{2x^2}\Big|_{1/3}^{1/2}\right]$
$$= 6\left(-2 + \frac{9}{2}\right) = 15$$

54. The sum is approximated by

$$\int_0^1 3e^{-x}\, dx = -3e^{-x}\Big|_0^1 = -3e^{-1} + 3 = 3\left(1 - e^{-1}\right).$$

55. $\int_a^c f(x)\, dx = .68 - .42 = .26$

$$\int_a^d f(x)\, dx = .68 - .42 + 1.7 = 1.96$$

56. $\int_0^1 \pi(1-x^2)^2\, dx = \int_0^1 \pi(x^4 - 2x^2 + 1)\, dx$
$$= \pi\left(\frac{x^5}{5} - \frac{2}{3}x^3 + x\right)\Big|_0^1$$
$$= \pi\left(\frac{1}{5} - \frac{2}{3} + 1\right) = \frac{8\pi}{15}$$

57. a. Inventory is decreasing, so the slope is $-\frac{Q}{A}$. From the graph we can see that
$$f(t) = Q - \frac{Q}{A}t\,.$$

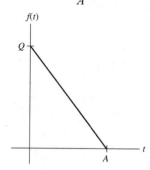

b. $\frac{1}{A}\int_0^A\left(Q - \frac{Q}{A}t\right)dt = \frac{1}{A}\left(Qt - \frac{Q}{2A}t^2\right)\Big|_0^A$
$$= \frac{Q}{A}A - \frac{QA^2}{2A^2} = \frac{Q}{2}$$

58. a. $f(t) = Q - \int_0^t rt\, dt = Q - \frac{rt^2}{2}$

b. $0 = Q - \frac{rA^2}{2} \Rightarrow r = \frac{2Q}{A^2}$

c. $f(t) = Q - \frac{\frac{2Q}{A^2}t^2}{2} = Q - \frac{Qt^2}{A^2}$

$$\frac{1}{A}\int_0^A\left(Q - \frac{Qt^2}{A^2}\right)dt = \frac{1}{A}\left(Qt - \frac{Qt^3}{3A^2}\right)\Big|_0^A$$
$$= Q - \frac{Q}{3} = \frac{2}{3}Q$$

59. a. $g(3)$ is the area under the curve $y = \frac{1}{1+t^2}$ from $t = 0$ to $t = 3$.

b. $g'(x) = \frac{1}{1+x^2}$

60. a. $h(0)$ is the area under one-quarter of the unit circle. $h(1)$ is the area under one-half of the unit circle.

b. $h'(x) = \sqrt{1-x^2}$

61. The sum is approximated by
$$\int_0^3 5000e^{-0.1t}\, dt = -50{,}000e^{-0.1t}\Big|_0^3 \approx 12{,}959$$
$$\approx 13{,}000$$

62. $\Delta x = \dfrac{1}{n}$, with left endpoints $t_i = i\Delta x$

$$\text{Sum} = \Delta x e^0 + \Delta x e^{\Delta x} + \Delta x e^{2\Delta x} + \cdots + \Delta x e^{(n-1)\Delta x}$$
$$= \Delta x \left[e^{t_0} + e^{t_1} + e^{t_2} + \cdots + e^{t_{n-1}} \right]$$
$$\approx \int_0^1 e^x dx = e - 1$$

63. $\Delta x = \dfrac{1}{n}$, with left endpoints $t_i = i\Delta x$

$$\text{Sum} = \Delta x \left[1 + (1+\Delta x)^3 + \cdots + (1+(n-1)\Delta x)^3 \right]$$
$$= \Delta x \left[1 + (1+t_1)^3 + \cdots + (1+t_{n-1})^3 \right]$$
$$\approx \int_0^1 (1+x)^3 dx = \left. \frac{(1+x)^4}{4} \right|_0^1 = \frac{15}{4}$$

64. Using the figure on the left, the average value of $f(x) = 4$ on $2 \le x \le 6$.

65. True; $3 \le f(x) \le 4$

$$\int_0^5 3 \, dx \le \int_0^5 f(x) \, dx \le \int_0^5 4 \, dx$$
$$15 \le \int_0^5 f(x) \, dx \le 20, \text{ so } 3 \le \frac{1}{5}\int_0^5 f(x)\, dx \le 4$$

66. a. $\displaystyle\int_{t_1}^{t_2} (20 - 4t_1) \, dt = (20 - 4t_1)t \Big|_{t_1}^{t_2}$
$$= (20 - 4t_1)t_2 - (20 - 4t_1)t_1$$
$$= (20 - 4t_1)\Delta t$$

b. Let $R(t) = $ the amount of water added up to time t. Then $R'(t) = r(t)$ and

$$\int_0^5 r(t) \, dt = R(5) - R(0) = \text{the total}$$
amount of water added to the tank from $t = 0$ to $t = 5$.

67. $\displaystyle\int_0^{35} 860 e^{0.04t} \, dt = 21{,}500 e^{0.04t} \Big|_0^{35}$
$$= 21{,}500(e^{1.4} - 1)$$
$$= 65{,}687 \text{ cubic kilometers}$$

68. $\displaystyle\int_0^1 4500 e^{0.09(1-t)} \, dt = -50{,}000 e^{0.09(1-t)} \Big|_0^1$
$$= \$4708.71$$

69. $\displaystyle\int (3x^2 - 2x + 1) dx = x^3 - x^2 + x + C$
Now $f(1) = 1$, so
$$(x^3 - x^2 + x + C)\Big|_{x=1} = 1 \Rightarrow$$
$$1 - 1 + 1 + C = 1 \Rightarrow C = 0$$
So, $f(x) = x^3 - x^2 + x$.

70. The slope of the line connecting $(0, 0)$ and (a, a^2) is $m = \dfrac{a^2 - 0}{a - 0} = a$. The equation of the line is $y - 0 = a(x - 0) \Rightarrow y = ax$.
The shaded area is equal to 1, so we have
$$\int_0^a (ax - x^2) dx = 1$$
$$\left(\frac{a}{2}x^2 - \frac{1}{3}x^3 \right)\Big|_0^a = 1$$
$$\frac{a^3}{2} - \frac{a^3}{3} - 0 = 1 \Rightarrow \frac{a^3}{6} = 1 \Rightarrow a = \sqrt[3]{6}$$

71. $\displaystyle\int_0^{b^2} \sqrt{x}\, dx + \int_0^b x^2 dx = \frac{2}{3}x^{3/2}\Big|_0^{b^2} + \frac{1}{3}x^3\Big|_0^b$
$$= \frac{2}{3}(b^2)^{3/2} - 0 + \frac{1}{3}b^3 - 0$$
$$= \frac{2}{3}|b|^3 + \frac{1}{3}b^3$$
$$\left(\text{since } (b^2)^{1/2} = |b| \right)$$
$$= \frac{2}{3}b^3 + \frac{1}{3}b^3$$
$$\left(\begin{array}{l} |b| = b \text{ since } b \\ \text{is positive.} \end{array} \right)$$
$$= b^3$$

72. $\displaystyle\int_0^{b^n} \sqrt[n]{x}\, dx + \int_0^b x^n dx$
$$= \left(\frac{n}{n+1}x^{(n+1)/n} \right)\Big|_0^{b^n} + \left(\frac{1}{n+1}x^{n+1} \right)\Big|_0^b$$
$$= \frac{n}{n+1}(b^n)^{(n+1)/n} - 0 + \frac{1}{n+1}b^{n+1} - 0$$
$$= \frac{n}{n+1}b^{n+1} + \frac{1}{n+1}b^{n+1} = b^{n+1}$$
$$\left[\text{Note: } (b^n)^{1/n} = \begin{cases} b, & n \text{ is odd} \\ |b|, & n \text{ is even} \end{cases}, \text{ but } |b| = b \right.$$
since b is positive. So, $(b^n)^{1/n} = b.$]

73. $\displaystyle\int_0^1 \left(\sqrt{x} - x^2 \right) dx = \int_0^1 \sqrt{x}\, dx - \int_0^1 x^2 dx$
$$= \left(\frac{2}{3}x^{3/2} - \frac{1}{3}x^3 \right)\Big|_0^1$$
$$= \left(\frac{2}{3} - \frac{1}{3} \right) - 0 = \frac{1}{3}$$

74. $\displaystyle\int_0^1 \left(\sqrt[n]{x} - x^n \right) dx = \left(\frac{n}{n+1}x^{\frac{n+1}{n}} - \frac{1}{n+1}x^{n+1} \right)\Big|_0^1$
$$= \frac{n}{n+1} - \frac{1}{n+1} - 0 = \frac{n-1}{n+1}$$

Chapter 7 Functions of Several Variables

7.1 Examples of Functions of Several Variables

1. $f(x, y) = x^2 - 3xy - y^2$

$f(5,0) = 5^2 - 3(5)(0) - 0^2 = 25$

$f(5,-2) = 5^2 - 3(5)(-2) - (-2)^2 = 51$

$f(a, b) = a^2 - 3ab - b^2$

3. $g(x, y, z) = \dfrac{x}{y - z}$

$g(2,3,4) = \dfrac{2}{3 - 4} = -2$

$g(7, 46, 44) = \dfrac{7}{46 - 44} = \dfrac{7}{2}$

5. $f(x,y) = xy \Rightarrow$

$f(2+h,3) = (2+h)3 = 6 + 3h$

$f(2,3) = (2)3 = 6$

$f(2+h,3) - f(2,3) = (6+3h) - 6 = 3h$

7. $C(x, y, z)$ is the cost of materials for the rectangular box with dimensions x, y, z in feet. The area of the top and the bottom together is $2xy$, so the cost is $3(2xy) = 6xy$. The area of the front and back together is $2xz$, so the cost is $5(2xz) = 10xz$. The area of the right and left side together is $2yz$, so the cost is $5(2yz) = 10yz$.
Thus, $C(x, y, z) = 6xy + 10xz + 10yz$.

9. $f(x, y) = 20x^{1/3}y^{2/3}$

$f(8,1) = 20\left(8^{1/3}\right)\left(1^{2/3}\right) = 40$

$f(1,27) = 20\left(1^{1/3}\right)\left(27^{2/3}\right) = 180$

$f(8,27) = 20\left(8^{1/3}\right)\left(27^{2/3}\right) = 360$

$f(8k, 27k)$

$= 20(8k)^{1/3}(27k)^{2/3}$

$= 20\left(8^{1/3}\right)\left(k^{1/3}\right)\left(27^{2/3}\right)\left(k^{2/3}\right)$

$= k(20)\left(8^{1/3}\right)\left(27^{2/3}\right) = kf(8, 27)$

11. $P(A, t) = Ae^{-0.05t}$

$P(100, 13.8) = 100e^{-0.05(13.8)} = 100e^{-0.69}$

≈ 50.16

$50 invested at 5% continuously compounded interest will yield $100 in 13.8 years.

13. $T = f(r,v,x) = \dfrac{r}{100}(0.40v - x)$

a. $v = 200,000, x = 5000, r = 2.5$

$T = \dfrac{r(0.4v - x)}{100}$

$= \dfrac{2.5(0.4(200,000) - 5000)}{100}$

$T = \$1875$

b. If $v = 200,000, x = 5000, r = 3$:

$T = \dfrac{r(0.4v - x)}{100} = \dfrac{3(0.4(200,000) - 5000)}{100}$

$= \$2250$

The tax due also increases by 20% since $1875 + (0.2)(1875) = \$2250$.

15. $C = 2x - y$, so $y = 2x - C$
The level curves are $y = 2x, \ y = 2x - 1,$ and $y = 2x - 2$.

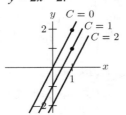

17. $C = x - y, \ y = x - C$
But $0 = 0 - C \Rightarrow C = 0$, so $y = x$.

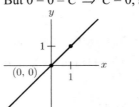

19. $y = 3x - 4 \Rightarrow y - 3x = -4$, so $y - 3x = C \Rightarrow f(x, y) = y - 3x$.

21. They correspond to the points having the same altitude above sea level.

23. (d)

25. (c)

7.2 Partial Derivatives

1. $f(x, y) = 5xy$

$\dfrac{\partial f}{\partial x} = 5(1)y = 5y; \quad \dfrac{\partial f}{\partial y} = 5(1)x = 5x$

3. $f(x, y) = 2x^2 e^y$

$$\frac{\partial f}{\partial x} = 2(2x)e^y = 4xe^y$$

$$\frac{\partial f}{\partial y} = 2x^2 e^y (1) = 2x^2 e^y$$

5. $f(x, y) = \dfrac{x}{y} + \dfrac{y}{x} = xy^{-1} + x^{-1}y$

$$\frac{\partial f}{\partial x} = \frac{1}{y} - \frac{y}{x^2}; \quad \frac{\partial f}{\partial y} = -\frac{x}{y^2} + \frac{1}{x}$$

7. $f(x, y) = (2x - y + 5)^2$

$$\frac{\partial f}{\partial x} = 2(2x - y + 5)(2) = 4(2x - y + 5)$$

$$\frac{\partial f}{\partial y} = 2(2x - y + 5)(-1) = -2(2x - y + 5)$$

9. $f(x, y) = xe^{x^2 y^2}$

$$\frac{\partial f}{\partial x} = x\left(\frac{\partial}{\partial x} e^{x^2 y^2}\right) + e^{x^2 y^2}\left(\frac{\partial}{\partial x} x\right)$$

$$= x\left(e^{x^2 y^2} \frac{\partial}{\partial x} x^2 y^2\right) + e^{x^2 y^2}$$

$$= 2x^2 y^2 e^{x^2 y^2} + e^{x^2 y^2}$$

$$= e^{x^2 y^2}\left(2x^2 y^2 + 1\right)$$

$$\frac{\partial f}{\partial y} = x\left(\frac{\partial}{\partial y} e^{x^2 y^2}\right) = xe^{x^2 y^2}\left(\frac{\partial}{\partial y} x^2 y^2\right)$$

$$= xe^{x^2 y^2}\left(2x^2 y\right) = 2x^3 ye^{x^2 y^2}$$

11. $f(x, y) = \dfrac{x - y}{x + y}$

$$\frac{\partial f}{\partial x} = \frac{1(x + y) - (x - y)(1)}{(x + y)^2} = \frac{2y}{(x + y)^2}$$

$$\frac{\partial f}{\partial y} = \frac{(-1)(x + y) - (x - y)(1)}{(x + y)^2} = -\frac{2x}{(x + y)^2}$$

13. $f(L, K) = 3\sqrt{LK}$

$$\frac{\partial f}{\partial L} = 3\left(\frac{1}{2}\right)(LK)^{-1/2}(K) = \frac{3}{2}\sqrt{\frac{K}{L}}$$

15. $f(x, y, z) = \dfrac{(1 + x^2 y)}{z} = z^{-1} + x^2 yz^{-1}$

$$\frac{\partial f}{\partial x} = 0 + 2xyz^{-1} = \frac{2xy}{z}$$

$$\frac{\partial f}{\partial y} = 0 + x^2 z^{-1} = \frac{x^2}{z}$$

$$\frac{\partial f}{\partial z} = -z^{-2} + (-z^{-2}x^2 y) = -\frac{1}{z^2} - \frac{x^2 y}{z^2}$$

$$= -\frac{1 + x^2 y}{z^2}$$

17. $f(x, y, z) = xze^{yz}$

$$\frac{\partial f}{\partial x} = (1)ze^{yz} = ze^{yz}$$

$$\frac{\partial f}{\partial y} = e^{yz}(z)xz = xz^2 e^{yz}$$

$$\frac{\partial f}{\partial z} = (1)xe^{yz} + e^{yz}(y)xz = xe^{yz}(1 + yz)$$

19. $f(x, y) = x^2 + 2xy + y^2 + 3x + 5y$

$$\frac{\partial f}{\partial x} = 2x + 2(y) + 0 + 3 + 0 = 2x + 2y + 3$$

$$\frac{\partial f}{\partial x}(2, -3) = 2(2) + 2(-3) + 3 = 1$$

$$\frac{\partial f}{\partial y} = 0 + 2x(1) + 2y + 0 + 5 = 2x + 2y + 5$$

$$\frac{\partial f}{\partial y}(2, -3) = 2(2) + 2(-3) + 5 = 3$$

21. $f(x, y) = xy^2 + 5$

$$\frac{\partial f}{\partial y} = 2xy$$

$$\frac{\partial f}{\partial y}(2, -1) = 2(2)(-1) = -4$$

This means that if x is kept constant at 2 and y is allowed to vary near -1, then $f(x, y)$ changes at a rate of -4 times the change in y.

23. $f(x, y) = x^3 y + 2xy^2$

$$\frac{\partial f}{\partial x} = 3x^2 y + 2y^2 \Rightarrow \frac{\partial^2 f}{\partial x^2} = 6xy$$

$$\frac{\partial^2 f}{\partial y \partial x} = 3x^2 + 4y$$

$$\frac{\partial f}{\partial y} = x^3 + 4xy \Rightarrow \frac{\partial^2 f}{\partial y^2} = 4x$$

$$\frac{\partial^2 f}{\partial x \partial y} = 3x^2 + 4y$$

25. $f(x, y) = 200\sqrt{6x^2 + y^2}$

a. $\dfrac{\partial f}{\partial x}$ is the marginal productivity of labor.

$$\frac{\partial f}{\partial x} = 200\left(\frac{1}{2}\right)(6x^2 + y^2)^{-1/2}(12x)$$

$$= 1200x(6x^2 + y^2)^{-1/2} = \frac{1200x}{\sqrt{6x^2 + y^2}}$$

When $x = 10$ and $y = 5$, the marginal productivity of labor is

$$\frac{\partial f}{\partial x}(10, 5) = \frac{1200(10)}{\sqrt{6(10)^2 + 5^2}} = 480 .$$

$\dfrac{\partial f}{\partial y}$ is the marginal productivity of capital.

$$\frac{\partial f}{\partial y} = 200\left(\frac{1}{2}\right)(6x^2 + y^2)^{-1/2}(2y)$$

$$= 200y(6x^2 + y^2)^{-1/2} = \frac{200y}{\sqrt{6x^2 + y^2}}$$

When $x = 10$ and $y = 5$, the marginal productivity of capital is

$$\frac{\partial f}{\partial y}(10, 5) = \frac{200(5)}{\sqrt{6(10)^2 + 5^2}} = 40 .$$

b. $f(10 + h, 5) - f(10, 5) \approx \dfrac{\partial f}{\partial x}(10, 5) \cdot h$
$$= 480h$$

c. Using part b, if $h = -.5$ then
$f(9.5, 5) \approx 480(-.5) = -240$. So, if capital is fixed at 5 units and labor decreased by .5 unit from 10 to 9.5 units, the number of goods produced will decrease by approximately 240 units.

27. As the price of a bus ride increases, fewer people will ride the bus if the train fare remains constant. An increase in train ticket prices, coupled with constant bus fare, should cause more people to ride the bus.

29. If the average price of MP3 players remains constant and the average price of audio files increases, people will purchase fewer MP3 players. An increase in the average price of the MP3 players, coupled with constant audio file prices, should cause a decline in the number of audio files purchased.

31. $V = .08\left(\dfrac{T}{P}\right) \Rightarrow \dfrac{\partial V}{\partial P} = \dfrac{-.08T}{P^2}$

When $P = 20$, $T = 300$,

$$\frac{\partial V}{\partial P} = \frac{-.08(300)}{400} = -.06 .$$

At this level, increasing the pressure by one unit will decrease the volume by approximately .06 unit.

$$\frac{\partial V}{\partial T} = \frac{.08}{P}$$

When $P = 20$, $T = 300$, $\dfrac{\partial V}{\partial T} = \dfrac{.08}{20} = .004$.

At this level, increasing the temperature by one unit will increase the volume by approximately .004 unit.

33. Assuming $m, p, r > 0$, $\dfrac{\partial f}{\partial m} > 0$, $\dfrac{\partial f}{\partial r} > 0$ and

$$\frac{\partial f}{\partial p} \approx -1.187m^{.595}r^{.922}p^{-1.543} < 0 .$$

Thus increases in aggregate income or retail prices of other goods (holding the other quantities constant) should cause an increase in the amount of food consumed; while an increase in the price of the food itself should cause the amount consumed to decrease.

35. $f(x, y) = 60x^{3/4}y^{1/4}$

$$\frac{\partial f}{\partial x} = 45y^{1/4}x^{-1/4} \Rightarrow \frac{\partial^2 f}{\partial x^2} = -\frac{45}{4}y^{1/4}x^{-5/4} < 0$$

for all $x, y > 0$. The fact that $\dfrac{\partial^2 f}{\partial x^2} < 0$ confirms the *law of diminishing returns*, which says that as additional units of a given productive input are added (holding other factors constant) production increases at a decreasing rate. In other words, the marginal productivity of labor is decreasing.

37. $f(x, y) = 3x^2 + 2xy + 5y$

$f(1 + h, 4) - f(1, 4)$
$$= \left[3(1 + h)^2 + 2(1 + h)(4) + 5(4)\right]$$
$$- \left[3(1)^2 + 2(1)(4) + 5(4)\right]$$
$$= 3h^2 + 14h$$

7.3 Maxima and Minima of Functions of Several Variables

1. $f(x, y) = x^2 - 3y^2 + 4x + 6y + 8$

$\dfrac{\partial f}{\partial x} = 2x + 4; \dfrac{\partial f}{\partial y} = -6y + 6$

$\left.\begin{array}{l} 2x + 4 = 0 \\ -6y + 6 = 0 \end{array}\right\} \begin{array}{l} x = -2 \\ y = 1 \end{array}$

The only possible extreme point is $(-2, 1)$.

3. $f(x, y) = x^2 - 5xy + 6y^2 + 3x - 2y + 4$

$\dfrac{\partial f}{\partial x} = 2x - 5y + 3; \dfrac{\partial f}{\partial y} = -5x + 12y - 2$

$\left.\begin{array}{l} 2x - 5y + 3 = 0 \\ -5x + 12y - 2 = 0 \end{array}\right\} \begin{array}{l} x = 26 \\ y = 11 \end{array}$

The only possible extreme point is $(26, 11)$.

5. $f(x, y) = 3x^2 + 8xy - 3y^2 - 2x + 4y - 1$

$\dfrac{\partial f}{\partial x} = 6x + 8y - 2; \quad \dfrac{\partial f}{\partial y} = 8x - 6y + 4$

$\left.\begin{array}{l} 6x + 8y - 2 = 0 \\ 8x - 6y + 4 = 0 \end{array}\right\} \begin{array}{l} x = -\frac{1}{5} \\ y = \frac{2}{5} \end{array}$

The only possible extreme point is $\left(-\frac{1}{5}, \frac{2}{5}\right)$.

7. $f(x, y) = x^3 + y^2 - 3x + 6y$

$\dfrac{\partial f}{\partial x} = 3x^2 - 3; \dfrac{\partial f}{\partial y} = 2y + 6$

$\left.\begin{array}{l} 3x^2 - 3 = 0 \\ 2y + 6 = 0 \end{array}\right\} \begin{array}{l} x = \pm 1 \\ y = -3 \end{array}$

The possible extreme points are $(1, -3)$ and $(-1, -3)$.

9. $f(x, y) = -8y^3 + 4xy + 9y^2 - 2y$

$\dfrac{\partial f}{\partial x} = 4y; \quad \dfrac{\partial f}{\partial y} = -24y^2 + 4x + 18y - 2$

$\left.\begin{array}{l} 4y = 0 \\ -24y^2 + 18y + 4x - 2 = 0 \end{array}\right\} \begin{array}{l} x = \frac{1}{2} \\ y = 0 \end{array}$

The only possible extreme point is $\left(\frac{1}{2}, 0\right)$.

11. $f(x, y) = 2x^3 + 2x^2y - y^2 + y$

$\dfrac{\partial f}{\partial x} = 6x^2 + 4xy; \quad \dfrac{\partial f}{\partial y} = 2x^2 - 2y + 1$

$\left.\begin{array}{ll} 6x^2 + 4xy = 0 & 3x^2 + 2xy = 0 \\ 2x^2 - 2y + 1 = 0 & y = x^2 + \frac{1}{2} \end{array}\right\}$

$3x^2 + 2x\left(x^2 + \frac{1}{2}\right) = 0 \Rightarrow 2x^3 + 3x^2 + x = 0 \Rightarrow$

$x\left(2x^2 + 3x + 1\right) = 0 \Rightarrow x(2x + 1)(x + 1) = 0 \Rightarrow$

$x = 0, -\dfrac{1}{2}, -1 \Rightarrow y = \dfrac{1}{2}, \dfrac{3}{4}, \dfrac{3}{2}$

The possible extreme points are $\left(0, \frac{1}{2}\right)$, $\left(-\frac{1}{2}, \frac{3}{4}\right)$, and $\left(-1, \frac{3}{2}\right)$.

13. $f(x, y) = \dfrac{1}{3}x^3 - 2y^3 - 5x + 6y - 5$

$\dfrac{\partial f}{\partial x} = x^2 - 5; \dfrac{\partial f}{\partial y} = -6y^2 + 6$

$\left.\begin{array}{l} x^2 - 5 = 0 \\ -6y^2 + 6 = 0 \end{array}\right\} \begin{array}{l} x = \pm\sqrt{5} \\ y = \pm 1 \end{array}$

There are four possible extreme points:
$\left(\sqrt{5}, 1\right), \left(\sqrt{5}, -1\right), \left(-\sqrt{5}, 1\right)$, and $\left(-\sqrt{5}, -1\right)$.

15. $f(x, y) = x^3 + x^2y - y$

$\dfrac{\partial f}{\partial x} = 3x^2 + 2xy; \quad \dfrac{\partial f}{\partial y} = x^2 - 1$

$\left.\begin{array}{ll} 3x^2 + 2xy = 0 & y = \dfrac{-3x^2}{2x} = -\dfrac{3}{2}x \\ x^2 - 1 = 0 & x^2 = 1 \end{array}\right\} \Rightarrow$

$x = \pm 1 \Rightarrow y = \mp\dfrac{3}{2}$

The possible extreme points are $\left(-1, \frac{3}{2}\right)$ and $\left(1, -\frac{3}{2}\right)$.

17. $f(x, y) = 2x + 3y + 9 - x^2 - xy - y^2$

$\dfrac{\partial f}{\partial x} = 2 - 2x - y; \dfrac{\partial f}{\partial y} = 3 - x - 2y$

$\left.\begin{array}{ll} 2 - 2x - y = 0 & y = 2 - 2x \\ 3 - x - 2y = 0 & x = \dfrac{1}{3} \end{array}\right\} \begin{array}{l} x = \dfrac{1}{3} \\ y = \dfrac{4}{3} \end{array}$

$\left(\dfrac{1}{3}, \dfrac{4}{3}\right)$ is the only point at which $f(x, y)$

can have a maximum, so the maximum value must occur at this point.

19. $f(x, y) = 3x^2 - 6xy + y^3 - 9y$

$\dfrac{\partial f}{\partial x} = 6x - 6y; \dfrac{\partial f}{\partial y} = -6x + 3y^2 - 9$

$\dfrac{\partial^2 f}{\partial x^2} = 6; \dfrac{\partial^2 f}{\partial y^2} = 6y; \dfrac{\partial^2 f}{\partial x \partial y} = -6$

$D(x, y) = \dfrac{\partial^2 f}{\partial x^2} \cdot \dfrac{\partial^2 f}{\partial y^2} - \left(\dfrac{\partial^2 f}{\partial x \partial y}\right)^2$

$\quad\quad = 6 \cdot 6y - (-6)^2 = 36y - 36$

$D(3, 3) = 36 \cdot 3 - 36 > 0$, and $\dfrac{\partial^2 f}{\partial x^2}(3, 3) > 0$

so $(3, 3)$ is a relative minimum of $f(x, y)$. $D(-1, -1) = 36(-1) - 36 < 0$, so $f(x, y)$ has neither a relative maximum nor a relative minimum at $(-1, -1)$.

21. $f(x, y) = 2x^2 - x^4 - y^2$

$\dfrac{\partial f}{\partial x} = 4x - 4x^3; \dfrac{\partial f}{\partial y} = -2y$

$\dfrac{\partial^2 f}{\partial x^2} = 4 - 12x^2; \dfrac{\partial^2 f}{\partial y^2} = -2; \dfrac{\partial^2 f}{\partial x \partial y} = 0$

$D(x, y) = \dfrac{\partial^2 f}{\partial x^2} \cdot \dfrac{\partial^2 f}{\partial y^2} - \left(\dfrac{\partial^2 f}{\partial x \partial y}\right)^2$

$\quad\quad = \left(4 - 12x^2\right)(-2) - (0)^2 = -8 + 24x^2$

$D(-1, 0) > 0$ and $\dfrac{\partial^2 f}{\partial x^2}(-1, 0) < 0$, so $f(x, y)$

has a relative maximum at $(-1, 0)$. $D(0, 0) < 0$, so $f(x, y)$ has neither a relative minimum nor a relative maximum at $(0, 0)$. $D(1, 0) > 0$ and

$\dfrac{\partial^2 f}{\partial x^2}(1, 0) < 0$, so $f(x, y)$ has a relative

maximum at $(1, 0)$.

23. $f(x, y) = ye^x - 3x - y + 5$

$\dfrac{\partial f}{\partial x} = ye^x - 3; \dfrac{\partial f}{\partial y} = e^x - 1$

$\dfrac{\partial^2 f}{\partial x^2} = ye^x; \dfrac{\partial^2 f}{\partial y^2} = 0; \dfrac{\partial^2 f}{\partial x \partial y} = e^x$

$D(x, y) = \dfrac{\partial^2 f}{\partial x^2} \cdot \dfrac{\partial^2 f}{\partial y^2} - \left(\dfrac{\partial^2 f}{\partial x \partial y}\right)^2$

$\quad\quad = ye^x(0) - \left(e^x\right)^2 = -e^{2x}$

$D(0, 3) < 0$, thus $f(x, y)$ has neither a maximum nor a minimum at $(0, 3)$.

25. $f(x, y) = -5x^2 + 4xy - 17y^2 - 6x + 6y + 2$;

$\dfrac{\partial f}{\partial x} = -10x + 4y - 6; \dfrac{\partial f}{\partial y} = -34y + 4x + 6$

$\dfrac{\partial^2 f}{\partial x^2} = -10; \dfrac{\partial^2 f}{\partial y^2} = -34; \dfrac{\partial^2 f}{\partial x \partial y} = 4$

$\left.\begin{array}{r} -10x + 4y - 6 = 0 \\ 4x - 34y + 6 = 0 \end{array}\right\} x = -\dfrac{5}{9}, \; y = \dfrac{1}{9}$

$D(x, y) = \dfrac{\partial^2 f}{\partial x^2} \cdot \dfrac{\partial^2 f}{\partial y^2} - \left(\dfrac{\partial^2 f}{\partial x \partial y}\right)^2$

$\quad\quad = -10(-34) - 4^2 = 324$

$D\left(-\dfrac{5}{9}, \dfrac{1}{9}\right) = -10(-34) - 4^2 = 324 > 0$ and

$\dfrac{\partial^2 f}{\partial x^2}\left(-\dfrac{5}{9}, \dfrac{1}{9}\right) < 0$, so $f(x, y)$ has a relative

maximum at $\left(-\dfrac{5}{9}, \dfrac{1}{9}\right)$.

27. $f(x, y) = 3x^2 + 8xy - 3y^2 + 2x + 6y$;

$\dfrac{\partial f}{\partial x} = 6x + 8y + 2; \dfrac{\partial f}{\partial y} = -6y + 8x + 6$

$\dfrac{\partial^2 f}{\partial x^2} = 6; \dfrac{\partial^2 f}{\partial y^2} = -6; \dfrac{\partial^2 f}{\partial x \partial y} = 8$

$\left.\begin{array}{r} 6x + 8y + 2 = 0 \\ 8x - 6y + 6 = 0 \end{array}\right\} x = -\dfrac{3}{5}, \; y = \dfrac{1}{5}$

$D(x, y) = \dfrac{\partial^2 f}{\partial x^2} \cdot \dfrac{\partial^2 f}{\partial y^2} - \left(\dfrac{\partial^2 f}{\partial x \partial y}\right)^2$

$\quad\quad = 6(-6) - 8^2 = -100$

$D\left(-\dfrac{3}{5}, \dfrac{1}{5}\right) = 6(-6) - 8^2 = -100 < 0$, so

$f(x, y)$ has neither a relative maximum nor a

relative minimum at $\left(-\dfrac{3}{5}, \dfrac{1}{5}\right)$.

29. $f(x, y) = x^4 - x^2 - 2xy + y^2 + 1$;

$\dfrac{\partial f}{\partial x} = 4x^3 - 2x - 2y; \quad \dfrac{\partial f}{\partial y} = 2y - 2x$

$\dfrac{\partial^2 f}{\partial x^2} = 12x^2 - 2; \quad \dfrac{\partial^2 f}{\partial y^2} = 2; \quad \dfrac{\partial^2 f}{\partial x \partial y} = -2$

Solving the system

$\left. \begin{array}{r} 4x^3 - 2x - 2y = 0 \\ -2x + 2y = 0 \end{array} \right\}$

yields the solutions $(0, 0)$, $(-1, -1)$, and $(1, 1)$.

$\begin{aligned} D(x, y) &= \dfrac{\partial^2 f}{\partial x^2} \cdot \dfrac{\partial^2 f}{\partial y^2} - \left(\dfrac{\partial^2 f}{\partial x \partial y} \right)^2 \\ &= \left(12x^2 - 2 \right) 2 - (-2)^2 \\ &= 24x^2 - 8 \end{aligned}$

$D(0, 0) = -8 < 0$, so $f(x, y)$ has neither a relative maximum nor a relative minimum at $(0, 0)$.

$D(-1, -1) = 16 > 0$ and $\dfrac{\partial^2 f}{\partial x^2} = 10 > 0$ so $f(x, y)$ has a relative minimum at $(-1, -1)$.

$D(1, 1) = 16 > 0$ and $\dfrac{\partial^2 f}{\partial x^2} = 10 > 0$ so $f(x, y)$ has a relative minimum at $(1, 1)$.

31. $f(x, y) = 6xy - 3y^2 - 2x + 4y - 1$;

$\dfrac{\partial f}{\partial x} = 6y - 2; \quad \dfrac{\partial f}{\partial y} = -6y + 6x + 4$

$\dfrac{\partial^2 f}{\partial x^2} = 0; \quad \dfrac{\partial^2 f}{\partial y^2} = -6; \quad \dfrac{\partial^2 f}{\partial x \partial y} = 6$

Solving the system

$\left. \begin{array}{r} 6y - 2 = 0 \\ 6x - 6y + 4 = 0 \end{array} \right\}$

yields the solution $\left(-\dfrac{1}{3}, \dfrac{1}{3} \right)$.

$\begin{aligned} D(x, y) &= \dfrac{\partial^2 f}{\partial x^2} \cdot \dfrac{\partial^2 f}{\partial y^2} - \left(\dfrac{\partial^2 f}{\partial x \partial y} \right)^2 \\ &= 0(-6) - 6^2 = -36 \end{aligned}$

$D\left(-\dfrac{1}{3}, \dfrac{1}{3} \right) = -36 < 0$, so $f(x, y)$ has neither a relative maximum nor a relative minimum at $\left(-\dfrac{1}{3}, \dfrac{1}{3} \right)$.

33. $f(x, y) = -2x^2 + 2xy - 25y^2 - 2x + 8y - 1$

$\dfrac{\partial f}{\partial x} = -4x + 2y - 2; \quad \dfrac{\partial f}{\partial y} = 2x - 50y + 8$

$\dfrac{\partial^2 f}{\partial x^2} = -4; \quad \dfrac{\partial^2 f}{\partial y^2} = -50; \quad \dfrac{\partial^2 f}{\partial x \partial y} = 2$

Solving the system

$\left. \begin{array}{r} -4x + 2y - 2 = 0 \\ 2x - 50y + 8 = 0 \end{array} \right\}$

yields the solution $\left(-\dfrac{3}{7}, \dfrac{1}{7} \right)$.

$\begin{aligned} D(x, y) &= \dfrac{\partial^2 f}{\partial x^2} \cdot \dfrac{\partial^2 f}{\partial y^2} - \left(\dfrac{\partial^2 f}{\partial x \partial y} \right)^2 \\ &= -4(-50) - 2^2 = 196 \end{aligned}$

$D\left(-\dfrac{3}{7}, \dfrac{1}{7} \right) = 196 > 0$ and

$\dfrac{\partial^2 f}{\partial x^2}\left(-\dfrac{3}{7}, \dfrac{1}{7} \right) = -4 < 0$, so $f(x, y)$ has a relative maximum at $\left(-\dfrac{3}{7}, \dfrac{1}{7} \right)$.

35. $f(x, y) = x^4 - 12x^2 - 4xy - y^2 + 16$

$\dfrac{\partial f}{\partial x} = 4x^3 - 24x - 4y; \quad \dfrac{\partial f}{\partial y} = -4x - 2y$

$\dfrac{\partial^2 f}{\partial x^2} = 12x^2 - 24; \quad \dfrac{\partial^2 f}{\partial y^2} = -2; \quad \dfrac{\partial^2 f}{\partial x \partial y} = -4$

Solving the system

$\left. \begin{array}{r} 4x^3 - 24x - 4y = 0 \\ -4x - 2y = 0 \end{array} \right\}$

yields the solutions $(0, 0)$, $(-2, 4)$, and $(2, -4)$.

$\begin{aligned} D(x, y) &= \dfrac{\partial^2 f}{\partial x^2} \cdot \dfrac{\partial^2 f}{\partial y^2} - \left(\dfrac{\partial^2 f}{\partial x \partial y} \right)^2 \\ &= \left(12x^2 - 24 \right)(-2) - (-4)^2 \\ &= -24x^2 + 32 \end{aligned}$

(*continued on next page*)

(continued)

$$D(0, 0) = 32 > 0 \text{ and } \frac{\partial^2 f}{\partial x^2}(0, 0) = -24 < 0,$$

so $f(x, y)$ has a relative maximum at $(0, 0)$.

$$D(-2, 4) = -24(-2)^2 + 32 = -64 < 0, \text{ so}$$

$f(x, y)$ has neither a relative maximum nor a relative minimum at $(-2, 4)$.

$$D(2, -4) = -24(2)^2 + 32 = -64 < 0, \text{ and}$$

$\frac{\partial^2 f}{\partial x^2} = 24 > 0$ so $f(x, y)$ has neither a relative maximum nor a relative minimum at $(2, -4)$.

37. $f(x, y) = x^2 - 2xy + 4y^2$

$$\frac{\partial f}{\partial x} = 2x - 2y; \frac{\partial f}{\partial y} = -2x + 8y$$

$$\frac{\partial^2 f}{\partial x^2} = 2; \frac{\partial^2 f}{\partial y^2} = 8; \frac{\partial^2 f}{\partial x \partial y} = -2$$

$$\left.\begin{matrix} 2x - 2y = 0 \\ -2x + 8y = 0 \end{matrix}\right\} \begin{matrix} x = 0 \\ y = 0 \end{matrix}$$

$$D(x, y) = \frac{\partial^2 f}{\partial x^2} \cdot \frac{\partial^2 f}{\partial y^2} - \left(\frac{\partial^2 f}{\partial x \partial y}\right)^2 = 2 \cdot 8 - (-2)^2$$
$$= 12$$

$$D(0, 0) = 2 \cdot 8 - (-2)^2 > 0 \text{ and } \frac{\partial^2 f}{\partial x^2}(0, 0) > 0,$$
so $f(x, y)$ has a relative minimum at $(0, 0)$.

39. $f(x, y) = -2x^2 + 2xy - y^2 + 4x - 6y + 5$

$$\frac{\partial f}{\partial x} = -4x + 2y + 4; \frac{\partial f}{\partial y} = 2x - 2y - 6$$

$$\frac{\partial^2 f}{\partial x^2} = -4; \frac{\partial^2 f}{\partial y^2} = -2; \frac{\partial^2 f}{\partial x \partial y} = 2$$

$$\left.\begin{matrix} -4x + 2y + 4 = 0 \\ 2x - 2y - 6 = 0 \end{matrix}\right\} \begin{matrix} x = -1 \\ y = -4 \end{matrix}$$

$$D(x, y) = \frac{\partial^2 f}{\partial x^2} \cdot \frac{\partial^2 f}{\partial y^2} - \left(\frac{\partial^2 f}{\partial x \partial y}\right)^2$$
$$= -4(-2) - 2^2 = 4$$

$$D(-1, -4) = (-4)(-2) - 2^2 > 0 \text{ and}$$

$\frac{\partial^2 f}{\partial x^2}(-1, -4) < 0$, so $f(x, y)$ has a relative maximum at $(-1, -4)$.

41. $f(x, y) = x^2 + 2xy + 5y^2 + 2x + 10y - 3$

$$\frac{\partial f}{\partial x} = 2x + 2y + 2; \frac{\partial f}{\partial y} = 2x + 10y + 10$$

$$\frac{\partial^2 f}{\partial x^2} = 2; \frac{\partial^2 f}{\partial y^2} = 10; \frac{\partial^2 f}{\partial x \partial y} = 2$$

$$\left.\begin{matrix} 2x + 2y + 2 = 0 \\ 2x + 10y + 10 = 0 \end{matrix}\right\} \begin{matrix} x = 0 \\ y = -1 \end{matrix}$$

$$D(x, y) = \frac{\partial^2 f}{\partial x^2} \cdot \frac{\partial^2 f}{\partial y^2} - \left(\frac{\partial^2 f}{\partial x \partial y}\right)^2$$
$$= 2(10) - 2^2 = 16$$

$$D(0, -1) = (2)(10) - 2^2 > 0, \frac{\partial^2 f}{\partial x^2} > 0, \text{ so}$$

$f(x, y)$ has a relative minimum at $(0, -1)$.

43. $f(x, y) = x^3 - y^2 - 3x + 4y$

$$\frac{\partial f}{\partial x} = 3x^2 - 3; \frac{\partial f}{\partial y} = -2y + 4$$

$$\frac{\partial^2 f}{\partial x^2} = 6x; \frac{\partial^2 f}{\partial y^2} = -2; \frac{\partial^2 f}{\partial x \partial y} = 0$$

$$\left.\begin{matrix} 3x^2 - 3 = 0 \\ -2y + 4 = 0 \end{matrix}\right\} \begin{matrix} x = \pm 1 \\ y = 2 \end{matrix}$$

$$D(x, y) = \frac{\partial^2 f}{\partial x^2} \cdot \frac{\partial^2 f}{\partial y^2} - \left(\frac{\partial^2 f}{\partial x \partial y}\right)^2$$
$$= 6x(-2) - 0^2 = -12x$$

$D(1, 2) < 0$, so $f(x, y)$ has neither a maximum nor a minimum at $(1, 2)$.

$$D(-1, 2) > 0 \text{ and } \frac{\partial^2 f}{\partial x^2}(-1, 2) = -6 < 0, \text{ so}$$

$f(x, y)$ has a relative maximum at $(-1, 2)$.

45. $f(x, y) = 2x^2 + y^3 - x - 12y + 7$

$$\frac{\partial f}{\partial x} = 4x - 1; \frac{\partial f}{\partial y} = 3y^2 - 12$$

$$\frac{\partial^2 f}{\partial x^2} = 4; \frac{\partial^2 f}{\partial y^2} = 6y; \frac{\partial^2 f}{\partial x \partial y} = 0$$

$$\left.\begin{matrix} 4x - 1 = 0 \\ 3y^2 - 12 = 0 \end{matrix}\right\} \begin{matrix} x = 1/4 \\ y = \pm 2 \end{matrix}$$

$$D(x, y) = \frac{\partial^2 f}{\partial x^2} \cdot \frac{\partial^2 f}{\partial y^2} - \left(\frac{\partial^2 f}{\partial x \partial y}\right)^2$$
$$= 4(6y) - 0^2 = 24y$$

(continued on next page)

(continued)

$$D\left(\frac{1}{4}, 2\right) = 48 > 0 \text{ and } \frac{\partial^2 f}{\partial x^2} > 0, \text{ so } f(x, y) \text{ has a relative minimum at } \left(\frac{1}{4}, 2\right).$$

$$D\left(\frac{1}{4}, -2\right) = -48 < 0, \text{ so } f(x, y) \text{ has neither a maximum nor a minimum at } \left(\frac{1}{4}, -2\right).$$

47. $f(x, y, z) = 2x^2 + 3y^2 + z^2 - 2x - y - z$

$$\frac{\partial f}{\partial x} = 4x - 2; \frac{\partial f}{\partial y} = 6y - 1; \frac{\partial f}{\partial z} = 2z - 1$$

$$\left.\begin{array}{r}4x - 2 = 0 \\ 6y - 1 = 0 \\ 2z - 1 = 0\end{array}\right\} x = \frac{1}{2}; y = \frac{1}{6}; z = \frac{1}{2}$$

$\left(\dfrac{1}{2}, \dfrac{1}{6}, \dfrac{1}{2}\right)$ is the only point at which $f(x, y, z)$ can have a relative minimum.

49.

Let x, y, and l be as shown in the figure.
Since $l = 84 - 2x - 2y$, the volume of the box may be written as

$$V(x, y) = xy(84 - 2x - 2y) = 84xy - 2x^2y - 2xy^2.$$

$$\frac{\partial V}{\partial x} = 84y - 4xy - 2y^2; \frac{\partial V}{\partial y} = 84x - 2x^2 - 4xy$$

$$\frac{\partial^2 V}{\partial x^2} = -4y; \frac{\partial^2 V}{\partial y^2} = -4x; \frac{\partial^2 V}{\partial x \partial y} = 84 - 4x - 4y$$

$$\left.\begin{array}{r}84y - 4xy - 2y^2 = 0 \\ 84x - 2x^2 - 4xy = 0\end{array}\right| \begin{array}{l}x = 0 \\ y = 0\end{array} \text{ or } \left.\begin{array}{r}84 - 4x - 2y = 0 \\ 84 - 4y - 2x = 0\end{array}\right| \begin{array}{l}x = 14 \\ y = 14\end{array}$$

Obviously, (0, 0) does not give the maximum value of $V(x, y)$. To verify that (14, 14) is the maximum, check

$$D = \frac{\partial^2 V}{\partial x^2} \cdot \frac{\partial^2 V}{\partial y^2} - \left(\frac{\partial^2 V}{\partial x \partial y}\right)^2 = -4y(-4x) - (84 - 4x - 4y)^2$$

$$D(14,14) = -4(14)(-4)(14) - [84 - 4(14) - 4(14)]^2 > 0 \text{ and } \frac{\partial^2 V}{\partial x^2} = -4(14) < 0.$$

Thus, the dimensions that give the maximum volume are $x = 14$, $y = 14$, $l = 84 - 56 = 28$; or $14 \times 14 \times 28$ in.

51. The revenue is $10x + 9y$, so the profit function is

$$P(x, y) = 10x + 9y - \left[400 + 2x + 3y + 0.01(3x^2 + xy + 3y^2)\right]$$

$$= 8x + 6y - 0.03x^2 - 0.01xy - 0.03y^2 - 400$$

$$\frac{\partial P}{\partial x} = 8 - 0.06x - 0.01y; \frac{\partial P}{\partial y} = 6 - 0.01x - 0.06y; \quad \frac{\partial^2 P}{\partial x^2} = -0.06; \frac{\partial^2 P}{\partial y^2} = -0.06; \frac{\partial^2 P}{\partial x \partial y} = -0.01$$

$$\left. \begin{array}{l} 8 - 0.06x - 0.01y = 0 \\ 6 - 0.01x - 0.06y = 0 \end{array} \right\} \begin{array}{l} x = 120 \\ y = 80 \end{array} \Rightarrow (120, 80) \text{ is a maximum. Verify this by checking}$$

$$D = \frac{\partial^2 P}{\partial x^2} \cdot \frac{\partial^2 P}{\partial y^2} - \left(\frac{\partial^2 P}{\partial x \partial y}\right)^2 = -0.06(-0.06) - (-0.01)^2 = 0.0035 \Rightarrow D(120, 80) > 0.$$

$$\frac{\partial^2 P}{\partial x^2}(120, 80) = -0.06 < 0.$$

Thus profit is maximized by producing 120 units of product I and 80 units of product II. **53.** Let $P(x, y)$ denote the company's profit from producing x units of product I and y units of product II. Then $P(x, y) = P_1 x + P_2 y - C(x, y)$. If (a, b) is the profit maximizing output combination, then

$$\frac{\partial P}{\partial x}(a, b) = \frac{\partial P}{\partial y}(a, b) = 0, \text{ so } P_1 - \frac{\partial C}{\partial x} = 0 \text{ and } P_2 - \frac{\partial C}{\partial y} = 0 \text{ or } \frac{\partial C}{\partial x} = P_1 \text{ and } \frac{\partial C}{\partial y} = P_2.$$

7.4   LaGrange Multipliers and Constrained Optimization

1. $F(x, y, \lambda) = x^2 + 3y^2 + 10 + \lambda(8 - x - y)$

$$\left. \begin{array}{l} \dfrac{\partial F}{\partial x} = 2x - \lambda = 0 \\[2mm] \dfrac{\partial F}{\partial y} = 6y - \lambda = 0 \\[2mm] \dfrac{\partial F}{\partial \lambda} = 8 - x - y = 0 \end{array} \right\} \begin{array}{l} \lambda = 2x \\[2mm] \lambda = 6y \\[2mm] 8 - x - y = 0 \end{array} \left| \begin{array}{l} 2x = 6y \Rightarrow x = 3y \\[2mm] 8 - 3y - y = 0 \end{array} \right\} \begin{array}{l} x = 6 \\[2mm] y = 2 \\[2mm] \lambda = 12 \end{array}$$

The minimum value is $6^2 + 3 \cdot 2^2 + 10 = 58.$

3. $F(x, y, \lambda) = x^2 + xy - 3y^2 + \lambda(2 - x - 2y)$

$$\left. \begin{array}{l} \dfrac{\partial F}{\partial x} = 2x + y - \lambda = 0 \\[2mm] \dfrac{\partial F}{\partial y} = x - 6y - 2\lambda = 0 \\[2mm] \dfrac{\partial F}{\partial \lambda} = 2 - x - 2y = 0 \end{array} \right| \begin{array}{l} \lambda = 2x + y \\[2mm] \lambda = \dfrac{1}{2}x - 3y \\[2mm] 2 - x - 2y = 0 \end{array} \left. \begin{array}{l} \dfrac{3}{2}x + 4y = 0 \\[2mm] x + 2y = 2 \end{array} \right| \begin{array}{l} x = 8 \\[2mm] y = -3 \\[2mm] \lambda = 13 \end{array}$$

The maximum value is $8^2 + 8(-3) - 3(-3)^2 = 13.$

5. $F(x, y, \lambda) = -2x^2 - 2xy - \dfrac{3}{2}y^2 + x + 2y + \lambda\left(x + y - \dfrac{5}{2}\right)$

$\dfrac{\partial F}{\partial x} = -4x - 2y + 1 + \lambda = 0$ $\left|\begin{array}{l}\lambda = 4x + 2y - 1\end{array}\right.$ $\left|\begin{array}{l}2x - y = -1\end{array}\right.$ $\left|\begin{array}{l}x = \dfrac{1}{2}\end{array}\right.$

$\dfrac{\partial F}{\partial y} = -2x - 3y + 2 + \lambda = 0$ $\left.\begin{array}{l}\lambda = 2x + 3y - 2\end{array}\right\}$ $x + y = \dfrac{5}{2}$ $\left.\begin{array}{l}\\ y = 2\end{array}\right.$

$\dfrac{\partial F}{\partial \lambda} = x + y - \dfrac{5}{2} = 0$ $\left.\begin{array}{l}x + y = \dfrac{5}{2}\end{array}\right.$

7. Minimize $xy + y^2 - x - 1$ subject to the constraint $x - 2y = 0$.

$F(x, y, \lambda) = xy + y^2 - x - 1 + \lambda(x - 2y)$

$\dfrac{\partial F}{\partial x} = y - 1 + \lambda = 0$ $\left|\begin{array}{l}\lambda = -y + 1\end{array}\right.$ $\left|\begin{array}{l}x + 4y = 2\end{array}\right\}$ $y = \dfrac{1}{3}$

$\dfrac{\partial F}{\partial y} = x + 2y - 2\lambda = 0$ $\left.\begin{array}{l}\lambda = \dfrac{x + 2y}{2}\end{array}\right\}$ $x = 2y$ $x = \dfrac{2}{3}$

$\dfrac{\partial F}{\partial \lambda} = x - 2y = 0$ $\left.\begin{array}{l}x = 2y\end{array}\right.$

9. Minimize $2x^2 + xy + y^2 - y$ subject to the constraint $x + y = 0$.

$F(x, y, \lambda) = 2x^2 + xy + y^2 - y + \lambda(x + y)$

$\dfrac{\partial F}{\partial x} = 4x + y + \lambda = 0$ $\left|\begin{array}{l}\lambda = -4x - y\end{array}\right.$ $\left|\begin{array}{l}-3x = -y + 1\end{array}\right\}$ $x = -\dfrac{1}{4}$

$\dfrac{\partial F}{\partial y} = x + 2y - 1 + \lambda = 0$ $\left.\begin{array}{l}\lambda = -x - 2y + 1\end{array}\right.$ $x = -y$ $y = \dfrac{1}{4}$

$\dfrac{\partial F}{\partial \lambda} = x + y = 0$ $\left.\begin{array}{l}x = -y\end{array}\right.$

11. Minimize $18x^2 + 12xy + 4y^2 + 6x - 4y + 5$ subject to the constraint $3x + 2y - 1 = 0$.

$F(x, y, \lambda) = 18x^2 + 12xy + 4y^2 + 6x - 4y + 5 + \lambda(3x + 2y - 1)$

$\dfrac{\partial F}{\partial x} = 36x + 12y + 6 + 3\lambda = 0$ $\left|\begin{array}{l}\lambda = -12x - 4y - 2\end{array}\right.$ $x = -\dfrac{2}{3}$ $x = -\dfrac{2}{3}$

$\dfrac{\partial F}{\partial y} = 12x + 8y - 4 + 2\lambda = 0$ $\left.\begin{array}{l}\lambda = -6x - 4y + 2\end{array}\right.$ $y = \dfrac{-3x + 1}{2}$ $y = \dfrac{3}{2}$

$\dfrac{\partial F}{\partial \lambda} = 3x + 2y - 1 = 0$ $y = \dfrac{-3x + 1}{2}$

13. Minimize $x - xy + 2y^2$ subject to the constraint $x - y + 1 = 0$.

$F(x, y, \lambda) = x - xy + 2y^2 + \lambda(x - y + 1)$

$\dfrac{\partial F}{\partial x} = 1 - y + \lambda = 0$ $\left|\begin{array}{l}\lambda = y - 1\end{array}\right.$ $x = 3y + 1$ $x = -2$

$\dfrac{\partial F}{\partial y} = -x + 4y - \lambda = 0$ $\left.\begin{array}{l}\lambda = -x + 4y\end{array}\right\}$ $y = x + 1$ $y = -1$

$\dfrac{\partial F}{\partial \lambda} = x - y + 1 = 0$ $y = x + 1$

15. Minimize $xy + xz - yz$ subject to the constraint $x + y + z = 1$.

$$F(x, y, z, \lambda) = xy + xz - yz + \lambda(x + y + z - 1)$$

$$\frac{\partial F}{\partial x} = y + z + \lambda = 0 \quad \left| \begin{array}{l} \lambda = -y - z \end{array} \right. \quad \left| \begin{array}{l} x - y = 2z \end{array} \right. \quad \left| \begin{array}{l} x = y + 2z \Rightarrow x = 3y \end{array} \right. \quad \left| \begin{array}{l} x = \dfrac{3}{5} \end{array} \right.$$

$$\frac{\partial F}{\partial y} = x - z + \lambda = 0 \quad \left| \begin{array}{l} \lambda = -x + z \end{array} \right. \quad \left| \begin{array}{l} y = z \end{array} \right. \quad \left| \begin{array}{l} y = z \end{array} \right.$$

$$\frac{\partial F}{\partial z} = x - y + \lambda = 0 \quad \left| \begin{array}{l} \lambda = -x + y \end{array} \right. \qquad\qquad\qquad\qquad\qquad \left| \begin{array}{l} y = \dfrac{1}{5} \end{array} \right.$$

$$\frac{\partial F}{\partial \lambda} = x + y + z - 1 = 0 \quad \left| \begin{array}{l} x + y + z = 1 \end{array} \right. \quad \left| \begin{array}{l} x + y + z = 1 \end{array} \right. \quad \left| \begin{array}{l} 3y + y + y = 1 \end{array} \right. \quad \left| \begin{array}{l} z = \dfrac{1}{5} \end{array} \right.$$

17. We want to minimize the function $x + y$ subject to the constraint $xy = 25$ or $xy - 25 = 0$.

$$F(x, y, \lambda) = x + y + \lambda(xy - 25)$$

$$\frac{\partial F}{\partial x} = 1 + \lambda y = 0 \quad \left| \begin{array}{l} \lambda = \dfrac{-1}{y} \end{array} \right. \quad \left| \begin{array}{l} x - y = 0 \end{array} \right. \quad \left| \begin{array}{l} x^2 = 25 \text{ or } x = \pm 5 \end{array} \right.$$

$$\frac{\partial F}{\partial y} = 1 + \lambda x = 0 \quad \left| \begin{array}{l} \lambda = \dfrac{-1}{x} \end{array} \right. \quad \left| \begin{array}{l} xy = 25 \end{array} \right.$$

$$\frac{\partial F}{\partial \lambda} = xy - 25 = 0 \quad \left| \begin{array}{l} xy - 25 = 0 \end{array} \right.$$

so $x = 5$, $y = 5$ (the positive numbers).

19. Let $x = $ length of a side of the base. Let $y = $ height of the box.

Area $= x^2 + 4xy = 300$

Maximize the volume $= x^2 y$ subject to $x^2 + 4xy - 300 = 0$.

$$F(x, y, \lambda) = x^2 y + \lambda(x^2 + 4xy - 300)$$

$$\frac{\partial F}{\partial x} = 2xy + 2x\lambda + 4y\lambda = 0 \quad \left| \begin{array}{l} \lambda = \dfrac{-2xy}{2(x + 2y)} \end{array} \right. \quad \left| \begin{array}{l} x - 2y = 0 \end{array} \right. \quad \left| \begin{array}{l} x = 10 \end{array} \right.$$

$$\frac{\partial F}{\partial y} = x^2 + 4x\lambda = 0 \quad \left| \begin{array}{l} \lambda = \dfrac{-x}{4x} = \dfrac{-x}{4} \end{array} \right. \quad \left| \begin{array}{l} x^2 + 4xy = 300 \end{array} \right. \quad \left| \begin{array}{l} y = 5 \end{array} \right.$$

$$\frac{\partial F}{\partial \lambda} = x^2 + 4xy - 300 = 0 \quad \left| \begin{array}{l} x^2 + 4xy = 300 \end{array} \right.$$

The sides of the base should be 10 in. and the height should be 5 in.

21. The length of the rectangle is $2x$, and the width of the rectangle is $2y$.

The problem is to maximize $4xy$ subject to $1 - x^2 - y^2 = 0$.

$$F(x, y, \lambda) = 4xy + \lambda(1 - x^2 - y^2)$$

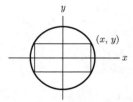

$$\frac{\partial F}{\partial x} = 4y - 2\lambda x = 0 \quad \left| \begin{array}{l} \lambda = \dfrac{2y}{x} \end{array} \right. \quad \left| \begin{array}{l} \text{Assuming } x \neq 0, \ y \neq 0 \\ \dfrac{2y}{x} = \dfrac{2x}{y} \end{array} \right. \quad \left| \begin{array}{l} x^2 = y^2 \\ 2x^2 = 1 \end{array} \right. \quad \left| \begin{array}{l} \text{Assuming } x > 0, \ y > 0, \\ x = \dfrac{\sqrt{2}}{2} \text{ and } y = \dfrac{\sqrt{2}}{2} \end{array} \right.$$

$$\frac{\partial F}{\partial y} = 4x - 2\lambda y = 0 \quad \left| \begin{array}{l} \lambda = \dfrac{2x}{y} \end{array} \right.$$

$$\frac{\partial F}{\partial \lambda} = 1 - x^2 - y^2 = 0 \quad \left| \begin{array}{l} x^2 + y^2 = 1 \end{array} \right. \quad \left| \begin{array}{l} x^2 + y^2 = 1 \end{array} \right.$$

The dimensions of the rectangle are $\sqrt{2} \times \sqrt{2}$.

23. The problem is to maximize $3x + 4y$ subject to $18,000 - 9x^2 - 4y^2 = 0$, $x \geq 0$, $y \geq 0$.

$$F(x, y, \lambda) = 3x + 4y + \lambda(18,000 - 9x^2 - 4y^2)$$

$$\frac{\partial F}{\partial x} = 3 - 18\lambda x = 0$$

$$\frac{\partial F}{\partial y} = 4 - 8\lambda x = 0$$

$$\frac{\partial F}{\partial \lambda} = 18,000 - 9x^2 - 4y^2 = 0$$

$$\left. \begin{array}{l} \lambda = \dfrac{1}{6x} \\[2mm] \lambda = \dfrac{1}{2y} \\[2mm] 9x^2 + 4y^2 = 18,000 \end{array} \right.$$

$$\left. \begin{array}{l} \text{If } x \neq 0 \text{ and } y \neq 0 \text{ then} \\ y = 3x. \\ 9x^2 + 36x^2 = 18,000 \end{array} \right\} \begin{array}{l} x = 20 \\ y = 60 \end{array}$$

Technically, we should also check the solutions $x = 0$, $y = \sqrt{\dfrac{18,000}{9}} \approx 44.7$ and $y = 0$, $x = \sqrt{\dfrac{18,000}{36}} \approx 22.4$.

These both give smaller values in the objective function $3x + 4y$ than does $(20, 60)$.

25. a. $F(x, y, \lambda) = 96x + 162y + \lambda\left(3456 - 64x^{3/4}y^{1/4}\right)$

Note that $3456 = 64x^{3/4}y^{1/4}$ implies $x \neq 0$, $y \neq 0$.

$$\frac{\partial F}{\partial x} = 96 - 48\lambda x^{-1/4}y^{1/4}$$

$$\frac{\partial F}{\partial y} = 162 - 16\lambda x^{3/4}y^{-3/4}$$

$$\frac{\partial F}{\partial \lambda} = 3456 - 64x^{3/4}y^{1/4}$$

$$\left. \begin{array}{l} \lambda = 2x^{1/4}y^{-1/4} \\[2mm] \lambda = \dfrac{81}{8}x^{-3/4}y^{3/4} \\[2mm] 3456 = 64x^{3/4}y^{1/4} \end{array} \right.$$

$$\left. \begin{array}{l} \text{Dividing 1st} \\ \text{equation by} \\ \text{2nd gives} \end{array} \right.$$

$$\left. \begin{array}{l} \dfrac{16}{81}xy^{-1} = 1 \\[2mm] y = \dfrac{16}{81}x \\[2mm] 3456 = 64x^{3/4}\left(\dfrac{16}{81}x\right)^{1/4} \end{array} \right.$$

$$\left. \begin{array}{l} x = 81 \\ y = 16 \end{array} \right.$$

b. $\lambda = 2(81)^{1/4}(16)^{-1/4} = 3$

c. The production function is $f(x, y) = 64x^{3/4}y^{1/4}$. Thus

$$\frac{\text{marginal productivity of labor}}{\text{marginal productivity of capital}} = \frac{\frac{\partial f}{\partial x}}{\frac{\partial f}{\partial y}} = \frac{48x^{-1/4}y^{1/4}}{16x^{3/4}y^{-3/4}} = \frac{48y}{16x}.$$

When $x = 81$ and $y = 16$, $\dfrac{48y}{16x} = \dfrac{48 \cdot 16}{16 \cdot 81} = \dfrac{96}{162}$, which is the ratio of the unit cost of labor and capital.

27. $F(x, y, z, \lambda) = xyz + \lambda(36 - x - 6y - 3z)$

$$\frac{\partial F}{\partial x} = yz - \lambda = 0$$

$$\frac{\partial F}{\partial y} = xz - 6\lambda = 0$$

$$\frac{\partial F}{\partial z} = xy - 3\lambda = 0$$

$$\frac{\partial F}{\partial \lambda} = 36 - x - 6y - 3z = 0$$

$$\left. \begin{array}{l} \lambda = yz \\[2mm] \lambda = \dfrac{xz}{6} \\[2mm] \lambda = \dfrac{xy}{3} \\[2mm] x + 6y + 3z = 36 \end{array} \right.$$

$$\left. \begin{array}{l} y = \dfrac{x}{6} \\[2mm] \dfrac{z}{6} = \dfrac{y}{3} \\[2mm] x + 6y + 3z = 36 \end{array} \right.$$

$$\left. \begin{array}{l} x = 6y \\ 3z = 6y \\ 3(6y) = 36 \end{array} \right.$$

$$\left. \begin{array}{l} x = 12 \\ y = 2 \\ z = 4 \end{array} \right.$$

29. $F(x, y, z, \lambda) = 3x + 5y + z - x^2 - y^2 - z^2 + \lambda(6 - x - y - z)$

$$\frac{\partial F}{\partial x} = 3 - 2x - \lambda = 0 \quad\Big| \quad \lambda = 3 - 2x \quad\Big| \quad -2x + 2y = 2 \quad\Big| \quad x = -1 + y \quad\Big| \quad x = 2$$

$$\frac{\partial F}{\partial y} = 5 - 2y - \lambda = 0 \quad\Big| \quad \lambda = 5 - 2y \quad\Big| \quad -2y + 2z = -4 \quad\Big| \quad z = -2 + y \quad\Big| \quad y = 3$$

$$\frac{\partial F}{\partial z} = 1 - 2z - \lambda = 0 \quad\Big| \quad \lambda = 1 - 2z$$

$$\frac{\partial F}{\partial \lambda} = 6 - x - y - z = 0 \quad\Big| \quad x + y + z = 6 \quad\Big| \quad x + y + z = 6 \quad\Big| \quad (-1 + y) + y + (-2 + y) = 6 \quad\Big| \quad z = 1$$

31.

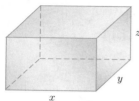

The problem is to minimize $3xy + 2xz + 2yz$ subject to the constraint $xyz = 12$ or $xyz - 12 = 0$.
$F(x, y, z, \lambda) = 3xy + 2xz + 2yz + \lambda(12 - xyz)$
(Note that $xyz = 12$ implies that $x \neq 0, y \neq 0, z \neq 0$.)

$$\frac{\partial F}{\partial x} = 3y + 2z - \lambda yz = 0 \quad\Big| \quad \lambda = \frac{3}{z} + \frac{2}{y} \quad\Big| \quad x = y \quad\Big| \quad x = 2$$

$$\frac{\partial F}{\partial y} = 3x + 2z - \lambda xz = 0 \quad\Big| \quad \lambda = \frac{3}{z} + \frac{2}{x} \quad\Big| \quad z = \frac{3}{2}y \quad\Big| \quad y = 2$$

$$\frac{\partial F}{\partial z} = 2x + 2y - \lambda xy = 0 \quad\Big| \quad \lambda = \frac{2}{y} + \frac{2}{x}$$

$$\frac{\partial F}{\partial \lambda} = 12 - xyz = 0 \quad\Big| \quad xyz = 12 \quad\Big| \quad y^2\left(\frac{3}{2}y\right) = 12 \quad\Big| \quad z = 3$$

The box is 2 ft $\times$ 2 ft $\times$ 3 ft.

33.

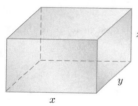

Let x, y, z be as shown in the figure. The problem is to minimize $xy + 2xz + 2yz$ subject to $xyz = 32$ or $32 - xyz = 0$.

$F(x, y, z, \lambda) = xy + 2xz + 2yz + \lambda(32 - xyz)$
(Note that $xyz = 32$ implies $x \neq 0, y \neq 0, z \neq 0$.)

$$\frac{\partial F}{\partial x} = y + 2z - \lambda yz = 0 \quad\Big| \quad \lambda = \frac{1}{z} + \frac{2}{y} \quad\Big| \quad \text{Equating} \quad\Big| \quad x = y \quad\Big| \quad x = 4$$

$$\frac{\partial F}{\partial y} = x + 2z - \lambda xz = 0 \quad\Big| \quad \lambda = \frac{1}{z} + \frac{2}{x} \quad\Big| \quad \substack{\text{expressions} \\ \text{for } \lambda \text{ gives}} \quad\Big| \quad z = \frac{1}{2}y \quad\Big| \quad y = 4$$

$$\frac{\partial F}{\partial z} = 2x + 2y - \lambda xy = 0 \quad\Big| \quad \lambda = \frac{2}{y} + \frac{2}{x} \quad\Big| \quad \frac{1}{2}y^3 = 32 \quad\Big| \quad z = 2$$

$$\frac{\partial F}{\partial \lambda} = xyz - 32 = 0 \quad\Big| \quad xyz = 32$$

The dimensions of the tank are 4 ft $\times$ 4 ft $\times$ 2 ft.

35. $F(x, y, \lambda) = f(x, y) + \lambda(c - ax - by)$

The values of x and y that minimize production subject to the cost constraint satisfy

$$\frac{\partial F}{\partial x}(x, y) = \frac{\partial f}{\partial x}(x, y) - \lambda a = 0$$

$$\frac{\partial F}{\partial y}(x, y) = \frac{\partial f}{\partial y}(x, y) - \lambda b = 0$$

$$\frac{\partial F}{\partial \lambda}(x, y) = c - ax - by = 0$$

$$\frac{\partial f}{\partial x}(x, y) = \lambda a$$

$$\frac{\partial f}{\partial y}(x, y) = \lambda b$$

Dividing the 1st equation by the second gives

$$\frac{\frac{\partial f}{\partial x}(x, y)}{\frac{\partial f}{\partial y}(x, y)} = \frac{a}{b}$$

7.5 The Method of Least Squares

1. The given points are (1, 3), (2, 6), (3, 8), and (4, 6) with straight line $y = 1.1x + 3$. When $x = 1, 2, 3, 4$ the corresponding y-coordinates are $1.1 + 3$, $2(1.1) + 3$, $3(1.1) + 3$, $4(1.1) + 3$ or 4.1, 5.2, 6.3, and 7.4, respectively. Then

$$E_1^2 = (4.1 - 3)^2 = (1.1)^2 = 1.21,$$

$$E_2^2 = (5.2 - 6)^2 = (-.8)^2 = .64,$$

$$E_3^2 = (6.3 - 8)^2 = (-1.7)^2 = 2.89,$$

$$E_4^2 = (7.4 - 6)^2 = (1.4)^2 = 1.96 \text{ so the least-}$$

squares error E is

$$E = E_1^2 + E_2^2 + E_3^2 + E_4^2$$
$$= 1.21 + .64 + 2.89 + 1.96 = 6.7.$$

3. $E = (2A + B - 6)^2 + (5A + B - 10)^2$
$$+ (9A + B - 15)^2$$

5. Let the straight line be $y = Ax + B$.

$$E_1^2 = (A + B - 2)^2, \ E_2^2 = (2A + B - 5)^2,$$

$$E_3^2 = (3A + B - 11)^2$$

$$f(A, B) = E_1^2 + E_2^2 + E_3^2$$
$$= (A + B - 2)^2 + (2A + B - 5)^2$$
$$+ (3A + B - 11)^2$$

$$\frac{\partial f}{\partial A} = 2(A + B - 2) + 2(2A + B - 5) \cdot 2$$
$$+ 2(3A + B - 11) \cdot 3$$
$$= 28A + 12B - 90$$

$$\frac{\partial f}{\partial B} = 2(A + B - 2) + 2(2A + B - 5)$$
$$+ 2(3A + B - 11)$$
$$= 12A + 6B - 36$$

Setting $\frac{\partial f}{\partial A}$ and $\frac{\partial f}{\partial B}$ equal to zero, we obtain the system

$$\begin{cases} 28A + 12B = 90 \\ 12A + 6B = 36 \end{cases}.$$

Then $A = 4.5$ and $B = -3$, so the line with the best least-squares fit to the data points is $y = 4.5x - 3$.

7. Let the straight line be $y = Ax + B$.

$$E_1^2 = (A + B - 9)^2, \ E_2^2 = (2A + B - 8)^2,$$

$$E_3^2 = (3A + B - 6)^2, \ E_4^2 = (4A + B - 3)^2$$

$$f(A, B) = E_1^2 + E_2^2 + E_3^2 + E_4^2$$
$$= (A + B - 9)^2 + (2A + B - 8)^2$$
$$+ (3A + B - 6)^2$$
$$+ (4A + B - 3)^2.$$

$$\frac{\partial f}{\partial A} = 2(A + B - 9) + 2(2A + B - 8) \cdot 2$$
$$+ 2(3A + B - 6) \cdot 3 + 2(4A + B - 3) \cdot 4$$
$$= 60A + 20B - 110$$

$$\frac{\partial f}{\partial B} = 2(A + B - 9) + 2(2A + B - 8)$$
$$+ 2(3A + B - 6) + 2(4A + B - 3)$$
$$= 20A + 8B - 52$$

Setting $\frac{\partial f}{\partial A}$ and $\frac{\partial f}{\partial B}$ equal to zero, we obtain the system

$$\begin{cases} 60A + 20B = 110 \\ 20A + 8B = 52 \end{cases} \Rightarrow A = -2, \ B = 11.5.$$

The line with the best least-squares fit to the data points is $y = -2x + 11.5$.

9.

x	y	xy	x^2
1	7	7	1
2	6	12	4
3	4	12	9
4	3	12	16
$\Sigma x = 10$	$\Sigma y = 20$	$\Sigma xy = 43$	$\Sigma x^2 = 30$

Let $y = Ax + B$. Then using

$$A = \frac{N \cdot \Sigma xy - \Sigma x \cdot \Sigma y}{N \cdot \Sigma x^2 - (\Sigma x)^2} \text{ and } B = \frac{\Sigma y - A \cdot \Sigma x}{N}$$

we have: $A = \dfrac{4 \cdot 43 - 10 \cdot 20}{4 \cdot 30 - 10^2} = -1.4$ and

$$B = \frac{20 - (-1.4) \cdot 10}{4} = 8.5.$$

The least-squares line is $y = -1.4x + 8.5$.

11. a.

x	y	xy	x^2
9	8175	73,575	81
10	8428	84,280	100
12	8996	107,952	144
13	9255	120,315	169
$\Sigma x = 44$	Σy $= 34{,}854$	Σxy $= 386{,}122$	Σx^2 $= 494$

Let $y = Ax + B$, where x = years after 2000 and y = dollars in thousands.

Then using $A = \dfrac{N \cdot \Sigma xy - \Sigma x \cdot \Sigma y}{N \cdot \Sigma x^2 - (\Sigma x)^2}$ and

$B = \dfrac{\Sigma y - A \cdot \Sigma x}{N}$

we have

$A = \dfrac{4 \cdot 386{,}122 - 44 \cdot 34{,}854}{4 \cdot 494 - (44)^2} = 272.8$

$B = \dfrac{34{,}854 - 272.8 \cdot 44}{4} = 5712.7$

The least-squares line is
$y = 272.8x + 5712.7$.

b. For $x = 16$,
$y = 272.8(16) + 5712.7 = 10{,}077.5$

c. $12{,}000 = 272.8x + 5712.7 \Rightarrow x \approx 23$
Per capita health care expenditures will reach \$12,000 sometime during 2023.

13. Given the table, convert *year* to *years after 2000*. Then we have the data points (0, 5.15), (5, 5.15), (10, 7.25), (16, 7.25).

a.

Using a graphing utility, we obtain
$y = .157x + 4.986$

By hand, we have
$N = 4$, $\Sigma x = 31$, $\Sigma y = 24.8$,

$\Sigma xy = 214.25$, and $\Sigma x^2 = 381$. The regression equation is the same.

b. $y = .157(8) + 4.986 = \$6.24$ per hour

c. $10 = .157x + 4.986 \Rightarrow x \approx 31.9$
The minimum wage will reach \$10 per hour in about 31.9 years, or in the year 2032.

15. a.

Using a graphing utility, we obtain
$y = -4.24x + 22.01$. By hand, we have
$N = 4$, $\Sigma x = 12$, $\Sigma y = 37.2$,

$\Sigma xy = 109.99$, and $\Sigma x^2 = 36.38$.
The regression equation is the same.

b. $y = -4.24(3.2) + 22.01 = 8.442°C$.

7.6 Double Integrals

1. $\displaystyle \int_0^1 \left(\int_0^1 e^{x+y}\,dy \right) dx = \int_0^1 \left(e^{x+y} \Big|_{y=0}^{1} \right) dx$

$= \displaystyle \int_0^1 (e^{x+1} - e^x)\,dx$

$= e^{x+1} - e^x \Big|_0^1$

$= e^2 - e - e + 1$

$= e^2 - 2e + 1$

3. $\displaystyle \int_{-1}^1 \left(\int_{-2}^0 xe^{xy}\,dy \right) dx = \int_{-1}^1 \left(e^{xy} \Big|_{y=-2}^{0} \right) dx$

$= \displaystyle \int_{-1}^1 \left(1 - e^{-2x} \right) dx$

$= x + \dfrac{e^{-2x}}{2} \Big|_{-1}^{1}$

$= 1 + \dfrac{e^{-2}}{2} - \left(-1 + \dfrac{e^2}{2} \right)$

$= 2 + \dfrac{e^{-2}}{2} - \dfrac{e^2}{2}$

5. $\int_1^4\left(\int_x^{x^2} xy\,dy\right)dx = \int_1^4\left(\frac{1}{2}xy^2\Big|_{y=x}^{x^2}\right)dx$

$\qquad = \int_1^4\left(\frac{1}{2}x^5 - \frac{1}{2}x^3\right)dx$

$\qquad = \frac{1}{12}x^6 - \frac{1}{8}x^4\Big|_1^4$

$\qquad = \frac{1}{12}4^6 - \frac{1}{8}4^4 - \frac{1}{12} + \frac{1}{8}$

$\qquad = 309\frac{3}{8}$

7. $\int_{-1}^1\left(\int_x^{2x}(x+y)\,dy\right)dx$

$\qquad = \int_{-1}^1\left(\left[xy + \frac{1}{2}y^2\right]_{y=x}^{2x}\right)dx$

$\qquad = \int_{-1}^1\left(2x^2 + 2x^2 - x^2 - \frac{1}{2}x^2\right)dx$

$\qquad = \int_{-1}^1\frac{5}{2}x^2dx = \frac{5}{6}x^3\Big|_{-1}^1 = \frac{5}{6} + \frac{5}{6} = \frac{5}{3}$

9. $\int_0^2\left(\int_2^3 xy^2dy\right)dx = \int_0^2\left(\frac{1}{3}xy^3\Big|_{y=2}^3\right)dx$

$\qquad = \int_0^2\left(9x - \frac{8}{3}x\right)dx$

$\qquad = \int_0^2\frac{19}{3}x\,dx = \frac{19}{6}x^2\Big|_0^2 = \frac{38}{3}$

11. $\int_0^2\left(\int_2^3 e^{-x-y}dy\right)dx = \int_0^2\left(-e^{-x-y}\Big|_{y=2}^3\right)dx$

$\qquad = \int_0^2(e^{-x-2} - e^{-x-3})\,dx$

$\qquad = -e^{-x-2} + e^{-x-3}\Big|_0^2$

$\qquad = -e^{-4} + e^{-5} + e^{-2} - e^{-3}$

13. $\int_1^3\left[\int_0^1(x^2+y^2)\,dy\right]dx$

$\qquad = \int_1^3\left[\left[x^2y + \frac{1}{3}y^3\right]_{y=0}^1\right]dx$

$\qquad = \int_1^3\left(x^2 + \frac{1}{3}\right)dx = \frac{1}{3}x^3 + \frac{1}{3}x\Big|_1^3$

$\qquad = 9 + 1 - \left(\frac{1}{3} + \frac{1}{3}\right) = \frac{28}{3}$

Chapter 7 Review Exercises

1. $f(x, y) = \frac{x\sqrt{y}}{1+x}$

$\qquad f(2, 9) = \frac{2\sqrt{9}}{1+2} = \frac{6}{3} = 2$

$\qquad f(5, 1) = \frac{5\sqrt{1}}{1+5} = \frac{5}{6};\quad f(0, 0) = \frac{0\sqrt{0}}{1+0} = 0$

2. $f(x, y, z) = x^2 e^{y/z}$

$\qquad f(-1, 0, 1) = (-1)^2 e^{0/1} = 1e^0 = 1$

$\qquad f(1, 3, 3) = 1^2 e^{3/3} = e$

$\qquad f(5, -2, 2) = 5^2 e^{-2/2} = \frac{25}{e}$

3. $f(A, t) = Ae^{0.06t}$

$\qquad f(10, 11.5) = 10e^{(0.06)(11.5)} \approx 19.94$

Ten dollars increases to approximately 20 dollars in 11.5 years.

4. $f(x, y, \lambda) = xy + \lambda(5 - x - y)$

$\qquad f(1, 2, 3) = (1)(2) + 3(5 - 1 - 2) = 2 + 3(2) = 8$

5. $f(x, y) = 3x^2 + xy + 5y^2$

$\qquad \frac{\partial f}{\partial x} = 6x + y;\quad \frac{\partial f}{\partial y} = x + 10y$

6. $f(x, y) = 3x - \frac{1}{2}y^4 + 1$

$\qquad \frac{\partial f}{\partial x} = 3;\quad \frac{\partial f}{\partial y} = -2y^3$

7. $f(x, y) = e^{x/y}$

$\qquad \frac{\partial f}{\partial x} = \frac{1}{y}e^{x/y};\quad \frac{\partial f}{\partial y} = -\frac{x}{y^2}e^{x/y}$

8. $f(x, y) = \frac{x}{x - 2y}$

$\qquad \frac{\partial f}{\partial x} = \frac{(1)(x-2y) - (x)(1)}{(x-2y)^2} = \frac{-2y}{(x-2y)^2}$

$\qquad \frac{\partial f}{\partial y} = \frac{(0)(x-2y) - (-2)(x)}{(x-2y)^2} = \frac{2x}{(x-2y)^2}$

9. $f(x, y, z) = x^3 - yz^2$

$\qquad \frac{\partial f}{\partial x} = 3x^2;\quad \frac{\partial f}{\partial y} = -z^2;\quad \frac{\partial f}{\partial z} = -2yz$

10. $f(x, y, \lambda) = xy + \lambda(5 - x - y)$
$$= xy + 5\lambda - x\lambda - y\lambda$$
$$\frac{\partial f}{\partial x} = y - \lambda; \frac{\partial f}{\partial y} = x - \lambda; \frac{\partial f}{\partial \lambda} = 5 - x - y$$

11. $f(x, y) = x^3 y + 8$
$$\frac{\partial f}{\partial x} = 3x^2 y; \ \frac{\partial f}{\partial x}(1, 2) = 3(1)^2(2) = 6$$
$$\frac{\partial f}{\partial y} = x^3; \ \frac{\partial f}{\partial y}(1, 2) = (1)^3 = 1$$

12. $f(x, y, z) = (x + y)z = xz + yz$
$$\frac{\partial f}{\partial y} = z; \ \frac{\partial f}{\partial y}(2, 3, 4) = 4$$

13. $f(x, y) = x^5 - 2x^3 y + \dfrac{y^4}{2}$
$$\frac{\partial f}{\partial x} = 5x^4 - 6x^2 y \Rightarrow \frac{\partial^2 f}{\partial x^2} = 20x^3 - 12xy$$
$$\frac{\partial f}{\partial y} = -2x^3 + 2y^3 \Rightarrow \frac{\partial^2 f}{\partial y^2} = 6y^2$$
$$\frac{\partial^2 f}{\partial x \partial y} = -6x^2; \frac{\partial^2 f}{\partial y \partial x} = -6x^2$$

14. $f(x, y) = 2x^3 + x^2 y - y^2$
$$\frac{\partial f}{\partial x} = 6x^2 + 2xy \Rightarrow \frac{\partial^2 f}{\partial x^2} = 12x + 2y$$
$$\frac{\partial f}{\partial y} = x^2 - 2y \Rightarrow \frac{\partial^2 f}{\partial y^2} = -2$$
$$\frac{\partial^2 f}{\partial x^2}(1, 2) = 16; \ \frac{\partial^2 f}{\partial y^2}(1, 2) = -2$$
$$\frac{\partial^2 f}{\partial x \partial y} = 2x; \ \frac{\partial^2 f}{\partial x \partial y}(1, 2) = 2$$

15. $f(p, t) = -p + 6t - 0.02pt$
$$\frac{\partial f}{\partial p} = -1 - 0.02t; \ \frac{\partial f}{\partial p}(25, 10,000) = -201$$
$$\frac{\partial f}{\partial t} = 6 - 0.02p; \ \frac{\partial f}{\partial t}(25, 10,000) = 5.5$$
At the level $p = 25$, $t = 10,000$, an increase in price of \$1 will result in a loss in sales of approximately 201 calculators, and an increase in advertising of \$1 will result in the sale of approximately 5.5 additional calculators.

16. The crime rate increases with increased unemployment and decreases with increased social services and police force size.

17. $f(x, y) = -x^2 + 2y^2 + 6x - 8y + 5$
$$\frac{\partial f}{\partial x} = -2x + 6; \ \frac{\partial f}{\partial y} = 4y - 8$$
$$-2x + 6 = 0 \Rightarrow x = 3; \ 4y - 8 = 0 \Rightarrow y = 2$$
The only possibility is $(x, y) = (3, 2)$.

18. $f(x, y) = x^2 + 3xy - y^2 - x - 8y + 4$
$$\frac{\partial f}{\partial x} = 2x + 3y - 1; \frac{\partial f}{\partial y} = 3x - 2y - 8$$
$$\left. \begin{array}{r} 2x + 3y = 1 \\ 3x - 2y = 8 \end{array} \right\} \begin{array}{l} x = 2 \\ y = -1 \end{array}$$
The only possibility is $(x, y) = (2, -1)$.

19. $f(x, y) = x^3 + 3x^2 + 3y^2 - 6y + 7$
$$\frac{\partial f}{\partial x} = 3x^2 + 6x; \frac{\partial f}{\partial y} = 6y - 6$$
$$3x^2 + 6x = 0 \Rightarrow 3x(x + 2) = 0 \Rightarrow x = 0, -2$$
$$6y - 6 = 0 \Rightarrow y = 1$$
$$(x, y) = (0, 1), (-2, 1)$$

20. $f(x, y) = \dfrac{1}{2}x^2 + 4xy + y^3 + 8y^2 + 3x + 2$
$$\frac{\partial f}{\partial x} = x + 4y + 3$$
$$\frac{\partial f}{\partial y} = 4x + 3y^2 + 16y$$
$$x + 4y + 3 = 0 \Rightarrow x = -4y - 3$$
$$4x + 3y^2 + 16y = 0 \Rightarrow$$
$$4(-4y - 3) + 3y^2 + 16y = 0 \Rightarrow$$
$$-16y - 12 + 3y^2 + 16y = 0 \Rightarrow y = \pm 2$$
$$x = -4(-2) - 3 = 5; \ x = -4(2) - 3 = -11;$$
$$(x, y) = (-11, 2), (5, -2)$$

21. $f(x, y) = x^2 + 3xy + 4y^2 - 13x - 30y + 12$
$$\frac{\partial f}{\partial x} = 2x + 3y - 13 \Rightarrow \frac{\partial^2 f}{\partial x^2} = 2$$
$$\frac{\partial f}{\partial y} = 3x + 8y - 30 \Rightarrow \frac{\partial^2 f}{\partial y^2} = 8; \ \frac{\partial^2 f}{\partial x \partial y} = 3$$
$$\left. \begin{array}{r} 2x + 3y = 13 \\ 3x + 8y = 30 \end{array} \right\} \begin{array}{l} x = 2 \\ y = 3 \end{array}$$
$$D(x, y) = \frac{\partial^2 f}{\partial x^2} \cdot \frac{\partial^2 f}{\partial y^2} - \left(\frac{\partial^2 f}{\partial x \partial y} \right)^2 \Rightarrow$$
$$D(2, 3) = 2 \cdot 8 - 3^2 > 0 \text{ and } \frac{\partial^2 f}{\partial x^2} > 0,$$
so $f(x, y)$ has a relative minimum at $(2, 3)$.

22. $f(x, y) = 7x^2 - 5xy + y^2 + x - y + 6$

$\dfrac{\partial f}{\partial x} = 14x - 5y + 1; \dfrac{\partial^2 f}{\partial x^2} = 14; \dfrac{\partial f}{\partial y} = -5x + 2y - 1; \dfrac{\partial^2 f}{\partial y^2} = 2; \dfrac{\partial^2 f}{\partial x \partial y} = -5$

$\left. \begin{array}{l} 14x - 5y = -1 \\ -5x + 2y = 1 \end{array} \right| \begin{array}{l} x = 1 \\ y = 3 \end{array}$

$D(x, y) = \dfrac{\partial^2 f}{\partial x^2} \cdot \dfrac{\partial^2 f}{\partial y^2} - \left(\dfrac{\partial^2 f}{\partial x \partial y} \right)^2 \Rightarrow D(1, 3) = 14 \cdot 2 - (-5)^2 > 0$ and $\dfrac{\partial^2 f}{\partial x^2} > 0$, so $f(x, y)$ has a relative

minimum at (1, 3).

23. $f(x, y) = x^3 + y^2 - 3x - 8y + 12$

$\dfrac{\partial f}{\partial x} = 3x^2 - 3; \dfrac{\partial^2 f}{\partial x^2} = 6x; \dfrac{\partial f}{\partial y} = 2y - 8; \dfrac{\partial^2 f}{\partial y^2} = 2; \dfrac{\partial^2 f}{\partial x \partial y} = 0$

$3x^2 = 3 \Rightarrow x = \pm 1$

$2y = 8 \Rightarrow y = 4$

$D(x, y) = \dfrac{\partial^2 f}{\partial x^2} \cdot \dfrac{\partial^2 f}{\partial y^2} - \left(\dfrac{\partial^2 f}{\partial x \partial y} \right)^2 \Rightarrow$

$D(1, 4) = 6(1)(2) - 0^2 > 0$ and $\dfrac{\partial^2 f}{\partial x^2}(1, 4) > 0$, so $f(x, y)$ has a relative minimum at (1, 4).

$D(-1, 4) = 6(-1)(2) - 0^2 < 0$, so $f(x, y)$ has neither a maximum nor a minimum at (−1, 4).

24. $f(x, y, z) = x^2 + 4y^2 + 5z^2 - 6x + 8y + 3$

$\dfrac{\partial f}{\partial x} = 2x - 6 = 0 \Rightarrow x = 3$

$\dfrac{\partial f}{\partial y} = 8y + 8 = 0 \Rightarrow y = -1$

$\dfrac{\partial f}{\partial z} = 10z = 0 \Rightarrow z = 0$

$f(x, y, z)$ must assume its minimum value at
(3, −1, 0).

25. $F(x, y, \lambda) = 3x^2 + 2xy - y^2 + \lambda(5 - 2x - y)$

$\dfrac{\partial F}{\partial x} = 6x + 2y - 2\lambda = 0 \left| \begin{array}{l} \lambda = 3x + y \\ \end{array} \right| \left. \begin{array}{l} x + 3y = 0 \\ 2x + y = 5 \end{array} \right| \begin{array}{l} x = 3 \\ y = -1 \end{array}$

$\dfrac{\partial F}{\partial y} = 2x - 2y - \lambda = 0 \left. \right\} \lambda = 2x - 2y$

$\dfrac{\partial F}{\partial \lambda} = 5 - 2x - y = 0 \quad \left| \quad 2x + y = 5 \right.$

The maximum value of $f(x, y)$ subject to the constraint is $3(3)^2 + 2(3)(-1) - (-1)^2 = 20$ occurs at
$(x, y) = (3, -1)$.

26. $F(x, y, \lambda) = -x^2 - 3xy - \dfrac{1}{2}y^2 + y + 10 + \lambda(10 - x - y)$

$$\begin{aligned}\dfrac{\partial F}{\partial x} &= -2x - 3y - \lambda = 0 \\[4pt] \dfrac{\partial F}{\partial y} &= -3x - y + 1 - \lambda = 0 \\[4pt] \dfrac{\partial F}{\partial \lambda} &= 10 - x - y = 0\end{aligned}\;\Bigg\}\;\begin{aligned}\lambda &= -2x - 3y \\[4pt] \lambda &= -3x - y + 1 \\[12pt] x &+ y = 10\end{aligned}\;\Bigg\}\;\begin{aligned}x - 2y &= 1 \\[2pt] x + y &= 10\end{aligned}\;\Big\}\;\begin{aligned}x &= 7 \\[2pt] y &= 3\end{aligned}$$

27. $F(x, y, z, \lambda) = 3x^2 + 2y^2 + z^2 + 4x + y + 3z + \lambda(4 - x - y - z)$

$$\begin{aligned}\dfrac{\partial F}{\partial x} &= 6x + 4 - \lambda = 0 \\[4pt] \dfrac{\partial F}{\partial y} &= 4y + 1 - \lambda = 0 \\[4pt] \dfrac{\partial F}{\partial z} &= 2z + 3 - \lambda = 0 \\[4pt] \dfrac{\partial F}{\partial \lambda} &= 4 - x - y - z = 0\end{aligned}\;\Bigg\}\;\begin{aligned}6x + 4 &= 4y + 1 \\[4pt] 4y + 1 &= 2z + 3 \\[8pt] x + y &+ z = 4\end{aligned}\;\Bigg\}\;\begin{aligned}x &= \dfrac{2}{3}y - \dfrac{1}{2} \\[4pt] z &= 2y - 1 \\[4pt] \dfrac{2}{3}y &- \dfrac{3}{2} + y + 2y = 4\end{aligned}\;\Bigg\}\;\begin{aligned}x &= \dfrac{1}{2} \\[4pt] y &= \dfrac{3}{2} \\[4pt] z &= 2\end{aligned}$$

28. The problem is to minimize $x + y + z$ subject to $xyz = 1000$.
$(x > 0, y > 0, z > 0)$
$F(x, y, z, \lambda) = x + y + z + \lambda(1000 - xyz)$
(Assuming $x \neq 0, y \neq 0, z \neq 0$)

$$\begin{aligned}\dfrac{\partial F}{\partial x} &= 1 - \lambda yz \\[4pt] \dfrac{\partial F}{\partial y} &= 1 - \lambda xz \\[4pt] \dfrac{\partial F}{\partial z} &= 1 - \lambda xy \\[4pt] \dfrac{\partial F}{\partial \lambda} &= 1000 - xyz\end{aligned}\;\Bigg|\;\begin{aligned}\lambda &= \dfrac{1}{yz} \\[4pt] \lambda &= \dfrac{1}{xz} \\[4pt] \lambda &= \dfrac{1}{xy} \\[4pt] xyz &= 1000\end{aligned}\;\Bigg|\;\begin{aligned}yz &= xz \\[4pt] xz &= xy \\[4pt] xyz &= 1000\end{aligned}\;\Bigg|\;\begin{aligned}x &= y = z \\[2pt] xyz &= 1000\end{aligned}\;\Big\}\;\begin{aligned}x &= 10 \\[2pt] y &= 10 \\[2pt] z &= 10\end{aligned}$$

The optimal dimensions are $10\text{ in.} \times 10\text{ in.} \times 10\text{ in.}$

29.

The problem is to maximize xy subject to $2x + y = 40$.
$F(x, y, \lambda) = xy + \lambda(40 - 2x - y)$

$$\begin{aligned}\dfrac{\partial F}{\partial x} &= y - 2\lambda = 0 \\[4pt] \dfrac{\partial F}{\partial y} &= x - \lambda = 0 \\[4pt] \dfrac{\partial F}{\partial \lambda} &= 40 - 2x - y = 0\end{aligned}\;\Bigg|\;\begin{aligned}\lambda &= \dfrac{1}{2}y \\[12pt] \lambda &= x \\[8pt] 2x &+ y = 40\end{aligned}\;\Bigg|\;\begin{aligned}x &= \dfrac{1}{2}y \\[12pt] 2y &= 40\end{aligned}\;\Bigg|\;\begin{aligned}x &= 10\text{ ft} \\[2pt] y &= 20\text{ ft}\end{aligned}$$

The dimensions of the garden should be $10\text{ ft} \times 20\text{ ft}$.

30. Maximize xy subject to $2x + y = 41$.

$$F(x, y, \lambda) = xy + \lambda(41 - 2x - y)$$

$$\left.\begin{array}{l} \dfrac{\partial F}{\partial x} = y - 2\lambda = 0 \\[2mm] \dfrac{\partial F}{\partial y} = x - \lambda = 0 \\[2mm] \dfrac{\partial F}{\partial \lambda} = 41 - 2x - y = 0 \end{array}\right\} \quad \left.\begin{array}{l} \lambda = \dfrac{1}{2}y \\[2mm] \lambda = x \\[2mm] 2x + y = 41 \end{array}\right\} \quad \left.\begin{array}{l} x = \dfrac{1}{2}y \\[2mm] 2y = 41 \end{array}\right\} \quad \left.\begin{array}{l} x = 10.25 \text{ ft} \\[2mm] y = 20.5 \text{ ft} \end{array}\right\}$$

The new area is $xy = (10.25)(20.5) = 210.125$ sq ft.
The increase in area (compared with the area in Exercise 29) is $210.125 - (10)(20) = 10.125$, which is approximately the value of λ.

31. Let the straight line be $y = Ax + B$.

$$E_1^2 = (A + B - 1)^2; \quad E_2^2 = (2A + B - 3)^2; \quad E_3^2 = (3A + B - 6)^2$$

Let $f(A, B) = E_1^2 + E_2^2 + E_3^2 = (A + B - 1)^2 + (2A + B - 3)^2 + (3A + B - 6)^2$.

$$\frac{\partial f}{\partial A} = 2(A + B - 1) + 2(2A + B - 3)(2) + 2(3A + B - 6)(3) = 28A + 12B - 50$$

$$\frac{\partial f}{\partial B} = 2(A + B - 1) + 2(2A + B - 3) + 2(3A + B - 6) = 12A + 6B - 20$$

Setting $\dfrac{\partial f}{\partial A}$ and $\dfrac{\partial f}{\partial B}$ equal to zero we obtain the system $\begin{cases} 28A + 12B = 50 \\ 12A + 6B = 20 \end{cases}$.

Then $A = \dfrac{5}{2}$ and $B = -\dfrac{5}{3}$ so the line with the best least-squares fit to the data points is $y = \dfrac{5}{2}x - \dfrac{5}{3}$.

32. Let the straight line be $y = Ax + B$.

$$E_1^2 = (A + B - 1)^2; \quad E_2^2 = (3A + B - 4)^2; \quad E_3^2 = (5A + B - 7)^2$$

Let $f(A, B) = E_1^2 + E_2^2 + E_3^2 = (A + B - 1)^2 + (3A + B - 4)^2 + (5A + B - 7)^2$.

$$\frac{\partial f}{\partial A} = 2(A + B - 1) + 2(3A + B - 4)(3) + 2(5A + B - 7)(5) = 70A + 18B - 96$$

$$\frac{\partial f}{\partial B} = 2(A + B - 1) + 2(3A + B - 4) + 2(5A + B - 7) = 18A + 6B - 24$$

Setting $\dfrac{\partial f}{\partial A}$ and $\dfrac{\partial f}{\partial B}$ equal to zero we obtain the system $\begin{cases} 70A + 18B = 96 \\ 18A + 6B = 24 \end{cases}$.

Then $A = \dfrac{3}{2}$ and $B = -\dfrac{1}{2}$ so the line with the best least-squares fit to the data points is $y = \dfrac{3}{2}x - \dfrac{1}{2}$.

33. Let the straight line be $y = Ax + B$.

$$E_1^2 = (0A + B - 1)^2; \quad E_2^2 = (A + B + 1)^2; \quad E_3^2 = (2A + B + 3)^2; \quad E_4^2 = (3A + B + 5)^2$$

Let $f(A, B) = E_1^2 + E_2^2 + E_3^2 + E_4^2 = (0A + B - 1)^2 + (A + B + 1)^2 + (2A + B + 3)^2 + (3A + B + 5)^2$.

$$\frac{\partial f}{\partial A} = 2(A + B + 1) + 2(2A + B + 3)(2) + 2(3A + B + 5)(3) = 28A + 12B + 44$$

$$\frac{\partial f}{\partial B} = 2(B - 1) + 2(A + B + 1) + 2(2A + B + 3) + 2(3A + B + 5) = 12A + 8B + 16$$

Setting $\dfrac{\partial f}{\partial A}$ and $\dfrac{\partial f}{\partial B}$ equal to zero we obtain the system: $\begin{cases} 28A + 12B = -44 \\ 12A + 8B = -16 \end{cases}$.

Then $A = -2$ and $B = 1$ so the line with the best least-squares fit to the data points is $y = -2x + 1$.

34. $\int_0^1 \left(\int_0^4 (x\sqrt{y} + y)\, dy \right) dx = \int_0^1 \left(\frac{2}{3} xy^{3/2} + \frac{1}{2} y^2 \Big|_{y=0}^{4} \right) dx = \int_0^1 \left(\frac{16}{3} x + 8 \right) dx = \frac{8}{3} x^2 + 8x \Big|_0^1 = \frac{8}{3} + 8 = \frac{32}{3}$

35. $\int_0^5 \left(\int_1^4 (2xy^4 + 3)\, dy \right) dx = \int_0^5 \left(\frac{2}{5} xy^5 + 3y \Big|_{y=1}^{4} \right) dx = \int_0^5 \left(\frac{2046}{5} x + 9 \right) dx = \frac{1023}{5} x^2 + 9x \Big|_0^5 = 5115 + 45 = 5160$

36. $\int_1^3 \left(\int_0^4 (2x + 3y)\, dx \right) dy = \int_1^3 \left(x^2 + 3xy \Big|_{x=0}^{4} \right) dy = \int_1^3 (16 + 12y)\, dy = 16y + 6y^2 \Big|_1^3 = 80$

37. $\iint_R 5\, dx\, dy$ represents the volume of the box with dimensions $(4-0) \times (3-1) \times 5$.

So $\iint_R 5\, dx\, dy = 4 \cdot 2 \cdot 5 = 40$.

38.

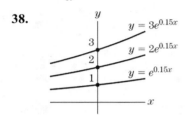

Chapter 8 The Trigonometric Functions

8.1 Radian Measure of Angles

1. $30° = 30 \cdot \dfrac{\pi}{180}$ radians $= \dfrac{\pi}{6}$ radian

$120° = 120 \cdot \dfrac{\pi}{180}$ radians $= \dfrac{2\pi}{3}$ radians

$315° = 315 \cdot \dfrac{\pi}{180}$ radians $= \dfrac{7\pi}{4}$ radians

3. $450° = 450 \cdot \dfrac{\pi}{180}$ radians $= \dfrac{5\pi}{2}$ radians

$-210° = -210 \cdot \dfrac{\pi}{180}$ radians $= -\dfrac{7\pi}{6}$ radians

$-90° = -90 \cdot \dfrac{\pi}{180}$ radians $= -\dfrac{\pi}{2}$ radians

5. $t = 8 \cdot \dfrac{\pi}{2} = 4\pi$ radians

7. $t = 7 \cdot \dfrac{\pi}{2} = \dfrac{7\pi}{2}$ radians

9. $t = -6 \cdot \dfrac{\pi}{2} = -3\pi$ radians

11. $t = 2 \cdot \dfrac{\pi}{3} = \dfrac{2\pi}{3}$ radians

13.

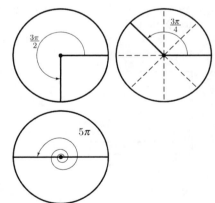

15.

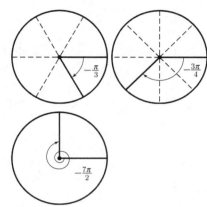

17.

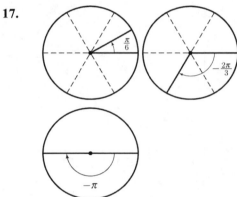

8.2 The Sine and the Cosine

1. $\sin t = \dfrac{1}{2};\ \cos t = \dfrac{\sqrt{3}}{2}$

3. $r = \sqrt{x^2 + y^2} = \sqrt{3^2 + 2^2} = \sqrt{13}$

$\sin t = \dfrac{2}{\sqrt{13}};\ \cos t = \dfrac{3}{\sqrt{13}}$

5. $r = \sqrt{x^2 + y^2}$, so

$x = \sqrt{r^2 - y^2} = \sqrt{13^2 - 12^2} = 5$

$\sin t = \dfrac{12}{13};\ \cos t = \dfrac{5}{13}$

7. $r = \sqrt{x^2 + y^2} = \sqrt{(-2)^2 + 1^2} = \sqrt{5}$

$\sin t = \dfrac{1}{\sqrt{5}};\ \cos t = -\dfrac{2}{\sqrt{5}}$

9. $r = \sqrt{x^2 + y^2} = \sqrt{(-2)^2 + (2)^2} = \sqrt{8}$

$$\sin t = \frac{2}{\sqrt{8}} = \frac{2}{2\sqrt{2}} = \frac{\sqrt{2}}{2}$$

$$\cos t = \frac{-2}{\sqrt{8}} = -\frac{2}{2\sqrt{2}} = -\frac{\sqrt{2}}{2}$$

11. $r = \sqrt{x^2 + y^2} = \sqrt{(-0.6)^2 + (-0.8)^2} = 1$

$$\sin t = \frac{-.8}{1} = -.8; \cos t = \frac{-.6}{1} = -.6$$

13. $a = 12$, $b = 5$, and $c = 13$

$$\sin t = \frac{b}{c} = \frac{5}{13} \approx .3846$$

$t = .4$ radians

15. $t = 1.1$ and $b = 3.2$

$$\sin t = \frac{b}{c} \Rightarrow \sin 1.1 = \frac{3.2}{c} \Rightarrow c \approx \frac{3.2}{.89121} = 3.6$$

17. $t = .4$, $a = 10.0$

$$\cos .4 = \frac{10.0}{c} \Rightarrow c \approx \frac{10.0}{.92106} = 10.9$$

19. $t = .5$, $a = 2.4$

$$\cos t = \frac{a}{c} \Rightarrow \cos .5 = \frac{2.4}{c} \Rightarrow$$

$$c \approx \frac{2.4}{.87758} = 2.7$$

$$\sin t = \frac{b}{c} \Rightarrow \sin .5 = \frac{b}{2.73} \Rightarrow$$

$$b \approx (.47943)(2.73) = 1.3$$

21. $\cos t = \cos\left(-\frac{\pi}{6}\right) \Rightarrow t = \frac{\pi}{6}$

23. $\cos t = \cos\left(\frac{5\pi}{4}\right) \Rightarrow t = \frac{3\pi}{4}$

25. $\cos t = \cos\left(-\frac{5\pi}{8}\right) \Rightarrow t = \frac{5\pi}{8}$

27. $\sin t = \sin\left(\frac{3\pi}{4}\right) \Rightarrow t = \frac{\pi}{4}$

29. $\sin t = \sin\left(-\frac{4\pi}{3}\right) \Rightarrow t = \frac{\pi}{3}$

31. $\sin t = -\sin\left(\frac{\pi}{6}\right) \Rightarrow t = -\frac{\pi}{6}$

33. $\sin t = \cos t \Rightarrow t = \frac{\pi}{4}$

35. $\cos t$ decreases from 1 to –1 as t increases from 0 to π.

37. $\sin 5\pi = 0$; $\sin(-2\pi) = 0$

$$\sin\left(\frac{17\pi}{2}\right) = 1; \sin\left(\frac{-13\pi}{2}\right) = -1$$

39. $\sin(.19) = \sqrt{1 - \cos^2 .19} = .2$
$\cos(.19 - 4\pi) = \cos(.19) = .98$
$\cos(-.19) = \cos(.19) = .98$
$\sin(-.19) = -\sin(.19) = -.2$

41. a.

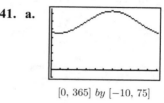

$[0, 365]$ *by* $[-10, 75]$

b. $59 + 14\cos\left[\frac{2\pi}{365}(45 - 208)\right]$

$\approx 59 + 14\cos(-2.81) \approx 45.76$

The temperature on February 14 is approximately 46°.

c. The coldest temperature will be when cosine has the value –1, and the warmest temperature will be when cosine has the value 1.

$59 + 14(-1) = 45°$ is the coldest tap water temperature

$59 + 14(1) = 73°$ is the warmest tap water temperature

d.

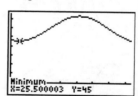

Estimate: $t = 25.5$

Since $\cos(-\pi) = -1$, $\frac{2\pi}{365}(t - 208) = -\pi$

$$t - 208 = -\frac{365}{2}$$

$$t = -\frac{365}{2} + 208 = 25.5$$

Tap water is the coldest on January 26.

e.

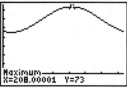

Estimate: $t = 208$

Since $\cos(0) = 1$, $\dfrac{2\pi}{365}(t - 208) = 0$

$t - 208 = 0; \; t = 208$

Tap water is the warmest on July 27.

f. $59 + 14\cos\left[\dfrac{2\pi}{365}(t - 208)\right] = 59$

$\cos\left[\dfrac{2\pi}{365}(t - 208)\right] = 0$

$\dfrac{2\pi}{365}(t - 208) = \dfrac{\pi}{2}$

$t - 208 = \dfrac{365}{4}$

$t = 299.25$

$\dfrac{2\pi}{365}(t - 208) = \dfrac{3\pi}{2}$

$t - 208 = \dfrac{1095}{4}$

$t = 481.75$, which is

equivalent to $481.75 - 365 = 116.75$.
Tap water is 59° on October 27 and April 27.

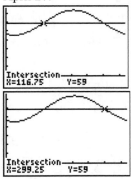

8.3 Differentiation and Integration of sin *t* and cos *t*

1. $\dfrac{d}{dt}\sin 4t = (\cos 4t)(4) = 4\cos 4t$

3. $\dfrac{d}{dt}4\sin t = 4(\cos t) = 4\cos t$

5. $\dfrac{d}{dt}2\cos 3t = 2(-\sin 3t)(3) = -6\sin 3t$

7. $\dfrac{d}{dt}(t + \cos \pi t) = 1 + (-\sin \pi t)(\pi) = 1 - \pi \sin \pi t$

9. $\dfrac{d}{dt}\sin(\pi - t) = \cos(\pi - t)(-1) = -\cos(\pi - t)$

11. $\dfrac{d}{dt}\cos^3 t = (3\cos^2 t)(-\sin t) = -3\cos^2 t \sin t$

13. $\dfrac{d}{dx}\sin \sqrt{x-1} = \cos(x-1)^{1/2} \cdot \dfrac{1}{2}(x-1)^{-1/2}(1)$

$= \dfrac{\cos \sqrt{x-1}}{2\sqrt{x-1}}$

15. $\dfrac{d}{dx}\sqrt{\sin(x-1)} = \dfrac{1}{2}\big(\sin(x-1)\big)^{-1/2}\big(\cos(x-1)\big)(1)$

$= \dfrac{\cos(x-1)}{2\sqrt{\sin(x-1)}}$

17. $\dfrac{d}{dx}(1 + \cos t)^8 = 8(1 + \cos t)^7(-\sin t)$

$= -8\sin t(1 + \cos t)^7$

19. $\dfrac{d}{dx}\cos^2 x^3 = 2\big(\cos x^3\big)\big(-\sin x^3\big)\big(3x^2\big)$

$= -6x^2 \sin x^3 \cos x^3$

21. $\dfrac{d}{dx}(e^x \sin x) = e^x \cos x + e^x \sin x$

$= e^x(\cos x + \sin x)$

23. $\dfrac{d}{dx}\Big[\sin(2x)\cos(3x)\Big]$

$= \sin(2x) \cdot \Big[-\sin(3x)\Big](3)$

$+ \Big[2\cos(2x)\Big]\cos(3x)$

$= 2\cos(2x)\cos(3x) - 3\sin(2x)\sin(3x)$

25. $\dfrac{d}{dt}\left(\dfrac{\sin t}{\cos t}\right) = \dfrac{\cos t(\cos t) - \sin t(-\sin t)}{\cos^2 t}$

$= \dfrac{\cos^2 t + \sin^2 t}{\cos^2 t} = \dfrac{1}{\cos^2 t}$

27. $\dfrac{d}{dt}\ln(\cos t) = \dfrac{1}{\cos t}(-\sin t) = -\dfrac{\sin t}{\cos t} = -\cot t$

29. $\dfrac{d}{dt}\sin(\ln t) = \cos(\ln t) \cdot \left(\dfrac{1}{t}\right) = \dfrac{\cos(\ln t)}{t}$

31. $y = \cos 3x$

slope $= \dfrac{dy}{dx} = -(\sin 3x)(3) = -3 \sin 3x$

When $x = \dfrac{13\pi}{6}$,

slope $= -3 \sin\left(3 \cdot \dfrac{13\pi}{6}\right) = -3(1) = -3.$

33. $y = 3 \sin x + \cos 2x$

When $x = \dfrac{\pi}{2}$,

$y = 3 \sin \dfrac{\pi}{2} + \cos 2\left(\dfrac{\pi}{2}\right) = 3 + (-1) = 2.$

slope $= \dfrac{dy}{dx} = 3 \cos x + (-\sin 2x)2$

$\qquad = 3 \cos x - 2 \sin 2x$

When $x = \dfrac{\pi}{2}$,

slope $= 3 \cos \dfrac{\pi}{2} - 2 \sin 2 \cdot \dfrac{\pi}{2} = 0 - 0 = 0.$

The equation of the tangent line is

$y - 2 = 0\left(x - \dfrac{\pi}{2}\right)$ or $y = 2.$

35. $\displaystyle\int \cos 2x \, dx = \dfrac{1}{2} \sin 2x + C$

37. $\displaystyle\int -\dfrac{1}{2} \cos \dfrac{x}{7} \, dx = -\dfrac{1}{2}\left(7 \sin \dfrac{x}{7}\right) + C$

$\qquad\qquad = -\dfrac{7}{2} \sin \dfrac{x}{7} + C$

39. $\displaystyle\int (\cos x - \sin x)dx = \sin x + \cos x + C$

41. $\displaystyle\int \left(-\sin x + 3 \cos(-3x)\right) dx$

$\qquad = \cos x - \sin(-3x) + C$

$\qquad = \cos x + \sin 3x + c$

43. $\displaystyle\int \sin(4x + 1) \, dx = -\dfrac{1}{4} \cos(4x + 1) + C$

45. $\displaystyle\int_0^{\pi/2} \cos t \, dt = \sin t \Big|_0^{\pi/2} = \sin \dfrac{\pi}{2} - \sin 0 = 1$

47. a. $P = 100 + 20 \cos 6t$

$\dfrac{dP}{dt} = (-20 \sin 6t)6 = -120 \sin 6t$

$\dfrac{d^2P}{dt^2} = (-120 \cos 6t)6 = -720 \cos 6t$

Setting $\dfrac{dP}{dt} = 0$ gives $-120 \sin 6t = 0 \Rightarrow$

$\sin 6t = 0 \Rightarrow 6t = 0, \pi, 2\pi, 3\pi, \ldots \Rightarrow$

$t = 0, \dfrac{\pi}{6}, \dfrac{\pi}{3}, \dfrac{\pi}{2}, \ldots$

When $t = 0$,

$\dfrac{d^2P}{dt^2} = -720 \cos 6(0) = -720.$

When $t = \dfrac{\pi}{6}$,

$\dfrac{d^2P}{dt^2} = -720 \cos 6\left(\dfrac{\pi}{6}\right) = -720(-1) = 720.$

When $t = \dfrac{\pi}{3}$,

$\dfrac{d^2P}{dt^2} = -720 \cos 6\left(\dfrac{\pi}{3}\right) = -720(1) = -720.$

When $t = \dfrac{\pi}{2}$,

$\dfrac{d^2P}{dt^2} = -720 \cos 6\left(\dfrac{\pi}{2}\right) = -720(-1) = 720.$

Thus $t = 0$ and $t = \dfrac{\pi}{3}$ give relative maximum values for P. The maximum value is

$P = 100 + 20 \cos 6\left(\dfrac{\pi}{3}\right)$

$\qquad = 100 + 20(1) = 120$

$t = \dfrac{\pi}{6}$ and $t = \dfrac{\pi}{2}$ give minimum values for P. The minimum value is

$P = 100 + 20 \cos 6\left(\dfrac{\pi}{6}\right) = 100 - 20 = 80.$

b. The length of time between two maximum values of P is $\dfrac{\pi}{3}$ seconds. The heart beats every $\dfrac{\pi}{3}$ seconds. Therefore the heart rate is $\dfrac{180}{\pi} \approx 57$ beats per minute.

49. Since $\sin\frac{\pi}{2} = 1$,

$$\lim_{h \to 0}\frac{\sin\left(\frac{\pi}{2}+h\right)-1}{h} = \lim_{h \to 0}\frac{\sin\left(\frac{\pi}{2}+h\right)-\sin\left(\frac{\pi}{2}\right)}{h}.$$

$$f'(a) \approx \frac{f(a+\Delta x)-f(a)}{\Delta x} \text{ so}$$

$$f(x) = \sin x \text{ and } \Delta x = h, \ a = \frac{\pi}{2}.$$

Therefore,

$$(\sin(a))' = \frac{\sin(a+h)-\sin(a)}{h}$$

$$= \frac{\sin\left(\frac{\pi}{2}+h\right)-\sin\left(\frac{\pi}{2}\right)}{h}$$

but $(\sin(a))' = \left(\sin\left(\frac{\pi}{2}\right)\right)' = \cos\left(\frac{\pi}{2}\right) = 0$.

51. a. $f(18) = 54 + 23\sin\left[\frac{2\pi}{52}(18-12)\right]$

$\approx 54 + 23\sin(.72) \approx 69.17 \approx 69°$

b. $f'(t) = 23\cos\left[\frac{2\pi}{52}(t-12)\right]\left(\frac{2\pi}{52}\right)$

$$f'(20) = 23\cos\left[\frac{2\pi}{52}(20-12)\right]\left(\frac{2\pi}{52}\right)$$

$$\approx 23\cos(.97)\left(\frac{2\pi}{52}\right) \approx 1.6$$

The temperature is increasing 1.6 degrees per week.

c. $54 + 23\sin\left[\frac{2\pi}{52}(t-12)\right] = 39$

$$\sin\left(\frac{2\pi}{52}(t-12)\right) = -0.65 \Rightarrow$$

$$\frac{2\pi}{52}(t-12) = -0.71 \text{ or } \frac{2\pi}{52}(t-12) = 0.71 + \pi$$

$$t - 12 = -5.88 \qquad\qquad t - 12 = 31.88$$

$$t = 6.12 \qquad\qquad\quad t = 43.88$$

Average weekly temperature is 39° during weeks 6 and 44.

d. $23\cos\left[\frac{2\pi}{52}(t-12)\right]\left(\frac{2\pi}{52}\right) = -1$

$$\cos\left[\frac{2\pi}{52}(t-12)\right] = -0.36$$

$$\frac{2\pi}{52}(t-12) = 1.94 \Rightarrow t - 12 = 16.06 \Rightarrow$$

$$t = 28.06 \text{ or}$$

$$\frac{2\pi}{52}(t-12) = 2\pi - 1.94 \Rightarrow$$

$$t - 12 = 35.94 \Rightarrow t = 47.94$$

Average weekly temperature is falling 1° per week during weeks 28 and 48.

e.   The average weekly temperature is greatest when sine is 1 and least when sine is −1. Since

$$\sin\left(\frac{\pi}{2}\right) = 1 \text{ and } \sin\left(\frac{3\pi}{2}\right) = -1,$$

$$\frac{2\pi}{52}(t-12) = \frac{\pi}{2} \quad \text{or} \quad \frac{2\pi}{52}(t-12) = \frac{3\pi}{2}$$

$$t - 12 = 13 \qquad\qquad t - 12 = 39$$

$$t = 25 \qquad\qquad\quad t = 51$$

The average weekly temperature is greatest at week 25 and least at week 51.

f.   The average weekly temperature is increasing fastest when cosine is 1 and decreasing fastest when cosine is −1. Since $\cos(0) = 1$ and $\cos(\pi) = -1$,

$$\frac{2\pi}{52}(t-12) = 0 \quad \text{or} \quad \frac{2\pi}{52}(t-12) = \pi$$

$$t = 12 \qquad\qquad\quad t - 12 = 26$$

$$\qquad\qquad\qquad\qquad\quad t = 38$$

The average weekly temperature is increasing fastest at week 12 and decreasing fastest at week 38.

8.4   The Tangent and Other Trigonometric Functions

1. $\sec t = \dfrac{1}{\cos t} = \dfrac{\text{hyp.}}{\text{adj.}}$

3. $(\text{adj.})^2 + 5^2 = 13^2 \Rightarrow \text{adj.} = 12$

$$\tan t = \frac{\sin t}{\cos t} = \frac{\text{opp.}}{\text{adj.}} = \frac{5}{12}$$

$$\sec t = \frac{1}{\cos t} = \frac{\text{hyp.}}{\text{adj.}} = \frac{13}{12}$$

5. $(-2)^2 + (1)^2 = (\text{hyp.})^2 \Rightarrow \text{hyp.} = \sqrt{5}$

$$\tan t = \frac{\sin t}{\cos t} = \frac{\text{opp.}}{\text{adj.}} = -\frac{1}{2}$$

$$\sec t = \frac{1}{\cos t} = \frac{\text{hyp.}}{\text{adj.}} = -\frac{\sqrt{5}}{2}$$

7. $(-2)^2 + 2^2 = (\text{hyp.})^2 \Rightarrow \text{hyp.} = \sqrt{8} = 2\sqrt{2}$

$\tan t = \dfrac{\sin t}{\cos t} = \dfrac{\text{opp.}}{\text{adj.}} = \dfrac{2}{-2} = -1; \quad \sec t = \dfrac{1}{\cos t} = \dfrac{\text{hyp.}}{\text{adj.}} = \dfrac{2\sqrt{2}}{-2} = -\sqrt{2}$

9. $(-.6)^2 + (-.8)^2 = (\text{hyp.})^2 \Rightarrow \text{hyp.} = 1$

$\tan t = \dfrac{\sin t}{\cos t} = \dfrac{\text{opp.}}{\text{adj.}} = \dfrac{-.8}{-.6} = \dfrac{4}{3}; \quad \sec t = \dfrac{1}{\cos t} = \dfrac{\text{hyp.}}{\text{adj.}} = \dfrac{1}{-.6} = -\dfrac{5}{3}$

11.

$40° = 40° \times \dfrac{\pi}{180°} = \dfrac{2\pi}{9} \approx 0.7$

$\tan(0.7) = \dfrac{\sin t}{\cos t} = \dfrac{\text{opp.}}{\text{adj.}} = \dfrac{AB}{75} \Rightarrow AB \approx 75 \tan(.7) \approx 63 \text{ feet}$

13. $\dfrac{d}{dt} \sec t = \dfrac{d}{dt}\left(\dfrac{1}{\cos t}\right) = \dfrac{0 - (-\sin t)(1)}{\cos^2 t} = \tan t \sec t$

15. $\dfrac{d}{dt} \cot t = \dfrac{d}{dt}\left(\dfrac{\cos t}{\sin t}\right) = \dfrac{-\sin t(\sin t) - \cos t(\cos t)}{\sin^2 t} = \dfrac{-\sin^2 t - \cos^2 t}{\sin^2 t} = \dfrac{-1}{\sin^2 t} = -\csc^2 t$

17. $\dfrac{d}{dt} \tan 4t = \dfrac{d}{dt}\left(\dfrac{\sin 4t}{\cos 4t}\right) = \dfrac{(\cos 4t)(4)\cos 4t - \sin 4t(-\sin 4t)(4)}{\cos^2 4t} = \dfrac{4(\cos^2 4t + \sin^2 4t)}{\cos^2 4t} = \dfrac{4(1)}{\cos^2 4t} = 4\sec^2 4t$

19. $\dfrac{d}{dx}\left[3\tan(\pi - x)\right] = 3\sec^2(\pi - x)(-1) = -3\sec^2(\pi - x)$

21. $\dfrac{d}{dx}\left[4\tan(x^2 + x + 3)\right] = 4\sec^2(x^2 + x + 3)(2x + 1) = (8x + 4)\sec^2(x^2 + x + 3)$

23. $\dfrac{d}{dx} \tan \sqrt{x} = \dfrac{d}{dx} \tan x^{1/2} = \sec^2 x^{1/2}\left(\dfrac{1}{2}x^{-1/2}\right) = \dfrac{\sec^2 \sqrt{x}}{2\sqrt{x}}$

25. $\dfrac{d}{dx}\left[x\tan x\right] = x\sec^2 x + (1)\tan x = x\sec^2 x + \tan x$

27. $\dfrac{d}{dx} \tan^2 x = \dfrac{d}{dx}(\tan x)^2 = 2(\tan x)\left(\sec^2 x\right) = 2\tan x \sec^2 x$

29. $\dfrac{d}{dt}(1 + \tan 2t)^3 = 3(1 + \tan 2t)^2\left(\sec^2 2t\right)(2) = 6\sec^2(2t)\left[1 + \tan(2t)\right]^2$

31. $\dfrac{d}{dx} \ln(\tan t + \sec t) = \dfrac{d}{dx}\ln\left(\tan t + \dfrac{1}{\cos t}\right) = \dfrac{1}{\tan t + \sec t}\left(\sec^2 t + \dfrac{\sin t}{\cos^2 t}\right)$

$= \dfrac{1}{\tan t + \sec t}\left(\sec^2 t + \dfrac{\sin t}{\cos t}\cdot\dfrac{1}{\cos t}\right) = \dfrac{1}{\tan t + \sec t}\left(\sec^2 t + \tan t \sec t\right)$

$= \dfrac{1}{\tan t + \sec t}\left(\sec t + \tan t\right)\sec t = \sec t$

33. a. $y = \tan x \Rightarrow \dfrac{dy}{dx} = \sec^2 x$

$$\left.\dfrac{dy}{dx}\right|_{x=\frac{\pi}{4}} = \sec^2 \dfrac{\pi}{4} = 2$$

$$y - 1 = 2\left(x - \dfrac{\pi}{4}\right)$$

b.

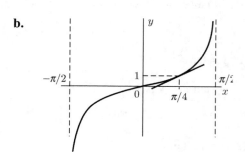

35. $\displaystyle\int \sec^2 3x\, dx = \dfrac{1}{3}\tan 3x + C$

37. $\displaystyle\int_{-\pi/4}^{\pi/4} \sec^2 x\, dx = (\tan x)\Big|_{-\pi/4}^{\pi/4} = 1 - (-1) = 2$

39. $\displaystyle\int \dfrac{1}{\cos^2 x}\, dx = \int \sec^2 x\, dx = \tan x + C$

Chapter 8 Review Exercises

1. $t = \dfrac{3\pi}{2}$

2. $t = -\dfrac{7\pi}{2}$

3. $t = -\dfrac{3\pi}{4}$

4.

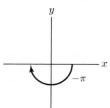

5.

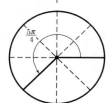

6.

7. $3^2 + 4^2 = (\text{hyp.})^2 \Rightarrow \text{hyp.} = 5$

$$\sin t = \dfrac{\text{opp.}}{\text{hyp.}} = \dfrac{4}{5};\ \ \cos t = \dfrac{\text{adj.}}{\text{hyp.}} = \dfrac{3}{5}$$

$$\tan t = \dfrac{\text{opp.}}{\text{adj.}} = \dfrac{4}{3}$$

8. $(-.6)^2 + (.8)^2 = (\text{hyp.})^2 \Rightarrow \text{hyp.} = 1$

$$\sin t = \dfrac{\text{opp.}}{\text{hyp.}} = .8;\ \ \cos t = \dfrac{\text{adj.}}{\text{hyp.}} = -.6$$

$$\tan t = \dfrac{\text{opp.}}{\text{adj.}} = \dfrac{.8}{-.6} = -\dfrac{4}{3}$$

9. $(-.6)^2 + (-.8)^2 = (\text{hyp.})^2 \Rightarrow \text{hyp.} = 1$

$$\sin t = \dfrac{\text{opp.}}{\text{hyp.}} = -.8; \cos t = \dfrac{\text{adj.}}{\text{hyp.}} = -.6$$

$$\tan t = \dfrac{\text{opp.}}{\text{adj.}} = \dfrac{-.8}{-.6} = \dfrac{4}{3}$$

10. $3^2 + (-4)^2 = (\text{hyp.})^2; \text{hyp.} = 5$

$$\sin t = \dfrac{\text{opp.}}{\text{hyp.}} = -\dfrac{4}{5}; \cos t = \dfrac{\text{adj.}}{\text{hyp.}} = \dfrac{3}{5}$$

$$\tan t = \dfrac{\text{opp.}}{\text{adj.}} = -\dfrac{4}{3}$$

11. $\sin t = \dfrac{1}{5}; (\text{opp.})^2 + (\text{adj.})^2 = (\text{hyp.})^2$

$$1 + (\text{adj.})^2 = 25; \text{adj.} = \pm\sqrt{24} = \pm 2\sqrt{6}$$

$$\cos t = \dfrac{\text{adj.}}{\text{hyp.}} = \pm\dfrac{2\sqrt{6}}{5}$$

12. $\cos t = -\dfrac{2}{3}; (\text{opp.})^2 + (\text{adj.})^2 = (\text{hyp.})^2$

$$(\text{opp.})^2 + 4 = 9; \text{opp.} = \pm\sqrt{5}; \sin t = \dfrac{\pm\sqrt{5}}{3}$$

13. $\dfrac{\pi}{4}, \dfrac{5\pi}{4}, -\dfrac{3\pi}{4}, -\dfrac{7\pi}{4}$

14. $\dfrac{3\pi}{4}, \dfrac{7\pi}{4}, -\dfrac{\pi}{4}, -\dfrac{5\pi}{4}$

15. negative

16. positive

17. Let r be the length of the rafter needed to support the roof.
$$r^2 = (15)^2 + \left[15(\tan 23°)\right]^2 \Rightarrow r \approx 16.3 \text{ ft}$$

18. Let t be the height of the tree.
$$t = 60(\tan 53°) \approx 79.62 \text{ feet}$$

19. $f(t) = 3 \sin t;\ \dfrac{d}{dt} 3 \sin t = 3 \cos t$

20. $f(t) = \sin 3t;\ \dfrac{d}{dt} \sin 3t = (\cos 3t)(3) = 3 \cos 3t$

21. $f(t) = \sin \sqrt{t} = \sin t^{1/2}$
$$\frac{d}{dt} \sin t^{1/2} = \left(\cos\left(t^{1/2}\right)\right)\left(\frac{1}{2}\right) t^{-1/2} = \frac{\cos \sqrt{t}}{2\sqrt{t}}$$

22. $f(t) = \cos t^3$
$$\frac{d}{dt} \cos t^3 = (-\sin t^3)(3t^2) = -3t^2 \sin t^3$$

23. $g(x) = x^3 \sin x$
$$\frac{d}{dx}(x^3 \sin x) = x^3 \cos x + 3x^2 \sin x$$

24. $g(x) = \sin(-2x) \cos 5x$
$$\frac{d}{dx}[\sin(-2x)\cos 5x]$$
$$= \sin(-2x)(-\sin 5x)(5) + \cos(-2x)(-2)\cos 5x$$
$$= -5 \sin 5x \sin(-2x) - 2 \cos(-2x)\cos 5x$$
$$= 5 \sin 5x \sin 2x - 2 \cos 5x \cos 2x$$

25. $f(x) = \dfrac{\cos 2x}{\sin 3x}$
$$\frac{d}{dx}\left(\frac{\cos 2x}{\sin 3x}\right)$$
$$= \frac{(-\sin 2x)(2)\sin 3x - (\cos 2x)(\cos 3x)(3)}{\sin^2 3x}$$
$$= -\frac{2\sin 2x \sin 3x + 3\cos 2x \cos 3x}{\sin^2 3x}$$

26. $f(x) = \dfrac{\cos x - 1}{x^3}$
$$\frac{d}{dx}\left(\frac{\cos x - 1}{x^3}\right) = \frac{(-\sin x)(x^3) - 3x^2(\cos x - 1)}{x^6}$$
$$= \frac{-x^3 \sin x - 3x^2(\cos x - 1)}{x^6}$$

27. $f(x) = \cos^3 4x$
$$\frac{d}{dx} \cos^3 4x = (3\cos^2 4x)(-\sin 4x)(4)$$
$$= -12 \cos^2 4x \sin 4x$$

28. $f(x) = \tan^3 2x$
$$\frac{d}{dx} \tan^3 2x = 3(\tan^2 2x)(\sec^2 2x)(2)$$
$$= 6\tan^2(2x)\sec^2(2x)$$

29. $y = \tan(x^4 + x^2)$
$$\frac{d}{dx} \tan(x^4 + x^2) = \left(\sec^2\left(x^4 + x^2\right)\right)\left(4x^3 + 2x\right)$$
$$= (4x^3 + 2x)\sec^2(x^4 + x^2)$$

30. $y = \tan e^{-2x}$
$$\frac{d}{dx} \tan e^{-2x} = (\sec^2 e^{-2x})e^{-2x}(-2)$$
$$= -2e^{-2x} \sec^2\left(e^{-2x}\right)$$

31. $y = \sin(\tan x)$
$$\frac{d}{dx} \sin(\tan x) = \cos(\tan x)\sec^2 x$$

32. $y = \tan(\sin x)$
$$\frac{d}{dx} \tan(\sin x) = \sec^2(\sin x)\cos x$$

33. $y = \sin x \tan x$
$$\frac{d}{dx}[\sin x \tan x] = \sin x \sec^2 x + \cos x \tan x$$
$$= \sin x \sec^2 x + \sin x$$

34. $y = (\ln x)\cos x$
$$\frac{d}{dx}[(\ln x)\cos x] = \ln x(-\sin x) + \frac{1}{x}\cos x$$
$$= \frac{\cos x}{x} - (\ln x)\sin x$$

35. $y = \ln(\sin x)$
$$\frac{d}{dx} \ln(\sin x) = \frac{1}{\sin x}(\cos x) = \cot x$$

36. $y = \ln(\cos x)$
$$\frac{d}{dx} \ln(\cos x) = \frac{1}{\cos x}(-\sin x) = -\tan x$$

37. $y = e^{3x} \sin^4 x$
$$\frac{d}{dx}[e^{3x} \sin^4 x]$$
$$= e^{3x}(4\sin^3 x)(\cos x) + e^{3x}(3)\sin^4 x$$
$$= 4e^{3x} \sin^3 x \cos x + 3e^{3x} \sin^4 x$$

38. $y = \sin^4 e^{3x}$
$$\frac{d}{dx} \sin^4 e^{3x} = \left(4\sin^3 e^{3x}\right)\left(\cos e^{3x}\right)\left(e^{3x}\right)(3)$$
$$= 12e^{3x}\left(\cos e^{3x}\right)\left(\sin^3 e^{3x}\right)$$

39. $f(t) = \dfrac{\sin t}{\tan 3t}$

$\dfrac{d}{dt}\left(\dfrac{\sin t}{\tan 3t}\right) = \dfrac{\cos t \tan 3t - (\sec^2 3t)(3)\sin t}{\tan^2 3t}$

$\qquad = \dfrac{\cos t \tan 3t - 3\sin t \sec^2 3t}{\tan^2 3t}$

40. $f(t) = \dfrac{\tan 2t}{\cos t}$

$\dfrac{d}{dt}\left(\dfrac{\tan 2t}{\cos t}\right) = \dfrac{(\sec^2 2t)(2)\cos t - (-\sin t)\tan 2t}{\cos^2 t}$

$\qquad = \dfrac{2\cos t \sec^2 2t + \sin t \tan 2t}{\cos^2 t}$

41. $f(t) = e^{\tan t}$

$\dfrac{d}{dt} e^{\tan t} = e^{\tan t}(\sec^2 t)$

42. $f(t) = e^t \tan t$

$\dfrac{d}{dt}[e^t \tan t] = e^t(\sec^2 t) + e^t \tan t$

$\qquad = e^t(\sec^2 t + \tan t)$

43. $f(t) = \sin^2 t$

$f'(t) = 2\sin t \cos t$

$f''(t) = 2[(\sin t)(-\sin t) + \cos t \cos t]$

$\qquad = 2(\cos^2 t - \sin^2 t)$

44. $y = 3\sin 2t + \cos 2t$

$y' = 3(\cos 2t)(2) + (-\sin 2t)(2)$

$\quad = 6\cos 2t - 2\sin 2t$

$y'' = 6[(-\sin 2t)(2)] - 2(\cos 2t)(2)$

$\quad = -12\sin 2t - 4\cos 2t$

$\quad = -4(3\sin 2t + \cos 2t)$

$-4y = -12\sin 2t - 4\cos 2t$

Therefore y'' and $-4y$ are equal.

45. $f(s, t) = \sin s \cos 2t;\ \dfrac{\partial f}{\partial s} = \cos s \cos 2t$

$\dfrac{\partial f}{\partial t} = \sin s\,(-\sin 2t)(2) = -2\sin s \sin 2t$

46. $z = \sin wt;\ \dfrac{\partial z}{\partial w} = t\cos wt;\ \dfrac{\partial z}{\partial t} = w\cos wt$

47. $f(s, t) = t\sin st;\ \dfrac{\partial f}{\partial s} = t(\cos st)(t) = t^2\cos st$

$\dfrac{\partial f}{\partial t} = t(\cos st)(s) + (1)\sin st = st\cos st + \sin st$

48. $\sin(s + t) = \sin s \cos t + \cos s \sin t$

$\dfrac{\partial}{\partial t}\sin(s + t) = \cos(s + t)$

$\dfrac{\partial}{\partial t}[\sin s \cos t + \cos s \sin t]$

$\qquad = \sin s(-\sin t) + \cos s \cos t$

$\qquad = \cos s \cos t - \sin s \sin t$

Thus, $\cos(s + t) = \cos s \cos t - \sin s \sin t$.

49. $y = \tan t = \dfrac{\sin t}{\cos t}$

When $t = \dfrac{\pi}{4}, y = \dfrac{\sin\frac{\pi}{4}}{\cos\frac{\pi}{4}} = 1$

$\text{slope} = y' = \sec^2 t = \sec^2\left(\dfrac{\pi}{4}\right) = 2$

The tangent line is $y - 1 = 2\left(t - \dfrac{\pi}{4}\right)$.

50. a. $f(t) = \sin t + \cos t$

$f'(t) = \cos t - \sin t$

$\cos t - \sin t = 0 \Rightarrow \cos t = \sin t \Rightarrow$

$t = \dfrac{\pi}{4}, \dfrac{5\pi}{4}, -\dfrac{3\pi}{4}, -\dfrac{7\pi}{4}$

b. $f''(t) = -\cos t - \sin t$

$f''\left(\dfrac{\pi}{4}\right) = -\sqrt{2}$, so the curve is concave

down at $t = \dfrac{\pi}{4}$. $f''\left(\dfrac{5\pi}{4}\right) = \sqrt{2}$, so the

curve is concave up at $t = \dfrac{5\pi}{4}$. Similarly,

$f''\left(-\dfrac{7\pi}{4}\right) = -\sqrt{2}$, so the curve is

concave down at $t = -\dfrac{7\pi}{4}$, and

$f''\left(-\dfrac{3\pi}{4}\right) = \sqrt{2}$, so the curve is concave

up at $t = \dfrac{3\pi}{4}$.

c. $f''(t) = -\cos t - \sin t = 0 \Rightarrow$

$-\cos t = \sin t \Rightarrow t = \dfrac{3\pi}{4}, \dfrac{7\pi}{4}, -\dfrac{\pi}{4}, -\dfrac{5\pi}{4}$

The inflection points are at $\left(\dfrac{3\pi}{4}, 0\right)$,

$\left(\dfrac{7\pi}{4}, 0\right), \left(-\dfrac{\pi}{4}, 0\right)$, and $\left(-\dfrac{5\pi}{4}, 0\right)$.

(*continued on next page*)

(*continued*)

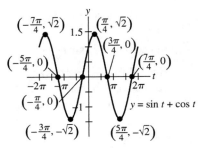

51.

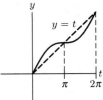

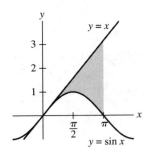

52. $y = 2 + \sin 3t$

Area under the curve is $\displaystyle\int_0^{\pi/2} (2 + \sin 3t)\, dt$.

$$\int_0^{\pi/2} (2 + \sin 3t)\, dt = \left[2t - (\cos 3t)\left(\frac{1}{3}\right) \right]\Bigg|_0^{\pi/2}$$

$$= \left[2t - \frac{1}{3}\cos 3t \right]\Bigg|_0^{\pi/2}$$

$$= \pi - \frac{1}{3}(0) - \left(0 - \frac{1}{3}\right) = \frac{1}{3} + \pi$$

53. The desired area is

$$\int_0^\pi \sin t\, dt + \int_\pi^{2\pi} (-\sin t)\, dt = (-\cos t)\Big|_0^\pi + \cos t\Big|_\pi^{2\pi}$$

$$= 2 + 2 = 4$$

54. The desired area is

$$\int_0^{\pi/2} \cos t\, dt + \int_{\pi/2}^{3\pi/2} (-\cos t)\, dt$$

$$= \sin t\Big|_0^{\pi/2} + (-\sin t)\Big|_{\pi/2}^{3\pi/2} = 1 + 2 = 3$$

55. It is easy to check that the line $y = x$ is tangent to the graph of $y = \sin x$ at $x = 0$. From the graph of $y = \sin x$, it is clear that $y = \sin x$ lies below $y = x$ for $x \geq 0$. So the area between these two curves from $x = 0$ to $x = \pi$ is given by

$$\text{Area} = \int_0^\pi (x - \sin x)\, dx = \left[\frac{x^2}{2} + \cos x \right]\Bigg|_0^\pi$$

$$= \left[\frac{\pi^2}{2} + \cos \pi \right] - [0 + \cos 0]$$

$$= \frac{\pi^2}{2} + (-1) - [0 + 1] = \frac{\pi^2}{2} - 2$$

Exercises 56–58 refer to the function

$V(t) = 3 + 0.05\sin\left(160\pi t - \dfrac{\pi}{2} \right)$, where $V(t)$ is the

lung volume in liters and t is measured in minutes.

56. a. $V(0) = 2.95$, $V\left(\dfrac{1}{320}\right) = 3$,

$V\left(\dfrac{1}{160}\right) = 3.05, V\left(\dfrac{1}{80}\right) = 2.95$

b. $V'(t) = 0.05\cos\left(160\pi t - \dfrac{\pi}{2} \right)(160\pi)$

Setting $V'(t) = 0$ gives

$$\cos\left(160\pi t - \frac{\pi}{2} \right) = 0 \Rightarrow$$

$$160\pi t - \frac{\pi}{2} = \frac{\pi}{2}, \frac{3\pi}{2}, \frac{5\pi}{2}, \dots \Rightarrow$$

$$160\pi t = \pi, 2\pi, 3\pi, \dots \Rightarrow$$

$$t = \frac{1}{160}, \frac{2}{160}, \frac{3}{160}, \dots$$

$$V''(t) = -8\pi \sin\left(160\pi t - \frac{\pi}{2} \right)(160\pi)$$

$$= -1280\pi^2 \sin\left(160\pi t - \frac{\pi}{2} \right)$$

At $t = \dfrac{1}{160}$, $V''(t) < 0$, so this value of t

gives a relative maximum, and the

maximum lung volume is $V\left(\dfrac{1}{160}\right) = 3.05$

liters.

57. a. $V'(t) = 8\pi\cos\left(160\pi t - \dfrac{\pi}{2} \right)$

b. Inspiration (in the first cycle) occurs from

$t = 0$ to $t = \dfrac{1}{160}$. To find the maximum

rate of inspiration, we need to find the

maximum of $V'(t)$ on $\left[0, \dfrac{1}{160} \right]$.

(*continued on next page*)

(continued)

$$V''(t) = -1280\pi^2 \sin\left(160\pi t - \frac{\pi}{2}\right)$$

Setting $V''(t) = 0$ gives $160\pi t - \dfrac{\pi}{2} = 0, \pi,$

$2\pi, \ldots \Rightarrow 160\pi t = \dfrac{\pi}{2}, \dfrac{3\pi}{2}, \dfrac{5\pi}{2}, \ldots \Rightarrow$

$t = \dfrac{1}{320}, \dfrac{3}{320}, \dfrac{5}{320}, \ldots$ Among these

values, only $\dfrac{1}{320}$ is within $\left[0, \dfrac{1}{160}\right]$.

Since $V'(t) = 0$ at the end points of the

interval, $t = \dfrac{1}{320}$ must give the maximum

value of $V'(t)$. Thus, the maximum rate of air flow is

$$V'\left(\frac{1}{320}\right) = 8\pi \text{ liters/min.}$$

c. The average value of $V'(t)$ on $\left[0, \dfrac{1}{160}\right]$

is

$$\frac{1}{\frac{1}{160}} \int_0^{1/160} V'(t)\, dt$$

$$= 160 \int_0^{1/160} 8\pi \cos\left(160\pi t - \frac{\pi}{2}\right) dt$$

$$= -8\sin\left(160\pi t - \frac{\pi}{2}\right)\Big|_0^{1/160}$$

$$= 8(1+1) = 16 \text{ liters/min}$$

58. During one minute, there will be 80 inspirations. Each inspiration represents

$$V\left(\frac{1}{160}\right) - V(0) = 3 + 0.05 - (3 - 0.05) = .1$$

liter. Therefore, the minute volume is $0.1(80)$ = 8 liters. In Exercise 57(b), we found that the peak respiratory flow was 8π liters/min. Thus the first statement is verified. In Exercise 57(c), we found the mean inspiratory flow to be $16 = 8 \cdot 2$ liters/min, verifying the second statement.

59. $\displaystyle\int \sin(\pi - x)\,dx = \cos(\pi - x) + C$

60. $\displaystyle\int (3\cos 3x - 2\sin 2x)\,dx = \sin 3x + \cos 2x + C$

61. $\displaystyle\int_0^{\pi/2} \cos 6x\,dx = \left(\frac{1}{6}\sin 6x\right)\Big|_0^{\pi/2} = 0 - 0 = 0$

62. $\displaystyle\int \cos(6 - 2x)\,dx = -\frac{1}{2}\sin(6 - 2x) + C$

63. $\displaystyle\int_0^\pi \left[x - 2\cos(\pi - 2x)\right]dx$

$$= \left[\frac{1}{2}x^2 + \sin(\pi - 2x)\right]\Big|_0^\pi$$

$$= \frac{\pi^2}{2} + 0 - (0) = \frac{\pi^2}{2}$$

64. $\displaystyle\int_{-\pi}^\pi (\cos 3x + 2\sin 7x)\,dx$

$$= \left(\frac{1}{3}\sin 3x - \frac{2}{7}\cos 7x\right)\Big|_{-\pi}^\pi$$

$$= 0 - \left(-\frac{2}{7}\right) - \left[0 - \left(-\frac{2}{7}\right)\right] = 0$$

65. $\displaystyle\int \sec^2\frac{x}{2}\,dx = 2\tan\frac{x}{2} + C$

66. $\displaystyle\int 2\sec^2 2x\,dx = \tan 2x + C$

67. $\displaystyle\int_0^{\pi/4} (\cos x - \sin x)\,dx = (\sin x + \cos x)\Big|_0^{\pi/4}$

$$= \frac{\sqrt{2}}{2} + \frac{\sqrt{2}}{2} - (0 + 1)$$

$$= \sqrt{2} - 1$$

68. $\displaystyle\int_0^{\pi/4} \sin x\,dx + \int_{\pi/4}^{\pi/2} \cos x\,dx$

$$= (-\cos x)\Big|_0^{\pi/4} + (\sin x)\Big|_{\pi/4}^{\pi/2}$$

$$= -\frac{\sqrt{2}}{2} - (-1) + 1 - \left(\frac{\sqrt{2}}{2}\right) = 2 - \sqrt{2}$$

69. $\displaystyle\int_{\pi/4}^{\pi/2} (\sin x - \cos x)\,dx + \int_{\pi/2}^\pi \sin x\,dx$

$$= (-\cos x - \sin x)\Big|_{\pi/4}^{\pi/2} + (-\cos x)\Big|_{\pi/2}^\pi$$

$$= 0 - 1 - \left(-\frac{\sqrt{2}}{2} - \frac{\sqrt{2}}{2}\right) + [-(-1)] - (0) = \sqrt{2}$$

70. $\displaystyle\int_{\pi/2}^\pi (0 - \cos x)\,dx = -\int_{\pi/2}^\pi \cos x\,dx = -(\sin x)\Big|_{\pi/2}^\pi$

$$= -[0 - (1)] = 1$$

71. $\text{Average} = \dfrac{1}{b-a}\displaystyle\int_a^b f(t)dt$

$\qquad = \dfrac{1}{2\pi-0}\displaystyle\int_0^{2\pi}\left(1+\sin 2t - \dfrac{1}{3}\cos 2t\right)dt$

$\qquad = \dfrac{1}{2\pi}\left[t - \dfrac{1}{2}\cos 2t - \dfrac{1}{6}\sin 2t\right]_0^{2\pi}$

$\qquad = \dfrac{1}{2\pi}\left[2\pi - \dfrac{1}{2} - 0 - \left(0 - \dfrac{1}{2} - 0\right)\right] = 1$

72. $\dfrac{1}{\pi-0}\displaystyle\int_0^{\pi}(t-\cos 2t)dt = \dfrac{1}{\pi}\left[\dfrac{1}{2}t^2 - \dfrac{1}{2}\sin 2t\right]_0^{\pi}$

$\qquad = \dfrac{1}{\pi}\left[\dfrac{\pi^2}{2} - 0 - (0-0)\right]$

$\qquad = \dfrac{\pi}{2}$

73. $\dfrac{1}{\frac{3\pi}{4}-0}\displaystyle\int_0^{3\pi/4}\left[1000 + 200\sin 2\left(t - \dfrac{\pi}{4}\right)\right]dt$

$\qquad = \dfrac{4}{3\pi}\left[1000t - 100\cos 2\left(t - \dfrac{\pi}{4}\right)\right]_0^{3\pi/4}$

$\qquad = \dfrac{4}{3\pi}\left[750\pi + 100 - (0-0)\right] = 1000 + \dfrac{400}{3\pi}$

74. $\dfrac{1}{0-(-\pi)}\displaystyle\int_{-\pi}^{0}(\cos t + \sin t)dt$

$\qquad = \dfrac{1}{\pi}\left[\sin t - \cos t\right]_{-\pi}^{0}$

$\qquad = \dfrac{1}{\pi}\left[0 - 1 - (0 - (-1))\right] = -\dfrac{2}{\pi}$

75. Substitute $\tan^2 x = \sec^2 x - 1$.

$\displaystyle\int \tan^2 x\, dx = \int(\sec^2 x - 1)dx = \tan x - x + C$

76. Substitute $\tan^2 3x = \sec^2 3x - 1$.

$\displaystyle\int \tan^2 3x\, dx = \int(\sec^2 3x - 1)dx$

$\qquad = \dfrac{1}{3}\tan 3x - x + C$

77. Substitute $1 + \tan^2 x = \sec^2 x$.

$\displaystyle\int(1 + \tan^2 x)dx = \int \sec^2 x\, dx = \tan x + C$

78. Substitute $1 + \tan^2 x = \sec^2 x$.

$\displaystyle\int(2 + \tan^2 x)dx = \int(1 + 1 + \tan^2 x)dx$

$\qquad = \displaystyle\int(1 + \sec^2 x)dx$

$\qquad = x + \tan x + C$

79. Substitute $\tan^2 x = \sec^2 x - 1$.

$\displaystyle\int_0^{\pi/4}\tan^2 x\, dx = \int_0^{\pi/4}(\sec^2 x - 1)dx$

$\qquad = (\tan x - x)\big|_0^{\pi/4}$

$\qquad = 1 - \dfrac{\pi}{4} - (0-0) = 1 - \dfrac{\pi}{4}$

80. Substitute $\tan^2 x = \sec^2 x - 1$.

$\displaystyle\int_0^{\pi/4}(2 + 2\tan^2 x)dx$

$\qquad = \displaystyle\int_0^{\pi/4}\left[2 + 2(\sec^2 x - 1)\right]dx$

$\qquad = \displaystyle\int_0^{\pi/4} 2\sec^2 x\, dx = 2\tan x\Big|_0^{\pi/4} = 2 - 0 = 2$

Chapter 9 Techniques of Integration

9.1 Integration by Substitution

1. $\int 2x(x^2+4)^5 dx$

Let $u = x^2 + 4$; then $du = 2x\,dx$.

$$\int 2x(x^2+4)^5 dx = \int u^5 \, du = \frac{1}{6}u^6 + C$$
$$= \frac{1}{6}(x^2+4)^6 + C$$

3. $\int \frac{2x+1}{\sqrt{x^2+x+3}}\,dx$

Let $u = x^2 + x + 3$; then $du = (2x+1)\,dx$.

$$\int \frac{2x+1}{\sqrt{x^2+x+3}}\,dx = \int \frac{1}{\sqrt{u}}\,du = 2u^{1/2} + C$$
$$= 2\sqrt{x^2+x+3} + C$$

5. $\int 3x^2 e^{(x^3-1)} dx$

Let $u = x^3 - 1$; then $du = 3x^2 dx$.

$$\int 3x^2 e^{(x^3-1)} dx = \int e^u du = e^u + C = e^{x^3-1} + C$$

7. $\int x\sqrt{4-x^2}\,dx$

Let $u = 4 - x^2$; then

$$du = -2xdx \Rightarrow -\frac{1}{2}du = xdx.$$

$$\int x\sqrt{4-x^2}\,dx = -\frac{1}{2}\int u^{1/2} du = -\frac{1}{3}u^{3/2} + C$$
$$= -\frac{1}{3}\left(4-x^2\right)^{3/2} + C$$

9. $\int \frac{1}{\sqrt{2x+1}}\,dx$

Let $u = 2x+1$; then $du = 2dx \Rightarrow \frac{1}{2}du = dx$.

$$\int \frac{1}{\sqrt{2x+1}}\,dx = \frac{1}{2}\int \frac{du}{\sqrt{u}} = \frac{1}{2}\int u^{-1/2} du$$
$$= \left(\frac{1}{2}\right)(2u^{1/2}) + C = u^{1/2} + C$$
$$= \sqrt{2x+1} + C$$

11. $\int xe^{x^2} dx$

Let $u = x^2$; then $du = 2xdx \Rightarrow \frac{1}{2}du = xdx$.

$$\int xe^{x^2} dx = \frac{1}{2}\int e^u du = \frac{e^u}{2} + C = \frac{1}{2}e^{x^2} + C$$

13. $\int \frac{\ln(2x)}{x}\,dx$

Let $u = \ln(2x)$; then $du = \frac{1}{2x}(2)\,dx = \frac{1}{x}\,dx$.

$$\int \frac{\ln(2x)}{x}\,dx = \int u\,du = \frac{u^2}{2} + C = \frac{(\ln(2x))^2}{2} + C$$

15. $\int \frac{x^4}{x^5+1}\,dx$

Let $u = x^5 + 1$; then

$$du = 5x^4 dx \Rightarrow \frac{1}{5}du = x^4 dx$$

$$\int \frac{x^4}{x^5+1}\,dx = \frac{1}{5}\int \frac{du}{u} = \frac{1}{5}\ln|u| + C$$
$$= \frac{1}{5}\ln|x^5+1| + C$$

17. $\int \frac{x-3}{(1-6x+x^2)^2}\,dx$

Let $u = 1 - 6x + x^2$; then

$$du = (-6+2x)dx \Rightarrow \frac{1}{2}du = (x-3)dx.$$

$$\int \frac{x-3}{(1-6x+x^2)^2}\,dx = \frac{1}{2}\int \frac{1}{u^2}\,du$$
$$= \left(\frac{1}{2}\right)\left(-\frac{1}{u}\right) + C$$
$$= -\frac{1}{2(1-6x+x^2)} + C$$
$$= -\frac{1}{2-12x+2x^2} + C$$

19. $\int \frac{\ln\sqrt{x}}{x}\,dx$

Let $u = \ln x$; then $du = \frac{1}{x}\,dx$.

$$\int \frac{\ln\sqrt{x}}{x}\,dx = \int \frac{\ln(x^{1/2})}{x}\,dx = \int \frac{\left(\frac{1}{2}\right)\ln x}{x}\,dx$$
$$= \frac{1}{2}\int \frac{\ln x}{x}\,dx = \frac{1}{2}\int u\,du$$
$$= \left(\frac{1}{2}\right)\frac{1}{2}u^2 + C = \frac{1}{4}u^2 + C$$
$$= \frac{1}{4}(\ln x)^2 + C = (\ln\sqrt{x})^2 + C$$

21. $\int \dfrac{x^2 - 2x}{x^3 - 3x^2 + 1}\,dx$

Let $u = x^3 - 3x^2 + 1$; then $du = (3x^2 - 6x)\,dx$

$\Rightarrow \dfrac{1}{3}du = (x^2 - 2x)dx$

$\int \dfrac{x^2 - 2x}{x^3 - 3x^2 + 1}\,dx = \dfrac{1}{3}\int \dfrac{du}{u}$

$\qquad = \dfrac{1}{3}\ln|u| + C$

$\qquad = \dfrac{1}{3}\ln\left|x^3 - 3x^2 + 1\right| + C$

23. $\int \dfrac{8x}{e^{x^2}}\,dx$

Let $u = x^2$; then $du = 2x\,dx \Rightarrow 4du = 8xdx$

$\int \dfrac{8x}{e^{x^2}}\,dx = 4\int e^{-u}\,du = -4e^{-u} + C$

$\qquad = -4e^{-x^2} + C$

25. $\int \dfrac{1}{x\ln x^2}\,dx$

Let $u = \ln x^2 = 2\ln x$; then

$du = \dfrac{2}{x}dx \Rightarrow \dfrac{1}{2}du = \dfrac{1}{x}dx$

$\int \dfrac{1}{x\ln x^2}\,dx = \dfrac{1}{2}\int \dfrac{1}{u}\,du = \dfrac{1}{2}\ln|u| + C$

$\qquad = \dfrac{1}{2}\ln\left|\ln x^2\right| + C$

27. $\int (3-x)(x^2 - 6x)^4\,dx$

Let $u = x^2 - 6x$; then $du = (2x - 6)dx \Rightarrow$

$-\dfrac{1}{2}du = (3 - x)dx$

$\int (3-x)(x^2 - 6x)^4\,dx = -\dfrac{1}{2}\int u^4\,du = -\dfrac{1}{10}u^5 + C$

$\qquad = -\dfrac{1}{10}(x^2 - 6x)^5 + C$

29. $\int e^x(1 + e^x)dx$

Let $u = 1 + e^x$; then $du = e^x\,dx$.

$\int e^x(1 + e^x)dx = \int u^5\,du = \dfrac{1}{6}u^6 + C$

$\qquad = \dfrac{1}{6}(1 + e^x)^6 + C$

31. $\int \dfrac{e^x}{1 + 2e^x}\,dx$

Let $u = 1 + 2e^x$; then $du = 2e^x dx \Rightarrow \dfrac{1}{2}du = e^x$

$\int \dfrac{e^x}{1 + 2e^x}\,dx = \dfrac{1}{2}\int \dfrac{1}{u}\,du = \dfrac{1}{2}\ln|u| + C$

$\qquad = \dfrac{1}{2}\ln\left|1 + 2e^x\right| + C$

33. $\int \dfrac{e^{-x}}{1 - e^{-x}}\,dx$

Let $u = 1 - e^{-x}$; then $du = e^{-x}dx$

$\int \dfrac{e^{-x}}{1 - e^{-x}}\,dx = \int \dfrac{1}{u}\,du = \ln|u| + C$

$\qquad = \ln\left|1 - e^{-x}\right| + C$

35. $\int \dfrac{1}{1 + e^x}\,dx$. Using hint,

$\int \dfrac{1}{1 + e^x}\,dx = \int \dfrac{e^{-x}}{e^{-x} + 1}\,dx$ Let $u = e^{-x} + 1$; then

$du = -e^{-x}dx \Rightarrow -du = e^{-x}dx$

$\int \dfrac{1}{1 + e^x}\,dx = -\int \dfrac{1}{u}\,du = -\ln|u| + C$

$\qquad = -\ln\left(e^{-x} + 1\right) + C$

37. $f'(x) = \dfrac{x}{\sqrt{x^2 + 9}}$, so $f(x) = \int \dfrac{x}{\sqrt{x^2 + 9}}\,dx$

Let $u = x^2 + 9$; then $du = 2xdx \Rightarrow \dfrac{1}{2}du = xdx$.

$\int \dfrac{x}{\sqrt{x^2 + 9}}\,dx = \dfrac{1}{2}\int u^{-1/2}du = \left(\dfrac{1}{2}\right)2u^{1/2} + C$

$\qquad = u^{1/2} + C = (x^2 + 9)^{1/2} + C$

But $f(4) = 8$, so $8 = (4^2 + 9)^{1/2} + C$, hence

$C = 3$, and $f(x) = (x^2 + 9)^{1/2} + 3$.

39. $\int (x + 5)^{-1/2}e^{\sqrt{x+5}}\,dx$

Let $u = \sqrt{x + 5}$; then $du = \dfrac{1}{2}(x + 5)^{-1/2}dx \Rightarrow$

$2du = (x + 5)^{-1/2}dx$

$\int (x + 5)^{-1/2}e^{(x+5)^{1/2}}\,dx = 2\int e^u\,du = 2e^u + C$

$\qquad = 2e^{(x+5)^{1/2}} + C$

41. $\int x\sec^2(x^2)\ dx$

Let $u = x^2$; then $du = 2x\ dx \Rightarrow \dfrac{1}{2}du = xdx$

$\int x\sec^2(x^2)\ dx = \dfrac{1}{2}\int\sec^2 u\ du = \dfrac{1}{2}\tan u + C$

$\qquad = \dfrac{1}{2}\tan x^2 + C$

43. $\int\sin x\cos x\ dx$

Let $u = \sin x$; then $du = \cos x\ dx$.

$\int\sin x\cos x\ dx = \int u\ du = \dfrac{u^2}{2} + C = \dfrac{\sin^2 x}{2} + C$

45. $\int\dfrac{\cos\sqrt{x}}{\sqrt{x}}\ dx$

Let $u = \sqrt{x} \Rightarrow du = \dfrac{1}{2}x^{-1/2}dx = \dfrac{1}{2\sqrt{x}}\ dx$

$\int\dfrac{\cos\sqrt{x}}{\sqrt{x}}\ dx = 2\int\cos u\ du = 2\sin u + C$

$\qquad = 2\sin\sqrt{x} + C$

47. $\int\cos^3 x\sin x\ dx$

Let $u = \cos x$; then $du = -\sin x\ dx \Rightarrow$
$-du = \sin xdx$

$\int\cos^3 x\sin x\ dx = -\int u^3 du = -\dfrac{1}{4}u^4 + C$

$\qquad = -\dfrac{1}{4}\cos^4 x + C$

49. $\int\dfrac{\cos 3x}{\sqrt{2-\sin 3x}}\ dx$

Let $u = 2 - \sin 3x$; then $du = -3\cos 3x\ dx \Rightarrow$
$-\dfrac{1}{3}du = \cos 3xdx$

$\int\dfrac{\cos 3x}{\sqrt{2-\sin 3x}}\ dx = -\dfrac{1}{3}\int\dfrac{1}{\sqrt{u}}\ du$

$\qquad = \left(-\dfrac{1}{3}\right)(2u^{1/2}) + C$

$\qquad = -\dfrac{2}{3}\sqrt{2-\sin 3x} + C$

51. $\int\dfrac{\sin x+\cos x}{\sin x-\cos x}dx$

Let $u = \sin x - \cos x$; then
$du = (\cos x + \sin x)dx$

$\int\dfrac{\sin x+\cos x}{\sin x-\cos x}dx = \int\dfrac{1}{u}\ du = \ln|u| + C$

$\qquad = \ln|\sin x - \cos x| + C$

53. $\int 2x(x^2 + 5)\ dx$

Let $u = x^2 + 5$; then $du = 2x\ dx$.

$\int 2x(x^2 + 5)\ dx = \int u\ du = \dfrac{1}{2}u^2 + C$

$\qquad = \dfrac{1}{2}(x^2 + 5)^2 + C$

$\qquad = \dfrac{1}{2}x^4 + 5x^2 + \dfrac{25}{2} + C$

$\int 2x(x^2 + 5)\ dx = \int(2x^3 + 10x)\ dx$

$\qquad = \dfrac{1}{2}x^4 + 5x^2 + C_1.$

The two methods differ by a constant.

9.2 Integration by Parts

1. $\int xe^{5x}dx$

Let $f(x) = x,\ g(x) = e^{5x}$.

Then $f'(x) = 1$ and $G(x) = \dfrac{1}{5}e^{5x}$.

$\int xe^{5x}dx = x\left(\dfrac{1}{5}e^{5x}\right) - \int(1)\dfrac{1}{5}e^{5x}$

$\qquad = \dfrac{1}{5}xe^{5x} - \dfrac{1}{25}e^{5x} + C$

3. $\int x(x+7)^4dx$

Let $f(x) = x,\ g(x) = (x+7)^4$.

Then $f'(x) = 1$ and $G(x) = \dfrac{1}{5}(x+7)^5$.

$\int x(x+7)^4dx$

$\qquad = x\left(\dfrac{1}{5}(x+7)^5\right) - \int(1)\dfrac{1}{5}(x+7)^5dx$

$\qquad = \dfrac{1}{5}x(x+7)^5 - \dfrac{1}{30}(x+7)^6 + C$

5. $\int\dfrac{x}{e^x}dx = \int xe^{-x}dx$

Let $f(x) = x,\ g(x) = e^{-x}$.

Then $f'(x) = 1$ and $G(x) = -e^{-x}$.

$\int xe^{-x}dx = x(-e^{-x}) - \int(1)(-e^{-x})\ dx$

$\qquad = -xe^{-x} - e^{-x} + C$

7. $\int \dfrac{x}{\sqrt{x+1}}\,dx$

Let $f(x) = x$, $g(x) = (x+1)^{-1/2}$.

Then $f'(x) = 1$ and $G(x) = 2(x+1)^{1/2}$.

$\int \dfrac{x}{\sqrt{x+1}}\,dx = x\left[2(x+1)^{1/2}\right] - \int (1)2(x+1)^{1/2}dx$

$\qquad = 2x(x+1)^{1/2} - \dfrac{4}{3}(x+1)^{3/2} + C$

9. $\int e^{2x}(1-3x)\,dx$

Let $f(x) = 1-3x$, $g(x) = e^{2x}$.

Then $f'(x) = -3$ and $G(x) = \dfrac{1}{2}e^{2x}$.

$\int e^{2x}(1-3x)\,dx$

$\qquad = (1-3x)\left(\dfrac{1}{2}e^{2x}\right) - \int (-3)\left(\dfrac{1}{2}e^{2x}\right)dx$

$\qquad = (1-3x)\left(\dfrac{1}{2}e^{2x}\right) + \left(\dfrac{3}{2}\right)\int e^{2x}dx$

$\qquad = (1-3x)\left(\dfrac{1}{2}e^{2x}\right) + \dfrac{3}{4}e^{2x} + C$

11. $\int \dfrac{6x}{e^{3x}}\,dx$

Let $f(x) = 6x$, $g(x) = e^{-3x}$.

Then $f'(x) = 6$ and $G(x) = -\dfrac{1}{3}e^{-3x}$.

$\int \dfrac{6x}{e^{3x}}\,dx = 6x\left(-\dfrac{1}{3}e^{-3x}\right) - \int 6\left(-\dfrac{1}{3}e^{-3x}\right)dx$

$\qquad = -2xe^{-3x} + (2)\int e^{-3x}dx$

$\qquad = -2xe^{-3x} - \dfrac{2}{3}e^{-3x} + C$

13. $\int x\sqrt{x+1}\,dx$

Let $f(x) = x$, $g(x) = (x+1)^{1/2}$.

Then $f'(x) = 1$ and $G(x) = \dfrac{2}{3}(x+1)^{3/2}$.

$\int x\sqrt{x+1}\,dx$

$\qquad = x\left[\dfrac{2}{3}(x+1)^{3/2}\right] - \int (1)\left(\dfrac{2}{3}(x+1)^{3/2}\right)dx$

$\qquad = \dfrac{2}{3}x(x+1)^{3/2} - \dfrac{4}{15}(x+1)^{5/2} + C$

15. $\int \sqrt{x}\,\ln\sqrt{x}\,dx$

Let $f(x) = \ln\sqrt{x}$, $g(x) = \sqrt{x}$.

Then $f'(x) = \dfrac{1}{\sqrt{x}}\left(\dfrac{1}{2}x^{-1/2}\right) = \dfrac{1}{2x}$ and

$G(x) = \dfrac{2}{3}x^{3/2}$.

$\int \sqrt{x}\,\ln\sqrt{x}\,dx$

$\qquad = \ln\sqrt{x}\left[\dfrac{2}{3}x^{3/2}\right] - \int \left(\dfrac{1}{2x}\right)\dfrac{2}{3}x^{3/2}dx$

$\qquad = \dfrac{2}{3}x^{3/2}\ln\sqrt{x} - \int \dfrac{1}{3}x^{1/2}dx$

$\qquad = \dfrac{2}{3}x^{3/2}\ln\sqrt{x} - \dfrac{2}{9}x^{3/2} + C$

$\qquad = \dfrac{1}{3}x^{3/2}\ln x - \dfrac{2}{9}x^{3/2} + C$

17. $\int x\cos x\,dx$

Let $f(x) = x$, $g(x) = \cos x$.

Then $f'(x) = 1$ and $G(x) = \sin x$.

$\int x\cos x\,dx = x\sin x - \int (1)(\sin x)\,dx$

$\qquad = x\sin x + \cos x + C$

19. $\int x\ln 5x\,dx$

Let $f(x) = \ln 5x$, $g(x) = x$.

Then $f'(x) = \dfrac{1}{5x}(5) = \dfrac{1}{x}$ and $G(x) = \dfrac{x^2}{2}$.

$\int x\ln 5x\,dx = (\ln 5x)\left(\dfrac{x^2}{2}\right) - \int \dfrac{1}{x}\left(\dfrac{x^2}{2}\right)dx$

$\qquad = \dfrac{1}{2}x^2\ln 5x - \dfrac{1}{4}x^2 + C$

21. $\int \ln x^4\,dx$

Let $f(x) = \ln x^4$, $g(x) = 1$.

Then $f'(x) = \dfrac{1}{x^4}(4x^3) = \dfrac{4}{x}$ and $G(x) = x$.

$\int \ln x^4\,dx = (\ln x^4)(x) - \int \dfrac{4}{x}(x)\,dx$

$\qquad = x\ln x^4 - 4x + C = 4x\ln x - 4x + C$

23. $\int x^2 e^{-x} dx$

Let $f(x) = x^2, g(x) = e^{-x}$.

Then $f'(x) = 2x$ and $G(x) = -e^{-x}$.

$\int x^2 e^{-x} dx = x^2(-e^{-x}) - \int 2x(-e^{-x}) \, dx$

$\qquad = -x^2 e^{-x} + \int 2x e^{-x} dx$

To evaluate $\int 2x e^{-x} dx$, use parts again:

Let $f(x) = 2x, \; g(x) = e^{-x}$.

Then $f'(x) = 2$ and $G(x) = -e^{-x}$.

$\int 2x e^{-x} dx = 2x(-e^{-x}) - \int 2(-e^{-x}) \, dx$

$\qquad = -2x e^{-x} - 2e^{-x} + C$

Therefore,

$\int x^2 e^{-x} dx = -x^2 e^{-x} - 2x e^{-x} - 2e^{-x} + C$

$\qquad = -e^{-x}\left(x^2 + 2x + 2\right) + C$

25. $\int x(x+5)^4 dx$

Use integration by parts: Let $f(x) = x$,

$g(x) = (x+5)^4$. Then $f'(x) = 1$ and

$G(x) = \dfrac{1}{5}(x+5)^5$.

$\int x(x+5)^4 dx = x\left[\dfrac{1}{5}(x+5)^5\right] - \int (1)\dfrac{1}{5}(x+5)^5 dx$

$\qquad = \dfrac{1}{5}x(x+5)^5 - \dfrac{1}{30}(x+5)^6 + C$

27. $\int x(x^2+5)^4 dx$

Use the substitution, $u = x^2 + 5$; then $du = 2x$

$dx \Rightarrow \dfrac{1}{2} du = x dx$

$\int x(x^2+5)^4 dx = \dfrac{1}{2}\int u^4 du = \left(\dfrac{1}{2}\right)\dfrac{1}{5}u^5 + C$

$\qquad = \dfrac{1}{10}u^5 + C = \dfrac{1}{10}(x^2+5)^5 + C$

29. $\int (3x+1)e^{x/3} dx$

Use integration by parts with $f(x) = 3x + 1$,

$g(x) = e^{x/3}$. Then $f'(x) = 3$ and $G(x) = 3e^{x/3}$.

$\int (3x+1)e^{x/3} dx$

$\quad = (3x+1)(3e^{x/3}) - \int (3)3e^{x/3} dx$

$\quad = (9x+3)e^{x/3} - 27e^{x/3} + C$

31. $\int x \sec^2(x^2+1) \, dx$

Use the substitution, $u = x^2 + 1$; then $du = 2x$

$dx \Rightarrow \dfrac{1}{2} du = x dx$

$\int x \sec^2(x^2+1) \, dx = \dfrac{1}{2}\int \sec^2 u \, du$

$\qquad = \dfrac{1}{2}\tan u + C$

$\qquad = \dfrac{1}{2}\tan\left(x^2+1\right) + C$

33. $\int (xe^{2x} + x^2) \, dx = \int xe^{2x} dx + \int x^2 dx$

$\int x^2 dx = \dfrac{1}{3}x^3 + C_1$

To evaluate $\int xe^{2x} dx$, use integration by parts

with $f(x) = x, \; g(x) = e^{2x}$.

Then $f'(x) = 1$ and $G(x) = \dfrac{1}{2}e^{2x}$.

$\int xe^{2x} dx = x\left(\dfrac{1}{2}e^{2x}\right) - \int \dfrac{1}{2}e^{2x} dx$

$\qquad = \dfrac{1}{2}xe^{2x} - \dfrac{1}{4}e^{2x} + C_2$

Therefore,

$\int (xe^{2x} + x^2) \, dx = \dfrac{1}{2}xe^{2x} - \dfrac{1}{4}e^{2x} + \dfrac{1}{3}x^3 + C$

35. $\int (xe^{x^2} - 2x) \, dx = \int xe^{x^2} dx - \int 2x \, dx$

$\int 2x \, dx = x^2 + C_1$

To evaluate $\int xe^{x^2} dx$, use the substitution,

$u = x^2; \, du = 2x \, dx \Rightarrow \dfrac{1}{2} du = x dx$

$\int xe^{x^2} dx = \dfrac{1}{2}\int e^{x^2} 2x \, dx = \dfrac{1}{2}\int e^u du$

$\qquad = \dfrac{1}{2}e^u + C_2 = \dfrac{1}{2}e^{x^2} + C_2$

Therefore, $\int (xe^{x^2} - 2x) \, dx = \dfrac{1}{2}e^{x^2} - x^2 + C$.

37. The slope is $\dfrac{x}{\sqrt{x+9}}$, so $f'(x) = \dfrac{x}{\sqrt{x+9}}$ and

thus $f(x) = \displaystyle\int \dfrac{x}{\sqrt{x+9}}\,dx$. Let $f(x) = x$ and

$g(x) = (x+9)^{-1/2}$.

Then $f'(x) = 1$ and $G(x) = 2(x+9)^{1/2}$.

$$\int \dfrac{x}{\sqrt{x+9}}\,dx = 2x(x+9)^{1/2} - \int 2(x+9)^{1/2}\,dx$$

$$= 2x(x+9)^{1/2} - \dfrac{4}{3}(x+9)^{3/2} + C \Rightarrow$$

$$f(x) = 2x(x+9)^{1/2} - \dfrac{4}{3}(x+9)^{3/2} + C$$

$$f(0) = 2 \Rightarrow$$

$$2(0)(0+9)^{1/2} - \dfrac{4}{3}(0+9)^{3/2} + C = 2 \Rightarrow$$

$$-36 + C = 2 \Rightarrow C = 38$$

Therefore,

$$f(x) = 2x(x+9)^{1/2} - \dfrac{4}{3}(x+9)^{3/2} + 38$$

39. $\displaystyle\int \dfrac{xe^x}{(x+1)^2}\,dx$

Use integration by parts with $f(x) = xe^x$ and

$g(x) = \dfrac{1}{(x+1)^2}$.

Then $f'(x) = e^x(x+1)$ and $G(x) = -\dfrac{1}{x+1}$.

$$\int \dfrac{xe^x}{(x+1)^2}\,dx$$

$$= xe^x\left(-\dfrac{1}{x+1}\right) - \int \left(-\dfrac{1}{x+1}\right)e^x(x+1)\,dx$$

$$= -\dfrac{xe^x}{x+1} - \int\left(-e^x\right)dx = -\dfrac{xe^x}{x+1} + e^x + C$$

$$= \dfrac{e^x}{x+1} + C$$

9.3 Evaluation of Definite Integrals

1. $\displaystyle\int_{5/2}^{3} 2(2x-5)^{14}\,dx$

Let $u = 2x - 5$; $du = 2\,dx$. When $x = \dfrac{5}{2}$,

$u = 2\left(\dfrac{5}{2}\right) - 5 = 0$.

When $x = 3$, $u = 2(3) - 5 = 1$.

$$\int_{5/2}^{3} 2(2x-5)^{14}\,dx = \int_0^1 u^{14}\,du = \dfrac{1}{15}u^{15}\Big|_0^1$$

$$= \dfrac{1}{15}(1)^{15} - 0 = \dfrac{1}{15}$$

3. $\displaystyle\int_0^2 4x(1+x^2)^3\,dx$

Let $u = 1 + x^2$; $du = 2x\,dx \Rightarrow 2\,du = 4x\,dx$

When $x = 0$, $u = 1 + 0^2 = 1$. When $x = 2$,

$u = 1 + 2^2 = 5$.

$$\int_0^2 4x\left(1+x^2\right)^3\,dx = 2\int_1^5 u^3\,du = \dfrac{1}{2}u^4\Big|_1^5$$

$$= \dfrac{625}{2} - \dfrac{1}{2} = 312$$

5. $\displaystyle\int_0^3 \dfrac{x}{\sqrt{x+1}}\,dx$

Use integration by parts with

$f(x) = x$, $g(x) = \dfrac{1}{\sqrt{x+1}}$. Then

$f'(x) = 1$, $G(x) = 2\sqrt{x+1}$

$$\int_0^3 \dfrac{x}{\sqrt{x+1}}\,dx = x\left(2\sqrt{x+1}\right)\Big|_0^3 - \int_0^3 2\sqrt{x+1}\,dx$$

$$= 12 - \dfrac{4}{3}(x+1)^{3/2}\Big|_0^3$$

$$= 12 - \left(\dfrac{32}{3} - \dfrac{4}{3}\right) = 12 - \dfrac{28}{3} = \dfrac{8}{3}$$

7. $\displaystyle\int_3^5 x\sqrt{x^2-9}\,dx$

Let $u = x^2 - 9$, $du = 2x\,dx \Rightarrow \dfrac{1}{2}du = x\,dx$

When $x = 3$, $u = 3^2 - 9 = 0$. When $x = 5$,

$u = 5^2 - 9 = 16$.

$$\int_2^5 x\sqrt{x^2-9}\,dx = \dfrac{1}{2}\int_2^5 \sqrt{x^2-9}\ 2x\,dx$$

$$= \dfrac{1}{2}\int_0^{16} u^{1/2}\,du = \left(\dfrac{1}{2}\right)\dfrac{2}{3}u^{3/2}\Big|_0^{16}$$

$$= \dfrac{1}{3}16^{3/2} - 0 = \dfrac{64}{3}$$

9. $\int_{-1}^{2}\left(x^2-1\right)\left(x^3-3x\right)^4 dx$

Let $u = x^3 - 3x$; $du = \left(3x^2 - 3\right)dx \Rightarrow$

$\frac{1}{3}du = \left(x^2 - 1\right)dx$

When $x = -1$, $u = (-1)^3 - 3(-1) = 2$.

When $x = 2$, $u = 2^3 - 3(2) = 2$.

$\int_{-1}^{2}\left(x^2-1\right)\left(x^3-3x\right)^4 dx = \frac{1}{3}\int_{2}^{2}u^4\,du$

$\qquad = \left(\frac{1}{15}u^5\right)\Big|_{2}^{2}$

$\qquad = \frac{32}{15} - \frac{32}{15} = 0$

11. $\int_{0}^{1}\frac{x}{x^2+3}dx$

Let $u = x^2 + 3$. Then $du = 2x\,dx \Rightarrow \frac{1}{2}du = x\,dx$.

When $x = 0$, $u = 3$. When $x = 1$, $u = 4$.

$\int_{0}^{1}\frac{x}{x^2+3}dx = \frac{1}{2}\int_{3}^{4}\frac{1}{u}du = \frac{1}{2}\ln u\Big|_{3}^{4}$

$\qquad = \frac{1}{2}\left(\ln 4 - \ln 3\right) = \frac{1}{2}\ln\frac{4}{3}$

13. $\int_{1}^{3}x^2 e^{x^3}dx$

Let $u = x^3$; $du = 3x^2 dx \Rightarrow \frac{1}{3}du = x^2 dx$.

When $x = 1$, $u = 1^3 = 1$. When $x = 3$,

$u = 3^3 = 27$.

$\int_{1}^{3}x^2 e^{x^3}dx = \frac{1}{3}\int_{1}^{3}e^{x^3}3x^2 dx$

$\qquad = \frac{1}{3}\int_{1}^{27}e^u du = \frac{1}{3}e^u\Big|_{1}^{27} = \frac{1}{3}e^{27} - \frac{1}{3}e$

15. $\int_{1}^{e}\frac{\ln x}{x}dx$

Let $u = \ln x$; $du = \frac{1}{x}dx$.

When $x = 1$, $u = \ln 1 = 0$. When $x = e$,

$u = \ln e = 1$.

$\int_{1}^{e}\frac{\ln x}{x}dx = \int_{0}^{1}u\,du = \frac{1}{2}u^2\Big|_{0}^{1} = \frac{1}{2}$

17. $\int_{0}^{\pi}e^{\sin x}\cos x\,dx$

Let $u = \sin x$; $du = \cos x\,dx$.

When $x = 0$, $u = \sin 0 = 0$. When $x = \pi$,

$u = \sin \pi = 0$.

$\int_{0}^{\pi}e^{\sin x}\cos x\,dx = \int_{0}^{0}e^u du = e^u\Big|_{0}^{0} = 1 - 1 = 0$

19. $\int_{0}^{1}x\sin \pi x\,dx$

Use integration by parts, with $f(x) = x$,

$g(x) = \sin \pi x$. Then $f'(x) = 1$ and

$G(x) = -\frac{1}{\pi}\cos \pi x$.

$\int_{0}^{1}x\sin \pi x\,dx$

$\qquad = x\left[-\frac{1}{\pi}\cos \pi x\right]\Big|_{0}^{1} - \int_{0}^{1}\left(-\frac{1}{\pi}\cos \pi x\right)dx$

$\qquad = \left[-\frac{1}{\pi}\cos \pi - 0\right] + \frac{1}{\pi^2}\sin \pi x\Big|_{0}^{1}$

$\qquad = \frac{1}{\pi} + \frac{1}{\pi^2}\sin \pi - \frac{1}{\pi^2}\sin 0 = \frac{1}{\pi}$

21. $\int_{-\pi/2}^{\pi/2}\sqrt{1-\sin^2 x}\cos x\,dx$

Let $u = \sin x$; $du = \cos x\,dx$.

When $x = -\frac{\pi}{2}$, $u = \sin\left(-\frac{\pi}{2}\right) = -1$.

When $x = \frac{\pi}{2}$, $u = \sin\left(\frac{\pi}{2}\right) = 1$.

$\int_{-\pi/2}^{\pi/2}\sqrt{1-\sin^2 x}\cos x\,dx = \int_{-1}^{1}\sqrt{1-u^2}\,du$

$=$ the area of the top half of the circle with

radius $1 = \frac{1}{2}\pi(1)^2 = \frac{\pi}{2}$

23. $\int_{-6}^{0}\sqrt{-x^2-6x}\,dx = \int_{-6}^{0}\sqrt{9-(x+3)^2}\,dx$

Let $u = (x+3)$; $du = dx$. When $x = -6$,

$u = -6 + 3 = -3$. When $x = 0$, $u = 0 + 3 = 3$.

$\int_{-6}^{0}\sqrt{9-(x+3)^2}\,dx = \int_{-3}^{3}\sqrt{9-u^2}\,du$

$=$ the area of the top half of the circle with

radius $3 = \frac{1}{2}\pi(3)^2 = \frac{9}{2}\pi$

25. $y = x\sqrt{4 - x^2}$

If $y = 0$, then $4 - x^2 = 0$ or $x = 0$. Thus,$\square x = -2, 0, 2$ are the intersection points with the x-axis. Area of the

portion from $x = 0$ to $x = 2$ is given by $\int_0^2 x\sqrt{4 - x^2}\,dx$. Let $u = 4 - x^2 \Rightarrow$

$du = -2x\,dx$.

$$\int x\sqrt{4 - x^2}\,dx = -\frac{1}{2}\int u^{1/2}\,du = -\frac{1}{3}u^{3/2} + C$$

$$= -\frac{1}{3}(4 - x^2)^{3/2} + C$$

$$\int_0^2 x\sqrt{4 - x^2}\,dx = -\frac{1}{3}(4 - x^2)^{3/2}\Big|_0^2 = \frac{8}{3}$$

By symmetry, area from $x = -2$ to $x = 0$ is also $\frac{8}{3}$, so the total area is $\frac{16}{3}$.

9.4 Approximation of Definite Integrals

1. $\Delta x = \dfrac{(5 - 3)}{5} = \dfrac{2}{5} = .4$

$a_0 = 3, \ a_1 = 3.4, \ a_2 = 3.8, \ a_3 = 4.2, \ a_4 = 4.6,$

$a_5 = 5$

3. $\Delta x = \dfrac{(1 - (-1))}{4} = .5$

$x_1 = -1 + \dfrac{.5}{2} = -.75, \ x_2 = -.25, \ x_3 = .25,$

$x_4 = .75$

5.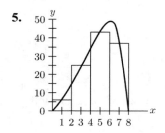

7. If $n = 2$, then $\Delta x = \dfrac{(4 - 0)}{2} = 2$. The first midpoint is $x_1 = 0 + \dfrac{2}{2} = 1$. The second midpoint is $x_2 = 1 + 2 = 3$.

Using the midpoint rule, we have $\int_0^4 (x^2 + 5)\,dx \approx [f(1) + f(3)] \cdot 2 = (6 + 14) \cdot 2 = 40$.

If $n = 4$, then $\Delta x = \dfrac{(4 - 0)}{4} = 1$, $x_1 = 0 + \dfrac{1}{2} = .5$, $x_2 = 1.5$, $x_3 = 2.5$, $x_4 = 3.5$. Hence,

$$\int_0^4 (x^2 + 5)\,dx \approx [f(x_1) + f(x_2) + f(x_3) + f(x_4)]\Delta x$$

$$= [[(.5)^2 + 5] + [(1.5)^2 + 5] + [(2.5)^2 + 5] + [(3.5)^2 + 5]](1) = 41$$

Evaluating the integral directly, we have

$$\int_0^4 (x^2 + 5)\,dx = \left(\frac{1}{3}x^3 + 5x\right)\Big|_0^4 = \frac{64}{3} + 20 = 41\frac{1}{3}.$$

9. $\int_0^1 e^{-x} dx$

Using the midpoint rule with $n = 5$, we have $\Delta x = \dfrac{(1-0)}{5} = .2$ and the midpoints are

$x_1 = 0 + \dfrac{.2}{2} = .1,\ x_2 = .3,\ x_3 = .5,\ x_4 = .7,\ x_5 = .9$.

$\int_0^1 e^{-x} dx \approx .2[e^{-0.1} + e^{-0.3} + e^{-0.5} + e^{-0.7} + e^{-0.9}] \approx 0 = .63107$

The exact value is $\int_0^1 e^{-x} dx = -e^{-x}\Big|_0^1 = -\dfrac{1}{e} + 1 \approx .63212$.

11. $\int_0^1 \left(x - \dfrac{1}{2}\right)^2 dx$

Using the trapezoidal rule with $n = 4$, we have $\Delta x = \dfrac{(1-0)}{4} = .25$ and the endpoints are

$a_0 = 0,\ a_1 = 0.25,\ a_2 = 0.5,\ a_3 = 0.75$ and $a_4 = 1$.

$\int_0^1 \left(x - \dfrac{1}{2}\right)^2 dx \approx \left[\left(0 - \dfrac{1}{2}\right)^2 + 2\left(.25 - \dfrac{1}{2}\right)^2 + 2\left(.5 - \dfrac{1}{2}\right)^2 + 2\left(.75 - \dfrac{1}{2}\right)^2 + \left(1 - \dfrac{1}{2}\right)^2\right]\dfrac{.25}{2} = .09375$

The exact value is $\int_0^1 \left(x - \dfrac{1}{2}\right)^2 dx = \dfrac{1}{3}\left(x - \dfrac{1}{2}\right)^3\Big|_0^1 = \dfrac{1}{3}[.5^3 - (-.5)^3] \approx .08333$.

13. $\int_1^5 \dfrac{1}{x^2} dx$

Using the trapezoidal rule with $n = 3$, we have $\Delta x = \dfrac{(5-1)}{3} \approx 1.33333$. The endpoints are

$a_0 = 1,\ a_1 = 2.33333,\ a_2 = 3.66667$, and $a_3 = 5$.

$\int_1^5 \dfrac{1}{x^2} dx \approx \left[\dfrac{1}{1^2} + (2)\dfrac{1}{2.33333^2} + (2)\dfrac{1}{3.66667^2} + \dfrac{1}{5^2}\right]\left(\dfrac{1.33333}{2}\right) \approx 1.03740$

The exact value is $\int_1^5 \dfrac{1}{x^2} dx = -\dfrac{1}{x}\Big|_1^5 = -\dfrac{1}{5} + 1 = .8$.

15. $\int_1^4 (2x - 3)^3 dx;\ n = 3$

$\Delta x = \dfrac{(4-1)}{3} = 1$

Midpoint rule: $x_1 = 1 + \dfrac{1}{2} = 1.5,\ x_2 = 2.5,\ x_3 = 3.5$

$\int_1^4 (2x-3)^3 dx \approx \left[(2(1.5) - 3)^3 + (2(2.5) - 3)^3 + (2(3.5) - 3)^3\right] = 72$

Trapezoidal rule: $a_0 = 1,\ a_1 = 2,\ a_2 = 3,\ a_3 = 4$

$\int_1^4 (2x-3)^3 dx \approx \left[(2(1) - 3)^3 + 2(2(2) - 3)^3 + 2(2(3) - 3)^3 + (2(4) - 3)^3\right]\dfrac{1}{2} = 90$

Simpson's rule: $\int_1^4 (2x-3)^3 dx \approx S = \dfrac{2M + T}{3} = \dfrac{2(72) + 90}{3} = 78$

Exact value: $\int_1^4 (2x-3)^3 dx = \dfrac{1}{8}(2x-3)^4\Big|_1^4 = \dfrac{1}{8}[(2(4) - 3)^4 - (2(1) - 3)^4] = 78$

17. $\int_0^2 2xe^{x^2} dx; \ n = 4$

$\Delta x = \dfrac{(2-0)}{4} = .5$

Midpoint rule: $x_1 = 0 + \dfrac{.5}{2} = .25, \ x_2 = .75, \ x_3 = 1.25, \ x_4 = 1.75$

$\int_0^2 2xe^{x^2} dx \approx 0.5[2(.25)e^{(0.25)^2} + 2(.75)e^{(0.75)^2} + 2(1.25)e^{(1.25)^2} + 2(1.75)e^{(1.75)^2}] \approx 44.96248$

Trapezoidal rule: $a_0 = 0, \ a_1 = .5, \ a_2 = 1, \ a_3 = 1.5, \ a_4 = 2$

$\int_0^2 2xe^{x^2} dx \approx \left[2(0)e^{0^2} + 2(2)(.5)e^{(0.5)^2} + 2(2)(1)e^{1^2} + 2(2)(1.5)e^{(1.5)^2} + 2(2)e^{2^2} \right]\dfrac{.5}{2} \approx 72.19005$

Simpson's rule: $\int_0^2 2xe^{x^2} dx \approx S = \dfrac{2M + T}{3} = \dfrac{2(44.96248) + 72.19005}{3} = 54.03834$

Exact value: $\int_0^2 2xe^{x^2} dx = e^{x^2} \Big|_0^2 = e^4 - 1 \approx 53.59815$

19. $\int_2^5 xe^x dx; \ n = 5$

$\Delta x = \dfrac{(5-2)}{5} = 0.6$

Midpoint rule: $x_1 = 2 + \dfrac{.6}{2} = 2.3, \ x_2 = 2.9, \ x_3 = 3.5, \ x_4 = 4.1, \ x_5 = 4.7$

$\int_2^5 xe^x dx \approx .6[(2.3)e^{2.3} + (2.9)e^{2.9} + (3.5)e^{3.5} + (4.1)e^{4.1} + (4.7)e^{4.7}] \approx 573.41797$

Trapezoidal rule: $a_0 = 2, \ a_1 = 2.6, \ a_2 = 3.2, \ a_3 = 3.8, \ a_4 = 4.4, \ a_5 = 5$

$\int_2^5 xe^x dx \approx [2e^2 + 2(2.6)e^{2.6} + 2(3.2)e^{3.2} + 2(3.8)e^{3.8} + 2(4.4)e^{4.4} + 5e^5]\dfrac{.6}{2} \approx 612.10806$

Simpson's rule: $\int_2^5 xe^x dx \approx \dfrac{2(573.41797) + 612.10806}{3} \approx 586.31466$

Exact value: $\int_2^5 xe^x dx$

Use integration by parts with $f(x) = x, \ g(x) = e^x$.

Then $f'(x) = 1$ and $G(x) = e^x$.

$\int_2^5 xe^x dx = xe^x \Big|_2^5 - \int_2^5 e^x dx = (xe^x - e^x) \Big|_2^5 = e^x (x-1) \Big|_2^5 = 4e^5 - e^2 \approx 586.26358$

21. $\int_0^2 \sqrt{1 + x^3} \, dx; \ n = 4$

$\Delta x = \dfrac{(2-0)}{4} = .5$

$a_0 = 0, \ a_1 = .5, \ a_2 = 1, \ a_3 = 1.5, \ a_4 = 2, \ x_1 = .25, \ x_2 = .75, \ x_3 = 1.25, \ x_4 = 1.75$

$\int_0^2 \sqrt{1 + x^3} \, dx$

$\approx \left[\sqrt{1 + 0^3} + 4\sqrt{1 + .25^3} + 2\sqrt{1 + .5^3} + 4\sqrt{1 + .75^3} + 2\sqrt{1 + 1^3} + 4\sqrt{1 + 1.25^3} \right.$

$\left. + 2\sqrt{1 + 1.5^3} + 4\sqrt{1 + 1.75^3} + \sqrt{1 + 2^3} \right]\dfrac{.5}{6}$

≈ 3.24124

23. $\int_0^2 \sqrt{\sin x}\, dx;\ n = 5$

$\Delta x = \dfrac{(2-0)}{5} = .4$

$a_0 = 0,\ a_1 = .4,\ a_2 = .8,\ a_3 = 1.2,\ a_4 = 1.6,\ a_5 = 2,\ x_1 = .2,\ x_2 = .6,\ x_3 = 1,\ x_4 = 1.4,\ x_5 = 1.8$

$\int_0^2 \sqrt{\sin x}\, dx$

$\approx \Big[\sqrt{\sin 0} + 4\sqrt{\sin(.2)} + 2\sqrt{\sin(.4)} + 4\sqrt{\sin(.6)} + 2\sqrt{\sin(.8)} + 4\sqrt{\sin 1} + 2\sqrt{\sin(1.2)} + 4\sqrt{\sin(1.4)}$

$\qquad\qquad\qquad\qquad + \ 2\sqrt{\sin(1.6)} + 4\sqrt{\sin(1.8)} + \sqrt{\sin(2)}\Big]\dfrac{.4}{6}$

≈ 1.61347

25. View the distance to the water as a function f of the position x of a point on the side. If a corresponds to the top of the diagram and b to the bottom, the area we wish to compute is $\int_a^b f(x)dx$. Because the measurements are taken 50 feet apart, $\Delta x = 50$. Thus the trapezoidal rule with $n = 4$ gives

$\int_a^b f(x)dx \approx [100 + 2(90) + 2(125) + 2(150) + 200]\dfrac{50}{2} = 25,750$ sq. ft .

27. (See hint.) using the trapezoidal rule with $\Delta t = 10,\ n = 10$,

$s(10) = \int_0^{10} v(t)dt \approx [0 + 2(30) + 2(75) + 2(115) + 2(155) + 2(200) + 2(250) + 2(300) + 2(360) + 2(420) + 490]\dfrac{1}{2}$

$= 2150$ ft

29. a.

$f(x) = \dfrac{1}{12}x^4 + 3x^2;\ \ f'(x) = \dfrac{1}{3}x^3 + 6x;\ \ f''(x) = x^2 + 6$

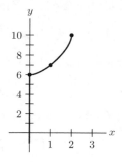

b. $f''(x) \le 10$ on $[0, 2]$, so $A = 10$

c. $a = 0,\ b = 2,\ n = 10,\ A = 10$

$\dfrac{A(b-a)^3}{24n^2} = \dfrac{10(2-0)^3}{24(10)^2} = \dfrac{8}{240} \approx .0333$

d. $8.5089 - 8.5333 = -.0244$, which satisfies the bound in part (c).

e. $a = 0,\ b = 2,\ n = 20,\ A = 10$

$\dfrac{A(b-a)^3}{24n^2} = \dfrac{10(2-0)^3}{24(20)^2} = \dfrac{10(8)}{24(400)} = \dfrac{1}{4}\left(\dfrac{8}{240}\right)$

The bound is quartered.

31. a. The area of the triangle on top is $\dfrac{1}{2}(k-h)l$ and the area of the rectangle on the bottom is lh. Thus the area of the trapezoid is $\dfrac{1}{2}(k-h)l + lh = \dfrac{1}{2}(h+k)l$.

b.

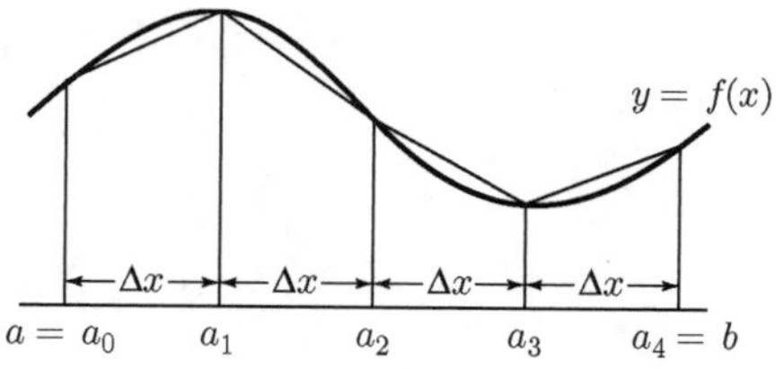

From part (a) with $f(a_0) = h$, $\Delta x = l$ and $f(a_1) = k$, we have area $= \dfrac{1}{2}[f(a_0) + f(a_1)]\Delta x$.

c. Applying the result in (a) as in part (b), we see that the sum of the areas of the 4 trapezoids in the figure is

$$\frac{1}{2}[f(a_0) + f(a_1)]\Delta x + \frac{1}{2}[f(a_1) + f(a_2)]\Delta x + \frac{1}{2}[f(a_2) + f(a_3)]\Delta x + \frac{1}{2}[f(a_3) + f(a_4)]\Delta x$$

$$= \left[f(a_0) + 2f(a_1) + 2f(a_2) + 2f(a_3) + f(a_4)\right]\frac{\Delta x}{2}$$

33. a.

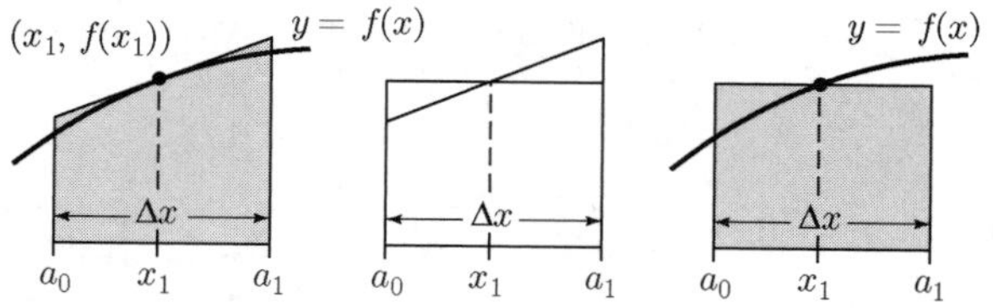

Let T_1 be the top triangle in Fig. 15(b) and let T_2 be the bottom one. Both are right triangles, their bases have the same length $\left(\dfrac{\Delta x}{2}\right)$, and the angle between the base and hypotenuse is the same in each.

Therefore the area of T_1, $A(T_1) = A(T_2)$, the area of T_2. The shaded area in Fig. 15(a) = area shaded in 15(c) + $A(T_1) - A(T_2)$.

b. $\displaystyle\int_a^b f(x)dx \le M$ follows from (a), using the fact that

$$\int_a^b f(x)dx = \int_a^{a_1} f(x)dx + \int_{a_1}^{a_2} f(x)dx \ldots + \int_{a_{n-1}}^b f(x)dx$$

Decomposing $\displaystyle\int_a^b f(x)dx$ as above, $T \le \displaystyle\int_a^b f(x)dx$ follows from the fact that if f is concave downward, then each of the trapezoids whose area is summed to obtain T is entirely *below* the graph of f. (See, for example, Fig. 13(b) in exercise 31, where the first two trapezoids are drawn over a region on which f is concave downward.) Thus, $T \le \displaystyle\int_a^b f(x)dx \le M$.

35. $f(x) = \dfrac{1}{25 - x^2}$,

$a = 3.1 - \dfrac{0.2}{2} = 3$, $b = 4.9 + \dfrac{0.2}{2} = 5$, $n = \dfrac{5-3}{0.2} = 10$

37. $\int_1^{11} \dfrac{1}{x}\,dx;\ \Delta x = \dfrac{(11-1)}{10} = 1$

Midpoint rule:

$x_1 = 1 + \dfrac{1}{2} = 1.5,\ x_2 = 2.5,\ x_3 = 3.5,\ x_4 = 4.5,\ x_5 = 5.5,\ x_6 = 6.5,\ x_7 = 7.5,\ x_8 = 8.5,\ x_9 = 9.5,\ x_{10} = 10.5$

$\int_1^{11} \dfrac{1}{x}\,dx \approx \left[\dfrac{1}{1.5} + \dfrac{1}{2.5} + \dfrac{1}{3.5} + \dfrac{1}{4.5} + \dfrac{1}{5.5} + \dfrac{1}{6.5} + \dfrac{1}{7.5} + \dfrac{1}{8.5} + \dfrac{1}{9.5} + \dfrac{1}{10.5}\right] \approx 2.361749156$

Trapezoidal rule: $a_0 = 1,\ a_1 = 2,\ a_2 = 3,\ a_3 = 4,\ a_4 = 5,\ a_5 = 6,\ a_6 = 7,\ a_7 = 8,\ a_8 = 9,\ a_9 = 10,\ a_{10} = 11$

$\int_1^{11} \dfrac{1}{x}\,dx \approx \left[\dfrac{1}{1} + (2)\dfrac{1}{2} + (2)\dfrac{1}{3} + (2)\dfrac{1}{4} + (2)\dfrac{1}{5} + (2)\dfrac{1}{6} + (2)\dfrac{1}{7} + (2)\dfrac{1}{8} + (2)\dfrac{1}{9} + (2)\dfrac{1}{10} + \dfrac{1}{11}\right]\dfrac{1}{2} \approx 2.474422799$

Simpson's rule:

$\int_1^{11} \dfrac{1}{x}\,dx \approx \Big[\dfrac{1}{1} + (4)\dfrac{1}{1.5} + (2)\dfrac{1}{2} + (4)\dfrac{1}{2.5} + (2)\dfrac{1}{3} + (4)\dfrac{1}{3.5} + (2)\dfrac{1}{4} + (4)\dfrac{1}{4.5} + (2)\dfrac{1}{5} + (4)\dfrac{1}{5.5} + (2)\dfrac{1}{6} + (4)\dfrac{1}{6.5}$

$\qquad + (2)\dfrac{1}{7} + (4)\dfrac{1}{7.5} + (2)\dfrac{1}{8} + (4)\dfrac{1}{8.5} + (2)\dfrac{1}{9} + (4)\dfrac{1}{9.5} + (2)\dfrac{1}{10} + (4)\dfrac{1}{10.5} + \dfrac{1}{11}\Big]\dfrac{1}{6}$

≈ 2.399307037

Exact value: $\int_1^{11} \dfrac{1}{x}\,dx = \ln|x|\Big|_1^{11} = \ln 11 = 2.397895273$

Error using midpoint rule = .036146117
Error using trapezoidal rule = .076527526
Error using Simpson's rule = .001411764

39. $\int_0^{\pi/4} \sec^2 x\,dx;\ \Delta x = \dfrac{\left(\frac{\pi}{4} - 0\right)}{10} = \dfrac{\pi}{40}$

Midpoint rule:

$x_1 = 0 + \dfrac{\left(\frac{\pi}{40}\right)}{2} = \dfrac{\pi}{80},\ x_2 = \dfrac{3\pi}{80},\ x_3 = \dfrac{5\pi}{80},\ x_4 = \dfrac{7\pi}{80},\ x_5 = \dfrac{9\pi}{80},\ x_6 = \dfrac{11\pi}{80},\ x_7 = \dfrac{13\pi}{80}\ x_8 = \dfrac{15\pi}{80},\ x_9 = \dfrac{17\pi}{80},$

$x_{10} = \dfrac{19\pi}{80}$

$\int_0^{\pi/4} \sec^2 x\,dx = \left[\sec^2\left(\dfrac{\pi}{80}\right) + \sec^2\left(\dfrac{3\pi}{80}\right) + \sec^2\left(\dfrac{5\pi}{80}\right) + \sec^2\left(\dfrac{7\pi}{80}\right) + \sec^2\left(\dfrac{9\pi}{80}\right) + \sec^2\left(\dfrac{11\pi}{80}\right)\right.$

$\qquad\left. + \sec^2\left(\dfrac{13\pi}{80}\right) + \sec^2\left(\dfrac{15\pi}{80}\right) + \sec^2\left(\dfrac{17\pi}{80}\right) + \sec^2\left(\dfrac{19\pi}{80}\right)\right]\dfrac{\pi}{40}$

≈ 0.9989755866

Trapezoidal rule: $a_0 = 0,\ a_1 = \dfrac{\pi}{40},\ a_2 = \dfrac{2\pi}{40},\ a_3 = \dfrac{3\pi}{40},\ a_4 = \dfrac{4\pi}{40},\ a_5 = \dfrac{5\pi}{40},\ a_6 = \dfrac{6\pi}{40},\ a_7 = \dfrac{7\pi}{40},\ a_8 = \dfrac{8\pi}{40},$

$a_9 = \dfrac{9\pi}{40},\ a_{10} = \dfrac{\pi}{4}$

$\int_0^{\pi/4} \sec^2 x\,dx \approx \left[\sec^2(0) + 2\sec^2\left(\dfrac{\pi}{40}\right) + 2\sec^2\left(\dfrac{2\pi}{40}\right) + 2\sec^2\left(\dfrac{3\pi}{40}\right) + 2\sec^2\left(\dfrac{4\pi}{40}\right) + 2\sec^2\left(\dfrac{5\pi}{40}\right)\right.$

$\qquad\left. + 2\sec^2\left(\dfrac{6\pi}{40}\right) + 2\sec^2\left(\dfrac{7\pi}{40}\right) + 2\sec^2\left(\dfrac{8\pi}{40}\right) + 2\sec^2\left(\dfrac{9\pi}{40}\right) + \sec^2\left(\dfrac{\pi}{4}\right)\right]\left(\dfrac{\pi}{40}\right)\left(\dfrac{1}{2}\right)$

≈ 1.00205197

(continued on next page)

(*continued*)

Simpson's rule:

$$\int_0^{\pi/4} \sec^2 x\, dx \approx \left[\sec^2(0) + 4\sec^2\left(\frac{\pi}{80}\right) + 2\sec^2\left(\frac{\pi}{40}\right) + 4\sec^2\left(\frac{3\pi}{80}\right) + 2\sec^2\left(\frac{2\pi}{40}\right) + 4\sec^2\left(\frac{5\pi}{80}\right) \right.$$

$$+ 2\sec^2\left(\frac{3\pi}{40}\right) + 4\sec^2\left(\frac{7\pi}{80}\right) + 2\sec^2\left(\frac{4\pi}{40}\right) + 4\sec^2\left(\frac{9\pi}{80}\right) + 2\sec^2\left(\frac{5\pi}{40}\right)$$

$$+ 4\sec^2\left(\frac{11\pi}{80}\right) + 2\sec^2\left(\frac{6\pi}{40}\right) + 4\sec^2\left(\frac{13\pi}{80}\right) + 2\sec^2\left(\frac{7\pi}{40}\right) + 4\sec^2\left(\frac{15\pi}{80}\right)$$

$$\left. + 2\sec^2\left(\frac{8\pi}{40}\right) + 4\sec^2\left(\frac{17\pi}{80}\right) + 2\sec^2\left(\frac{9\pi}{40}\right) + 4\sec^2\left(\frac{19\pi}{80}\right) + \sec^2\left(\frac{\pi}{4}\right) \right]\left(\frac{\pi}{40}\right)\left(\frac{1}{6}\right)$$

$$\approx 1.000001048$$

Exact value: $\int_0^{\pi/4} \sec^2 x\, dx = \tan x \Big|_0^{\pi/4} = \tan\left(\frac{\pi}{4}\right) - \tan(0) = 1$

Error using midpoint rule = .0010244134; error using trapezoidal rule = .00205197
Error using Simpson's rule = .000001048

41. $f(x) = \dfrac{4}{1+x^2} = 4(1+x^2)^{-1}$

$f'(x) = -4(1+x^2)^{-2}(2x) = -8x(1+x^2)^{-2}$

$f''(x) = -8x(-2)(1+x^2)^{-3}(2x) + (1+x^2)^{-2}(-8)$

$\qquad = \dfrac{32x^2}{(1+x^2)^3} - \dfrac{8}{(1+x^2)^2}$

Let $A = 8$, $a = 0$, $b = 1$, $n = 20$.

A bound on the error of the estimate is $\dfrac{A(b-a)^3}{24n^2} = \dfrac{8(1-0)^3}{24(20)^2} = .0008333$.

9.5   Some Applications of the Integral

1. Present value $= \displaystyle\int_{T_1}^{T_2} K(t)e^{-rt}\, dt$;

$K(t) = 35{,}000$, $T_1 = 0$, $T_2 = 5$, $r = .07$.
So

$\text{P.V.} = \displaystyle\int_0^5 35{,}000e^{-.07t}\, dt$

$\qquad = -\dfrac{100}{7}(35{,}000)e^{-.07t}\Big|_0^5$

$\qquad = -500{,}000\left[e^{-.07(5)} - e^0\right] \approx \$147{,}656$

3. $K(t) = 12{,}000$, $T_1 = 1$, $T_2 = 5$, $r = .1$
So

$\text{P.V.} = \displaystyle\int_1^5 12{,}000e^{-0.1t}\, dt = -10(12{,}000)e^{-0.1t}\Big|_1^5$

$\qquad \approx \$35{,}797$

5. $K(t) = 80e^{-0.08t}$, $T_1 = 0$, $T_2 = 3$, $r = 0.11$
The present value is

$\displaystyle\int_0^3 (80e^{-0.08t})e^{-0.11t}\, dt = \int_0^3 80e^{-0.19t}\, dt$

$\qquad = -\dfrac{100}{19}(80)e^{-0.19t}\Big|_0^3$

$\qquad = -\dfrac{8000}{19}\left[e^{-0.57} - 1\right]$

$\qquad \approx \$182{,}937$

7. a. $K(t) = 30 + 5t$, $T_1 = 0$, $T_2 = 2$, $r = 0.1$.

So $\text{P.V.} = \displaystyle\int_0^2 (30 + 5t)e^{-0.10t}\, dt$.

b. Use integration by parts with $f(t) = 30 + 5t$ and $g(t) = e^{-0.1t}$. Then $f'(t) = 5$ and $G(t) = -10e^{-0.1t}$.

$$\int_0^2 (30 + 5t)e^{-0.1t} dt$$

$$= (30 + 5t)(-10)e^{-0.1t}\Big|_0^2 - \int_0^2 5(-10)e^{-0.1t} dt$$

$$= \left(-400e^{-0.2} + 300\right) - 500e^{-0.1t}\Big|_0^2$$

$$= \left(-400e^{-0.2} + 300\right) - 500e^{-0.2} + 500$$

$$= -900e^{-0.2} + 800$$

$$\approx 63.1 \text{ million dollars}$$

9. a. $D(t) = 120e^{-0.65t}$

$$\text{Population} = \int_0^5 (2\pi t)120e^{-0.65t} dt$$

$$= 240\pi \int_0^5 te^{-0.65t} dt$$

b. Use integration by parts with $f(t) = t$, $g(t) = e^{-0.65t}$. Then $f'(t) = 1$ and $G(t) = \dfrac{-100}{65}e^{-0.65t} = -\dfrac{20}{13}e^{-0.65t}$.

$$240\pi \int_0^5 te^{-0.65t} dt$$

$$= 240\pi \left[t\left(-\frac{20}{13}\right)e^{-0.65t}\Big|_0^5 \right.$$

$$\left. - \int_0^5 \left(-\frac{20}{13}\right)e^{-0.65t} dt \right]$$

$$\approx 1490.493 \text{ thousand or } 1,490,493 \text{ people}$$

11. $D(t) = 60e^{-0.4t}$

$$\text{Population} = \int_0^5 (2\pi t)60e^{-0.4t} dt$$

$$= 120\pi \int_0^5 te^{-0.4t} dt$$

Use integration by parts with $f(t) = t$, $g(t) = e^{-0.4t}$. Then $f'(t) = 1$ and $G(t) = -\dfrac{10}{4}e^{-0.4t} = -\dfrac{5}{2}e^{-0.4t}$.

$$120\pi \int_0^5 te^{-0.4t} dt$$

$$= 120\pi \left[t\left(-\frac{5}{2}\right)e^{-0.4t}\Big|_0^5 - \int_0^5 \left(-\frac{5}{2}\right)e^{-0.4t} dt \right]$$

$$\approx 120\pi(3.712)$$

$$\approx 1399.391$$

About 1,400,000 people

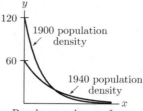

People moved away from the center of the city

The graphs show that people moved away from the center of the city.

13. a. The area of the ring is $2\pi t(\Delta t)$. The population density is $40e^{-0.5t}$. So the population is

$$2\pi t(\Delta t)40e^{-0.5t} = 80\pi t(\Delta t)e^{-0.5t}$$

thousand.

b. $\dfrac{dP}{dt}$ or $P'(t)$

c. It represents the number of people who live between $(5 + \Delta t)$ miles from the city center and 5 miles from the city center.

d. $P(t + \Delta t) - P(t) = 80\pi t(\Delta t)e^{-0.5t}$ from (a).

So, $\dfrac{P(t + \Delta t) - P(t)}{\Delta t} \approx P'(t) = 80\pi t e^{-0.5t}$

e. $\int_a^b P'(t) dt = P(b) - P(a) = \int_a^b 80\pi t e^{-0.5t} dt$

9.6 Improper Integrals

1. As $b \to \infty$, $\dfrac{5}{b}$ approaches 0.

3. As $b \to \infty$, $-3e^{2b}$ decreases without bound.

5. As $b \to \infty$, $\dfrac{1}{4} - \dfrac{1}{b^2}$ approaches $\dfrac{1}{4} - 0 = \dfrac{1}{4}$.

7. As $b \to \infty$, $2 - (b+1)^{-1/2}$ approaches $2 - 0 = 2$.

9. As $b \to \infty$, $5(b^2 + 3)^{-1}$ approaches 0.

11. As $b \to \infty$, $e^{-b/2} + 5$ approaches $0 + 5 = 5$.

13. The given area is $\int_2^\infty \frac{1}{x^2}\,dx$.

$$\int_2^b \frac{1}{x^2}\,dx = -\frac{1}{x}\Big|_2^b = -\frac{1}{b} + \frac{1}{2}$$

As $b \to \infty$, $-\frac{1}{b} + \frac{1}{2}$ approaches $\frac{1}{2}$.

Thus, $\int_2^\infty \frac{1}{x^2}\,dx = \lim_{b\to\infty}\int_2^b \frac{1}{x^2}\,dx = \frac{1}{2}$.

15. The given area is $\int_0^\infty e^{-x/2}\,dx$.

$$\int_0^b e^{-x/2}\,dx = -2e^{-x/2}\Big|_0^b = -2e^{-b/2} + 2$$

As $b \to \infty$, $-2e^{-b/2} + 2$ approaches 2.

Thus, $\int_0^\infty e^{-x/2}\,dx = \lim_{b\to\infty}\int_0^b e^{-x/2}\,dx = 2$.

17. The given area is $\int_3^\infty (x+1)^{-3/2}\,dx$.

$$\int_3^b (x+1)^{-3/2}\,dx = -2(x+1)^{-1/2}\Big|_3^b$$
$$= -2(b+1)^{-1/2} + 1$$

As $b \to \infty$, $-2(b+1)^{-1/2} + 1$ approaches 1. So,

$$\int_3^\infty (x+1)^{-3/2}\,dx = \lim_{b\to\infty}\int_3^b (x+1)^{-3/2}\,dx = 1.$$

19. $\int_1^b (14x+18)^{-4/5}\,dx = \frac{5}{14}(14x+18)^{1/5}\Big|_1^b$

$$= \frac{5}{14}(14b+18)^{1/5} - \frac{5}{7}$$

As $b \to \infty$, $\frac{5}{14}(14b+18)^{1/5} - \frac{5}{7}$ increases

without bound. Thus, $\int_1^\infty (14x+18)^{-4/5}\,dx$

diverges and the area under the curve

$y = (14x+18)^{-4/5}$ for $x \geq 1$ cannot be

assigned any finite number.

21. $\int_1^\infty \frac{1}{x^3}\,dx = \lim_{b\to\infty}\int_1^b \frac{1}{x^3}\,dx = \lim_{b\to\infty}\left[-\frac{1}{2}x^{-2}\Big|_1^b\right]$

$$= \lim_{b\to\infty}\left[-\frac{1}{2}b^{-2} + \frac{1}{2}\right] = \frac{1}{2}$$

23. $\int_0^\infty \frac{1}{(2x+3)^2}\,dx = \lim_{b\to\infty}\int_0^b \frac{1}{(2x+3)^2}\,dx$

$$= \lim_{b\to\infty}\left[-\frac{1}{2}(2x+3)^{-1}\Big|_0^b\right]$$

$$= \lim_{b\to\infty}\left[-\frac{1}{2}(2b+3)^{-1} + \frac{1}{6}\right] = \frac{1}{6}$$

25. $\int_0^\infty e^{2x}\,dx = \lim_{b\to\infty}\int_0^b e^{2x}\,dx = \lim_{b\to\infty}\left[\frac{1}{2}e^{2x}\Big|_0^b\right]$

$$= \lim_{b\to\infty}\left[\frac{1}{2}e^{2b} - \frac{1}{2}\right]$$

As $b \to \infty$, $\frac{1}{2}e^{2b} - \frac{1}{2}$ increases without

bound. Therefore, $\int_0^\infty e^{2x}\,dx$ diverges.

27. $\int_2^\infty \frac{1}{(x-1)^{5/2}}\,dx = \lim_{b\to\infty}\int_2^b \frac{1}{(x-1)^{5/2}}\,dx$

$$= \lim_{b\to\infty}\left[-\frac{2}{3}(x-1)^{-3/2}\Big|_2^b\right]$$

$$= \lim_{b\to\infty}\left[-\frac{2}{3}(b-1)^{-3/2} + \frac{2}{3}\right] = \frac{2}{3}$$

29. $\int_0^\infty 0.01e^{-0.01x}\,dx = \lim_{b\to\infty}\int_0^b 0.01e^{-0.01x}\,dx$

$$= \lim_{b\to\infty}\left[-e^{-0.01x}\Big|_0^b\right]$$

$$= \lim_{b\to\infty}\left[-e^{-0.01b} + 1\right] = 1$$

31. $\int_0^\infty 6e^{1-3x}\,dx = \lim_{b\to\infty}\int_0^b 6e^{1-3x}\,dx$

$$= \lim_{b\to\infty}\left[-2e^{1-3x}\Big|_0^b\right]$$

$$= \lim_{b\to\infty}\left[-2e^{1-3b} + 2e\right] = 2e$$

33. $\int_3^\infty \frac{x^2}{\sqrt{x^3-1}}\,dx = \lim_{b\to\infty}\int_3^b \frac{x^2}{\sqrt{x^3-1}}\,dx$

To evaluate $\int_3^b \frac{x^2}{\sqrt{x^3-1}}\,dx$, use the substitution

$u = x^3 - 1$, $du = 3x^2\,dx$. When $x = 3$,

$u = 3^3 - 1 = 26$; when $x = b$, $u = b^3 - 1$. Thus,

$$\int_3^b \frac{x^2}{\sqrt{x^3-1}}\,dx = \frac{1}{3}\int_{26}^{b^3-1} \frac{1}{\sqrt{u}}\,du = \left(\frac{1}{3}\right)2u^{1/2}\Big|_{26}^{b^3-1}$$

$$= \frac{2}{3}(b^3-1) - \frac{2}{3}\sqrt{26}$$

As $b \to \infty$, $\frac{2}{3}(b^3-1)^{1/2} - \frac{2}{3}\sqrt{26}$ increases

without bound. Therefore, $\int_3^\infty \frac{x^2}{\sqrt{x^3-1}}\,dx$

diverges.

35. $\int_0^\infty xe^{-x^2}\,dx = \lim_{b\to\infty}\int_0^b xe^{-x^2}\,dx$

To evaluate $\int_0^b xe^{-x^2}\,dx,$ use the substitution

$u = -x^2,\ du = -2x\,dx.$ When $x = 0,\ u = 0;$

when $x = b,\ u = -b^2.$

$\int_0^b xe^{-x^2}\,dx = -\frac{1}{2}\int_0^{-b^2} e^u\,du = -\frac{1}{2}e^u\Big|_0^{-b^2}$

$\qquad = -\frac{1}{2}e^{-b^2} + \frac{1}{2}$

Thus, $\int_0^\infty xe^{-x^2}\,dx = \lim_{b\to\infty}\left[-\frac{1}{2}e^{-b^2} + \frac{1}{2}\right] = \frac{1}{2}.$

37. $\int_0^\infty 2x(x^2+1)^{-3/2}\,dx = \lim_{b\to\infty}\int_0^b 2x(x^2+1)^{-3/2}\,dx$

To evaluate $\int_0^b 2x(x^2+1)^{-3/2}\,dx,$ use the

substitution $u = x^2+1,\ du = 2x\,dx.$ When $x = 0,$

$u = 1;$ when $x = b,\ u = b^2+1.$

$\int_0^b 2x(x^2+1)^{-3/2}\,dx = \int_1^{b^2+1} u^{-3/2}\,du$

$\qquad = -2u^{-1/2}\Big|_1^{b^2+1}$

$\qquad = -2(b^2+1)^{-1/2} + 2$

Thus,

$\int_0^\infty 2x(x^2+1)^{-3/2}\,dx = \lim_{b\to\infty}\left[-2(b^2+1)^{-1/2}+2\right]$

$\qquad = 2$

39. $\int_{-\infty}^0 e^{4x}\,dx = \lim_{b\to-\infty}\int_b^0 e^{4x}\,dx = \lim_{b\to-\infty}\left[\frac{1}{4}e^{4x}\Big|_b^0\right]$

$\qquad = \lim_{b\to-\infty}\left[\frac{1}{4} - \frac{1}{4}e^{4b}\right] = \frac{1}{4}$

41. $\int_{-\infty}^0 \frac{6}{(1-3x)^2}\,dx = \lim_{b\to-\infty}\int_b^0 \frac{6}{(1-3x)^2}\,dx$

$\qquad = \lim_{b\to-\infty}\left[\frac{2}{(1-3x)}\Big|_b^0\right]$

$\qquad = \lim_{b\to-\infty}\left[2 - \frac{2}{1-3b}\right] = 2$

43. $\int_0^\infty \frac{e^{-x}}{(e^{-x}+2)^2}\,dx = \lim_{b\to\infty}\int_0^b \frac{e^{-x}}{(e^{-x}+2)^2}\,dx$

To evaluate $\int_0^b \frac{e^{-x}}{(e^{-x}+2)^2}\,dx,$ use the

substitution $u = e^{-x}+2,\ du = -e^{-x}dx.$

When $x = 0,\ u = 3;$ when $x = b,\ u = e^{-b}+2.$

$\int_0^b \frac{e^{-x}}{(e^{-x}+2)^2}\,dx = -\int_3^{e^{-b}+2}\frac{du}{u^2} = u^{-1}\Big|_3^{e^{-b}+2}$

$\qquad = \frac{1}{e^{-b}+2} - \frac{1}{3}$

Thus,

$\int_0^b \frac{e^{-x}}{(e^{-x}+2)^2}\,dx = \lim_{b\to\infty}\left[\frac{1}{e^{-b}+2} - \frac{1}{3}\right].$

$\qquad = \frac{1}{2} - \frac{1}{3} = \frac{1}{6}$

45. $\int_0^\infty ke^{-kx}\,dx = \lim_{b\to\infty}\int_0^b ke^{-kx}\,dx$

$\qquad = \lim_{b\to\infty}\left[-e^{-kx}\Big|_0^b\right]$

$\qquad = \lim_{b\to\infty}\left[-e^{-kb}+1\right]$

If $k > 0,$ as $b\to\infty,\ -e^{-kb}$ approaches 0. Thus,

in this case $\int_0^\infty ke^{-kx}\,dx = 1.$

47. $\int_e^\infty \frac{k}{x(\ln x)^{k+1}}\,dx = \lim_{b\to\infty}\int_e^b \frac{k}{x(\ln x)^{k+1}}\,dx$

$\qquad = \lim_{b\to\infty}\left[(k)\frac{(\ln x)^{-k}}{-k}\Big|_e^b\right]$

$\qquad = \lim_{b\to\infty}\left[-(\ln b)^{-k}+1\right]$

If $k > 0,$ as $b\to\infty,\ (\ln b)^{-k}$ approaches 0.

Thus, in this case $\int_e^\infty \frac{k}{x(\ln x)^{k+1}}\,dx = 1.$

49. Capital value $= \int_0^\infty Ke^{-rt}\,dt$

$\qquad = \lim_{b\to\infty} K\left(-\frac{1}{r}\right)\int_0^b -re^{-rt}\,dt$

$\qquad = -\frac{K}{r}\lim_{b\to\infty}\left[e^{-rt}\Big|_0^b\right]$

$\qquad = -\frac{K}{r}\lim_{b\to\infty}[e^{-rb}-1] = \frac{K}{r}$

Chapter 9 Review Exercises

1. $\int x \sin 3x^2 \, dx$

Let $u = 3x^2$, $du = 6x \, dx$.
Then

$$\int x \sin 3x^2 \, dx = \frac{1}{6} \int \sin u \, du = -\frac{1}{6} \cos u + C$$

$$= -\frac{1}{6} \cos 3x^2 + C$$

2. $\int \sqrt{2x+1} \, dx = \frac{1}{3}(2x+1)^{3/2} + C$

3. $\int x(1-3x^2)^5 \, dx$

Let $u = 1 - 3x^2$, $du = -6x \, dx$. Then

$$\int x(1-3x^2)^5 \, dx = -\frac{1}{6} \int u^5 \, du = -\frac{1}{36} u^6 + C$$

$$= -\frac{1}{36}(1-3x^2)^6 + C$$

4. $\int \frac{(\ln x)^5}{x} \, dx$

Let $u = \ln x$, $du = \frac{1}{x} \, dx$. Then

$$\int \frac{(\ln x)^5}{x} \, dx = \int u^5 \, du = \frac{1}{6} u^6 + C$$

$$= \frac{1}{6}(\ln x)^6 + C$$

5. $\int \frac{(\ln x)^2}{x} \, dx$

Let $u = \ln x$, $du = \frac{1}{x} \, dx$. Then

$$\int \frac{(\ln x)^2}{x} \, dx = \int u^2 \, du = \frac{1}{3} u^3 + C$$

$$= \frac{1}{3}(\ln x)^3 + C$$

6. $\int \frac{1}{\sqrt{4x+3}} \, dx = \frac{1}{2}(4x+3)^{1/2} + C$

7. $\int x \sqrt{4-x^2} \, dx$

Let $u = 4 - x^2$, $du = -2x \, dx$. Then

$$\int x \sqrt{4-x^2} \, dx = -\frac{1}{2} \int \sqrt{u} \, du = -\frac{1}{3} u^{3/2} + C$$

$$= -\frac{1}{3}(4-x^2)^{3/2} + C$$

8. $\int x \sin 3x \, dx$

Use integration by parts with $f(x) = x$,
$g(x) = \sin 3x$. Then

$$f'(x) = 1, \; G(x) = -\frac{1}{3} \cos 3x \text{ and}$$

$$\int x \sin 3x \, dx = -\frac{1}{3} x \cos 3x + \frac{1}{3} \int \cos 3x \, dx$$

$$= -\frac{1}{3} x \cos 3x + \frac{1}{9} \sin 3x + C$$

9. $\int x^2 e^{-x^3} \, dx$

Let $u = -x^3$, $du = -3x^2 \, dx$. Then

$$\int x^2 e^{-x^3} \, dx = -\frac{1}{3} \int e^u \, du = -\frac{1}{3} e^u + C$$

$$= -\frac{1}{3} e^{-x^3} + C$$

10. $\int \frac{x \ln(x^2+1)}{x^2+1} \, dx$

Let $u = \ln(x^2+1)$, $du = \frac{2x}{x^2+1} \, dx$.

Then

$$\int \frac{x \ln(x^2+1)}{x^2+1} \, dx = \frac{1}{2} \int u \, du = \frac{1}{4} u^2 + C$$

$$= \frac{1}{4}(\ln(x^2+1))^2 + C$$

11. $\int x^2 \cos 3x \, dx$

Use integration by parts with $f(x) = x^2$,
$g(x) = \cos 3x$. Then

$$f'(x) = 2x, \; G(x) = \frac{1}{3} \sin 3x \text{ and}$$

$$\int x^2 \cos 3x \, dx = \frac{1}{3} x^2 \sin 3x - \frac{2}{3} \int x \sin 3x \, dx.$$

To evaluate $\int x \sin 3x \, dx$ integrate by parts
again:

$$\int x \sin 3x \, dx = -\frac{1}{3} x \cos 3x + \frac{1}{3} \int \cos 3x \, dx$$

$$= -\frac{1}{3} x \cos 3x + \frac{1}{9} \sin 3x + C_1.$$

Thus,

$$\int x^2 \cos 3x \, dx$$

$$= \frac{1}{3} x^2 \sin 3x + \frac{2}{9} x \cos 3x - \frac{2}{27} \sin 3x + C$$

12. $\displaystyle\int \frac{\ln(\ln x)}{x \ln x} dx$

Let $u = \ln(\ln x)$, $du = \dfrac{1}{x \ln x} dx$.

$\displaystyle\int \frac{\ln(\ln x)}{x \ln x} dx = \int u\, du = \frac{u^2}{2} + C$

$\qquad = \dfrac{1}{2}(\ln(\ln x))^2 + C$

13. $\displaystyle\int \ln x^2 dx = \int 2 \ln x\, dx = 2\int \ln x\, dx$

To evaluate $\displaystyle\int \ln x\, dx$, use integration by parts

with $f(x) = \ln x$, $g(x) = 1$. Then $f'(x) = \dfrac{1}{x}$,

$G(x) = x$ and

$\displaystyle\int \ln x\, dx = x \ln x - \int dx = x \ln x - x + C_1$.

Thus, $\displaystyle\int \ln x^2 dx = 2x \ln x - 2x + C$.

14. $\displaystyle\int x\sqrt{x+1}\, dx$

Use integration by parts with $f(x) = x$,
$g(x) = \sqrt{x+1}$. Then

$f'(x) = 1$, $G(x) = \dfrac{2}{3}(x+1)^{3/2}$ and

$\displaystyle\int x\sqrt{x+1}\, dx = \frac{2}{3}x(x+1)^{3/2} - \frac{2}{3}\int (x+1)^{3/2} dx$

$\qquad = \dfrac{2}{3}x(x+1)^{3/2} - \dfrac{4}{15}(x+1)^{5/2} + C$.

15. $\displaystyle\int \frac{x}{\sqrt{3x-1}} dx$

Use integration by parts with $f(x) = x$,
$g(x) = (3x-1)^{-1/2}$. Then

$f'(x) = 1$, $G(x) = \dfrac{2}{3}(3x-1)^{1/2}$ and

$\displaystyle\int \frac{x}{\sqrt{3x-1}} dx$

$\qquad = \dfrac{2}{3}x(3x-1)^{1/2} - \dfrac{2}{3}\int (3x-1)^{1/2} dx$

$\qquad = \dfrac{2}{3}x(3x-1)^{1/2} - \dfrac{4}{27}(3x-1)^{3/2} + C$

16. $\displaystyle\int x^2 \ln x^2 dx$

Use integration by parts with
$f(x) = \ln x^2$, $g(x) = x^2$. Then

$f'(x) = \dfrac{2x}{x^2} = \dfrac{2}{x}$, and $G(x) = \dfrac{x^3}{3}$.

$\displaystyle\int x^2 \ln x^2 dx = \frac{x^3}{3}\ln x^2 - \int \frac{2x^2}{3} dx$

$\qquad = \dfrac{x^3}{3}\ln x^2 - \dfrac{2}{9}x^3 + C$.

17. $\displaystyle\int \frac{x}{(1-x)^5} dx$

Use integration by parts with $f(x) = x$,
$g(x) = (1-x)^{-5}$. Then $f'(x) = 1$,

$G(x) = \dfrac{1}{4}(1-x)^{-4}$ and

$\displaystyle\int \frac{x}{(1-x)^5} dx = \frac{1}{4}x(1-x)^{-4} - \frac{1}{4}\int (1-x)^{-4} dx$

$\qquad = \dfrac{1}{4}x(1-x)^{-4} - \dfrac{1}{12}(1-x)^{-3} + C$.

18. $\displaystyle\int x(\ln x)^2 dx$

Use integration by parts with $f(x) = (\ln x)^2$,

$g(x) = x$. Then $f'(x) = \dfrac{2\ln x}{x}$, $G(x) = \dfrac{x^2}{2}$ and

$\displaystyle\int x(\ln x)^2 dx = \frac{x^2(\ln x)^2}{2} - \int x \ln x\, dx$

$\qquad = \dfrac{x^2(\ln x)^2}{2} - \left[\dfrac{x^2}{2}\ln x - \int \dfrac{x}{2} dx\right]$

(using parts again)

$\qquad = \dfrac{x^2(\ln x)^2}{2} - \dfrac{x^2}{2}\ln x + \dfrac{x^2}{4} + C$

$\qquad = \dfrac{x^2}{2}\left[(\ln x)^2 - \ln x + \dfrac{1}{2}\right] + C$.

19. Integration by parts: $f(x) = x$, $g(x) = e^{2x}$

20. Integration by parts: $f(x) = x - 3$, $g(x) = e^{-x}$

21. Substitution: $u = \sqrt{x+1}$

22. Substitution: $u = x^3 - 1$

23. Substitution: $u = x^4 - x^2 + 4$

24. Integration by parts:

$f(x) = \ln\sqrt{5-x} = \dfrac{1}{2}\ln(5-x)$, $g(x) = 1$

25. Repeated integration by parts, starting with
$f(x) = (3x-1)^2$, $g(x) = e^{-x}$

26. Substitution: $u = 3 - x^2$

27. Integration by parts: $f(x) = 500 - 4x$, $g(x) = e^{-x/2}$

28. Integration by parts: $f(x) = \ln x$, $g(x) = x^{5/2}$

29. Integration by parts: $f(x) = \ln(x + 2)$, $g(x) = \sqrt{x + 2}$

30. Repeated integration by parts, starting with $f(x) = (x+1)^2$, $g(x) = e^{3x}$

31. Substitution: $u = x^2 + 6x$

32. Substitution: $u = \sin x$

33. Substitution: $u = x^2 - 9$

34. Integration by parts: $f(x) = 3 - x$, $g(x) = \sin 3x$

35. Substitution: $u = x^3 - 6x$

36. Substitution: $u = \ln x$

37. $\int_0^1 \dfrac{2x}{(x^2+1)^3}\,dx$

    Let $u = x^2 + 1$, $du = 2x\,dx$. When $x = 0$, $u = 1$; when $x = 1$, $u = 2$.

    $\int_0^1 \dfrac{2x}{(x^2+1)^3}\,dx = \int_1^2 \dfrac{du}{u^3} = -\dfrac{1}{2}u^{-2}\Big|_1^2 = -\dfrac{1}{8} + \dfrac{1}{2} = \dfrac{3}{8}$

38. $\int_0^{\pi/2} x \sin 8x\,dx$

    Using integration by parts with $f(x) = x$,

    $g(x) = \sin 8x$, we have $f'(x) = 1$, $G(x) = -\dfrac{1}{8}\cos 8x$ and

    $\int_0^{\pi/2} x \sin 8x\,dx = -\dfrac{1}{8}x\cos 8x\Big|_0^{\pi/2} + \dfrac{1}{8}\int_0^{\pi/2}\cos 8x\,dx = \dfrac{1}{8}\left[-x\cos 8x + \dfrac{1}{8}\sin 8x\right]\Big|_0^{\pi/2} = \dfrac{1}{8}\left[-\dfrac{\pi}{2} - 0\right] = -\dfrac{\pi}{16}$

39. $\int_0^2 xe^{-(1/2)x^2}\,dx$

    Let $u = -\dfrac{1}{2}x^2$, $du = -x\,dx$. When $x = 0$, $u = 0$; when $x = 2$, $u = -2$.

    $\int_0^2 xe^{-(1/2)x^2}\,dx = -\int_0^{-2} e^u\,du = -e^u\Big|_0^{-2} = 1 - e^{-2}$

40. $\int_{1/2}^1 \dfrac{\ln(2x+3)}{2x+3}\,dx$

    Let $u = \ln(2x + 3)$, $du = \dfrac{2}{2x+3}\,dx$. When $x = \dfrac{1}{2}$, $u = \ln(4)$, when $x = 1$, $u = \ln 5$.

    $\int_{1/2}^1 \dfrac{\ln(2x+3)}{2x+3}\,dx = \dfrac{1}{2}\int_{\ln 4}^{\ln 5} u\,du = \dfrac{u^2}{4}\Big|_{\ln 4}^{\ln 5} = \dfrac{1}{4}\left[(\ln 5)^2 - (\ln 4)^2\right]$

41. $\int_1^2 xe^{-2x}dx$

Use integration by parts with $f(x) = x$, $g(x) = e^{-2x}$. Then $f'(x) = 1$, $G(x) = -\frac{1}{2}e^{-2x}$

$$\int_1^2 xe^{-2x}dx = -\frac{1}{2}xe^{-2x}\Big|_1^2 + \frac{1}{2}\int_1^2 e^{-2x}dx = \left(-\frac{1}{2}xe^{-2x} - \frac{1}{4}e^{-2x}\right)\Big|_1^2 = -\frac{1}{2}\left[e^{-2x}\left(x+\frac{1}{2}\right)\Big|_1^2\right]$$

$$= -\frac{1}{2}\left[\frac{5}{2}e^{-4} - \frac{3}{2}e^{-2}\right] = \frac{3}{4}e^{-2} - \frac{5}{4}e^{-4}$$

42. $\int_1^2 x^{-3/2}\ln x\, dx$

Use integration by parts with $f(x) = \ln x$, $g(x) = x^{-3/2}$. Then $f'(x) = \frac{1}{x}$, $G(x) = -2x^{-1/2}$ and

$$\int_1^2 x^{-3/2}\ln x\, dx = -2x^{-1/2}\ln x\Big|_1^2 + \int_1^2 2x^{-3/2}dx = \left(-2x^{-1/2}\ln x - 4x^{-1/2}\right)\Big|_1^2 = -2\left[x^{-1/2}(\ln x + 2)\Big|_1^2\right]$$

$$= -2\left[\frac{\ln 2 + 2}{\sqrt{2}} - 2\right] = -\sqrt{2}\ln 2 - 2\sqrt{2} + 4$$

43. $\int_1^9 \frac{1}{\sqrt{x}}dx$; $n = 4$, $\Delta x = \frac{(9-1)}{4} = 2$

Midpoint rule: $x_1 = 1 + \frac{2}{2} = 2$, $x_2 = 4$, $x_3 = 6$, $x_4 = 8$

$$\int_1^9 \frac{1}{\sqrt{x}}dx \approx \left[\frac{1}{\sqrt{2}} + \frac{1}{\sqrt{4}} + \frac{1}{\sqrt{6}} + \frac{1}{\sqrt{8}}\right](2) \approx 3.93782$$

Trapezoidal rule: $a_0 = 1$, $a_1 = 3$, $a_2 = 5$, $a_3 = 7$, $a_4 = 9$

$$\int_1^9 \frac{1}{\sqrt{x}}dx \approx \left[\frac{1}{\sqrt{1}} + \frac{2}{\sqrt{3}} + \frac{2}{\sqrt{5}} + \frac{2}{\sqrt{7}} + \frac{1}{\sqrt{9}}\right]\frac{2}{2} \approx 4.13839$$

Simpson's rule: $\int_1^9 \frac{1}{\sqrt{x}}dx \approx \frac{2(3.93782) + 4.13839}{3} \approx 4.00468$

44. $\int_0^{10} e^{\sqrt{x}}dx$; $n = 5$, $\Delta x = \frac{(10-0)}{5} = 2$

Midpoint rule: $x_1 = 1$, $x_2 = 3$, $x_3 = 5$, $x_4 = 7$, $x_5 = 9$

$$\int_0^{10} e^{\sqrt{x}}dx \approx \left[e^{\sqrt{1}} + e^{\sqrt{3}} + e^{\sqrt{7}} + e^{\sqrt{9}}\right]2 \approx 103.81310$$

Trapezoidal rule: $a_0 = 0$, $a_1 = 2$, $a_2 = 4$, $a_3 = 6$, $a_4 = 8$, $a_5 = 10$

$$\int_0^{10} e^{\sqrt{x}}dx \approx \left[e^{\sqrt{0}} + 2e^{\sqrt{2}} + 2e^{\sqrt{4}} + 2e^{\sqrt{6}} + 2e^{\sqrt{8}} + e^{\sqrt{10}}\right]\frac{2}{2} \approx 104.63148$$

Simpson's rule: $\int_0^{10} e^{\sqrt{x}}dx \approx \frac{2(103.8130) + 104.63148}{3} \approx 104.08589$

45. $\int_1^4 \dfrac{e^x}{x+1}\,dx$; $n=5$, $\Delta x = \dfrac{3}{5} = 0.6$

Midpoint rule: $x_1 = 1.3$, $x_2 = 1.9$, $x_3 = 2.5$, $x_4 = 3.1$, $x_5 = 3.7$

$$\int_1^4 \frac{e^x}{x+1}\,dx \approx \left[\frac{e^{1.3}}{2.3} + \frac{e^{1.9}}{2.9} + \frac{e^{2.5}}{3.5} + \frac{e^{3.1}}{4.1} + \frac{e^{3.7}}{4.7}\right](0.6) \approx 12.84089$$

Trapezoidal rule: $a_0 = 1$, $a_1 = 1.6$, $a_2 = 2.2$, $a_3 = 2.8$, $a_4 = 3.4$, $a_5 = 4$

$$\int_1^4 \frac{e^x}{x+1}\,dx \approx \left[\frac{e}{2} + \frac{2e^{1.6}}{2.6} + \frac{2e^{2.2}}{3.2} + \frac{2e^{2.8}}{3.8} + \frac{2e^{3.4}}{4.4} + \frac{e^4}{5}\right](0.3) \approx 13.20137$$

Simpson's rule: $\int_1^4 \dfrac{e^x}{x+1}\,dx \approx \dfrac{2(12.84089)+13.20137}{3} \approx 12.96105$

46. $\int_{-1}^1 \dfrac{1}{1+x^2}\,dx$; $n=5$, $\Delta x = \dfrac{2}{5} = 0.4$

Midpoint rule: $x_1 = -0.8$, $x_2 = -0.4$, $x_3 = 0$, $x_4 = 0.4$, $x_5 = 0.8$

$$\int_{-1}^1 \frac{1}{1+x^2}\,dx \approx \left[\frac{1}{1+(-0.8)^2} + \frac{1}{1+(-0.4)^2} + \frac{1}{1+0^2} + \frac{1}{1+(0.4)^2} + \frac{1}{1+(0.8)^2}\right](0.4) \approx 1.57746$$

Trapezoidal rule: $a_0 = -1$, $a_1 = -0.6$, $a_2 = -0.2$, $a_3 = 0.2$, $a_4 = 0.6$, $a_5 = 1$

$$\int_{-1}^1 \frac{1}{1+x^2}\,dx \approx \left[\frac{1}{1+(-1)^2} + \frac{2}{1+(-0.6)^2} + \frac{2}{1+(-0.2)^2} + \frac{2}{1+(0.2)^2} + \frac{2}{1+(0.6)^2} + \frac{1}{1+1^2}\right](0.2) \approx 1.55747$$

Simpson's rule:

$$\int_{-1}^1 \frac{1}{1+x^2}\,dx \approx \frac{2(1.57746)+1.55747}{3} \approx 1.57080$$

47. $\int_0^\infty e^{6-3x}\,dx = \lim_{b\to\infty} \int_0^b e^{6-3x}\,dx$

Let $u = 6 - 3x$, $du = -3dx$. When $x = 0$, $u = 6$;
when $x = b$, $u = 6 - 3b$.

$$\int_0^b e^{6-3x}\,dx = -\frac{1}{3}\int_6^{6-3b} e^u\,du = -\frac{1}{3}e^u\Big|_6^{6-3b}$$

$$= -\frac{1}{3}e^{6-3b} + \frac{1}{3}e^6$$

Thus $\int_0^\infty e^{6-3x}\,dx = \lim_{b\to\infty}\left[-\frac{1}{3}e^{6-3b} + \frac{e^6}{3}\right] = \frac{e^6}{3}$.

48. $\int_1^\infty x^{-2/3}\,dx = \lim_{b\to\infty} \int_1^b x^{-2/3}\,dx$

$$= \lim_{b\to\infty}\left[3x^{1/3}\Big|_1^b\right] = \lim_{b\to\infty}[3b^{1/3} - 3]$$

As $b \to \infty$, $3b^{1/3}$ increases without bound.

Thus, $\int_1^\infty x^{-2/3}\,dx$ diverges.

49. $\int_1^\infty \dfrac{x+2}{x^2+4x-2}\,dx = \lim_{b\to\infty} \int_1^b \dfrac{x+2}{x^2+4x-2}\,dx$

Let $u = x^2 + 4x - 2$,
$du = (2x+4)dx = 2(x+2)dx$. When $x = 1$,
$u = 3$; when $x = b$, $u = b^2 + 4b - 2$.

$$\int_1^b \frac{x+2}{x^2+4x-2}\,dx = \frac{1}{2}\int_3^{b^2+4b-2} \frac{du}{u}$$

$$= \frac{1}{2}\ln|u|\Big|_3^{b^2+4b-2}$$

$$= \frac{1}{2}\left[\ln|b^2+4b-2| - \ln 3\right]$$

As $b \to \infty$, $\ln|b^2+4b-2|$ increases without

bound. Thus $\int_1^\infty \dfrac{x+2}{x^2+4x-2}\,dx$ diverges.

50. $\int_0^\infty x^2 e^{-x^3} dx = \lim_{b\to\infty} \int_0^b x^2 e^{-x^3} dx$

Let $u = -x^3$, then $du = -3x^2 dx$.

When $x = 0$, $u = 0$; when $x = b$, $u = -b^3$.

$\int_0^b x^2 e^{-x^3} dx = -\frac{1}{3}\int_0^{-b^3} e^u du = -\frac{1}{3}\left[e^{-b^3} - 1\right]$

$\qquad\qquad = \frac{1}{3} - \frac{e^{-b^3}}{3}$

Thus $\int_0^\infty x^2 e^{-x^3} dx = \lim_{b\to\infty}\left[\frac{1}{3} - \frac{e^{-b^3}}{3}\right] = \frac{1}{3}$.

51. $\int_{-1}^\infty (x+3)^{-5/4} dx = \lim_{b\to\infty}\int_{-1}^b (x+3)^{-5/4} dx$

$\qquad = \lim_{b\to\infty}\left[-4(x+3)^{-1/4}\Big|_{-1}^b\right]$

$\qquad = \lim_{b\to\infty}\left[2^{7/4} - 4(b+3)^{-1/4}\right]$

$\qquad = 2^{7/4}$

52. $\int_{-\infty}^0 \frac{8}{(5-2x)^3} dx = \lim_{b\to-\infty}\int_b^0 \frac{8}{(5-2x)^3} dx$

$\qquad = \lim_{b\to-\infty}\left[2(5-2x)^{-2}\Big|_b^0\right]$

$\qquad = \lim_{b\to-\infty}\left[\frac{2}{25} - 2(5-2b)^{-2}\right]$

$\qquad = \frac{2}{25}$

53. $\int_1^\infty xe^{-x} dx = \lim_{b\to\infty}\int_1^b xe^{-x} dx$

Using integration by parts with $f(x) = x$,

$g(x) = e^{-x}$; $f'(x) = 1$, $G(x) = -e^{-x}$ gives

$\int_1^b xe^{-x} dx = -xe^{-x}\Big|_1^b + \int_1^b e^{-x} dx$

$\qquad = (-xe^{-x} - e^{-x})\Big|_1^b$

$\qquad = 2e^{-1} - be^{-b} - e^{-b}$

Thus

$\int_1^\infty xe^{-x} dx = \lim_{b\to\infty}\left[\frac{2}{e} - be^{-b} - e^{-b}\right] = \frac{2}{e}$

$\left(\text{since } \lim_{b\to\infty} be^{-b} = 0\right)$.

54. $\int_0^\infty xe^{-kx} dx = \lim_{b\to\infty}\int_0^b xe^{-kx} dx$

Using integration by parts with $f(x) = x$,

$g(x) = e^{-kx}$; $f'(x) = 1$, $G(x) = -\frac{1}{k}e^{-kx}$, we

have

$\int_0^b xe^{-kx} dx = \frac{-x}{k}e^{-kx}\Big|_0^b + \frac{1}{k}\int_0^b e^{-kx} dx$

$\qquad = \left(-\frac{x}{k}e^{-kx} - \frac{1}{k^2}e^{-kx}\right)\Big|_0^b$

$\qquad = \frac{1}{k^2} - \frac{b}{k}e^{-kb} - \frac{1}{k^2}e^{-kb}$

Thus

$\int_0^\infty xe^{-kx} dx = \lim_{b\to\infty}\left[\frac{1}{k}\left(\frac{1}{k} - be^{-kb} - \frac{1}{k}e^{-kb}\right)\right]$

$\qquad = \frac{1}{k^2}\left(\text{since } \lim_{b\to\infty} be^{-kb} = 0\right)$.

55. The present value is

$\int_0^4 50e^{-0.08t}e^{-0.12t} dt = \int_0^4 50e^{-0.2t} dt$

$\qquad = -250e^{-0.2t}\Big|_0^4$

$\qquad = 250 - 250e^{-0.8}$

$\qquad \approx \$137,668$

56. Using the method of Example 3, section 9.5, the total tax revenue is

$\int_0^{10} (2\pi t)50e^{-t/20} dt$

$= 100\pi\int_0^{10} te^{-t/20} dt$

$= 100\pi\left[-20te^{-t/20}\Big|_0^{10} + \int_0^{10} 20e^{-t/20} dt\right]$

$= 100\pi\left[-20te^{-t/20} - 400e^{-t/20}\Big|_0^{10}\right]$

$= 100\pi[400 - e^{-0.5}(600)]$

$\approx \$11,335$ thousand

57. a. $M(t_1)\Delta t + \cdots + M(t_n)\Delta t \approx \int_0^2 M(t) dt$

b. $M(t_1)e^{-0.1t_1}\Delta t + \cdots + M(t_n)e^{-0.1t_n}\Delta t$

$\qquad\qquad \approx \int_0^2 M(t)e^{-0.1t} dt$

58. $80,000 + \int_0^\infty 50,000e^{-rt} dt$

Chapter 10 Differential Equations

10.1 Solutions of Differential Equations

1. The differential equation says that $y' - 2ty = t$ for all values of t. We must show that this result holds if y is replaced by $\frac{3}{2}e^{t^2} - \frac{1}{2}$ and y' is replaced by $\left(\frac{3}{2}e^{t^2} - \frac{1}{2}\right)'$.

$$y = f(t) = \frac{3}{2}e^{t^2} - \frac{1}{2} \Rightarrow y' = 3te^{t^2}$$

$$y' - 2ty = 3te^{t^2} - 2t\left[\frac{3}{2}e^{t^2} - \frac{1}{2}\right] = t$$

Thus, $f(t) = \frac{3}{2}e^{t^2} - \frac{1}{2}$ is a solution of the differential equation.

3. The differential equation says that $y'' - 3y' + 2y = 0$, $y(0) = 5$, $y'(0) = 10$. We must show that this result holds if y is replaced by $5e^{2t}$, y' is replaced by $\left(5e^{2t}\right)'$, and y'' is replaced by $\left(5e^{2t}\right)''$.

$$y = f(t) = 5e^{2t} \Rightarrow y' = 10e^{2t} \Rightarrow y'' = 20e^{2t}$$

$$y'' - 3y' + 2y = 20e^{2t} - 3(10e^{2t}) + 2(5e^{2t}) = 0$$

$$y(0) = 5e^0 = 5, \; y'(0) = 10e^0 = 10$$

Thus, $f(t) = 5e^{2t}$ is a solution of the differential equation.

5. The highest derivative is the second derivative y'', so the differential equation is of second order.
$$y(t) = t \Rightarrow y'(t) = 1 \Rightarrow y''(t) = 0$$
$$(1 - t^2)y'' - 2ty' + 2y = (1 - t^2)(0) - 2t(1) + 2(t)$$
$$= 0$$

7. Yes. Given $y = f(t) = 3$, $y' = 0 = 6 - 2(3)$.

9. $y' = t^2y - 5t^2 = t^2(y - 5)$; $y = 5$ is a constant solution. $\dfrac{dy}{dt} = 0$, and $t^2(y - 5) = 0$ at $y = 5$.

11. $f(0) = y(0) = 4$. Since $y' = 2y - 3$, we have in particular $y'(0) = 2y(0) - 3 = 2(4) - 3 = 5$.

13. $y' = .2(160 - y)$. When $y = 60$,
$$y' = .2(160 - 60) = .2(100) = 20 \text{ ft/sec/sec}$$

15. $y' = 0.05y - 10,000$.

a. When $t = 1$, $y = 150,000$.
$$y' = .05(150,000) - 10,000 = -2500$$
So the balance is decreasing at a rate of $2500 per year after 1 year.

b. $y' = .05(y - 200,000)$

c. The rate of change of the savings account balance is proportional to the difference between the balance at the end of t years and $200,000.

17. The number of people who have heard the news broadcast after t hours is increasing at a rate that is proportional to the difference between that number and 200,000. At the beginning of the broadcast there are 10 people tuned in.

19. $y' = k(C - y)$, $k > 0$

21. $y' = k(P_b - y)$, $y(0) = P_0$, $k > 0$

23. $y = f(t) = 2e^{-t} + t - 1 \Rightarrow y' = -2e^{-t} + 1$
$t - y = t - (2e^{-t} + t - 1) = -2e^{-t} + 1 = y'$, so $f(t)$ is a solution of the differential equation.
$y(0) = f(0) = 2e^0 + 0 - 1 = 1$, so $f(t)$ satisfies the initial value condition.

25. The solution to $y' = .0002y(5000 - y)$, $y(0) = 1500$ approaches $y = 5000$ asymptotically from below. So $f(t)$ will never exceed 5000.

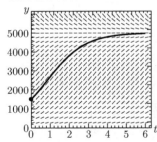

27. The slope field indicates that $y = 0$ and $y = 1$ may be the constant solutions.

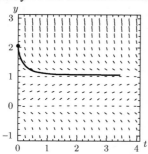

Check: $y = 0$, $y' = 0$.

So $2y(1 - y) = 2(0)(1 - 0) = 0 = y'$.

Thus $y = 0$ is a solution.

Check: $y = 1$, $y' = 0$.

So $2y(1 - y) = 2(1)(1 - 1) = 0 = y'$.

Thus $y = 1$ is also a solution.

29. Yes. The solution will approach $y = 1$ asymptotically from above, and is decreasing for all $t > 0$.

31. a.

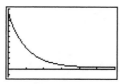

$[0, 30]$ *by* $[-75, 550]$

b. Let $y_1 = f(t)$.

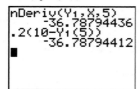

10.2 Separation of Variables

1. $\dfrac{dy}{dt} = \dfrac{5 - t}{y^2} \Rightarrow y^2 \dfrac{dy}{dt} = 5 - t$

$\displaystyle \int y^2 dy = \int (5 - t)\, dt$

$\dfrac{y^3}{3} + C_1 = 5t - \dfrac{t^2}{2} + C_2$

$\qquad y = \sqrt[3]{15t - \dfrac{3}{2}t^2 + C}$

3. $\dfrac{dy}{dt} = \dfrac{e^t}{e^y} \Rightarrow \dfrac{dy}{dt} e^y = e^t \Rightarrow \displaystyle \int e^y dy = \int e^t dt \Rightarrow$

$e^y + C_1 = e^t + C_2 \Rightarrow y = \ln \left| e^t + C \right|$

5. First check for constant solutions; if $y = 0$,

then $\dfrac{dy}{dt} = 0$ and $t^{1/2} y^2 = 0$; otherwise assume

$y \neq 0$.

$y' = t^{1/2} y^2 \Rightarrow y' y^{-2} = t^{1/2}$ (Assuming $y \neq 0$)

$\displaystyle \int y^{-2} dy = \int t^{1/2} dt \Rightarrow -\dfrac{1}{y} + C_1 = \dfrac{2}{3} t^{3/2} + C_2 \Rightarrow$

$y = \dfrac{1}{C - \left(\frac{2}{3}\right) t^{3/2}}$

The solutions are $y = \dfrac{1}{C - \left(\frac{2}{3}\right) t^{3/2}}$ or $y = 0$.

7. $\qquad y' = \dfrac{e^{t^3} t^2}{y^2} \Rightarrow y^2 \dfrac{dy}{dt} = t^2 e^{t^3}$

$\displaystyle \int y^2 dy = \int t^2 e^{t^3} dt$

$\dfrac{y^3}{3} + C_1 = \dfrac{e^{t^3}}{3} + C_2$

$\qquad y = \sqrt[3]{e^{t^3} + C}$

9. $\dfrac{dy}{dt} = \sqrt{\dfrac{y}{t}} = \dfrac{\sqrt{y}}{\sqrt{t}} \Rightarrow y^{-1/2} \dfrac{dy}{dt} = t^{-1/2}$

(Assuming $y \neq 0$)

$\displaystyle \int y^{-1/2} dy = \int t^{-1/2} dt$

$2y^{1/2} + C_1 = 2t^{1/2} + C_2 \Rightarrow y = (\sqrt{t} + C)^2$

$y = 0$ is also a solution.

11. $y' = 3t^2 y^2 \Rightarrow y^{-2} \dfrac{dy}{dt} = 3t^2$ (assuming $y \neq 0$)

$\displaystyle \int \dfrac{1}{y^2} dy = \int 3t^2 dt \Rightarrow -\dfrac{1}{y} + C_1 = t^3 + C_2 \Rightarrow$

$y = -\dfrac{1}{t^3 + C}$

$y = 0$ is also a solution.

13. $y' e^y = t e^{t^2} \Rightarrow e^y \dfrac{dy}{dt} = t e^{t^2}$

$\displaystyle \int e^y dy = \int t e^{t^2} dt \Rightarrow e^y + C_1 = \dfrac{1}{2} e^{t^2} + C_2 \Rightarrow$

$y = \ln \left(\dfrac{1}{2} e^{t^2} + C \right)$

15. $y' = \dfrac{\ln t}{ty} \Rightarrow y \dfrac{dy}{dt} = \dfrac{\ln t}{t} \Rightarrow \displaystyle \int y\, dy = \int \dfrac{\ln t}{t} dt \Rightarrow$

$\dfrac{y^2}{2} + C_1 = \dfrac{(\ln t)^2}{2} + C_2 \Rightarrow y = \pm\sqrt{(\ln t)^2 + C}$

17. $y' = (y-3)^2 \ln t \Rightarrow (y-3)^{-2} \dfrac{dy}{dt} = \ln t$

    (Assuming $y \neq 3$)

    $\displaystyle \int (y-3)^{-2} dy = \int \ln t\ dt$

    $-(y-3)^{-1} + C_1 = t \ln t - t + C_2$

    $y = 3 + \dfrac{1}{t - t \ln t + C}$

    $y = 3$ is also a solution.

19. $y' = 2te^{-2y} - e^{-2y}$, $y(0) = 3$

    $e^{2y} \dfrac{dy}{dt} = 2t - 1$

    $\displaystyle \int e^{2y} dy = \int (2t - 1)\ dt$

    $\dfrac{1}{2} e^{2y} + C_1 = t^2 - t + C_2$

    $e^{2y} = 2t^2 - 2t + C$

    $y = \dfrac{1}{2} \ln(2t^2 - 2t + C)$

    $y(0) = 3 = \dfrac{1}{2} \ln C$, so $C = e^6$ and the solution

    is $y = \dfrac{1}{2} \ln(2t^2 - 2t + e^6)$.

21. $y^2 y' = t \cos t$, $y(0) = 2$

    $\displaystyle \int y^2 dy = \int t \cos t\ dt$

    $\dfrac{y^3}{3} + C_1 = t \sin t + \cos t + C_2$

    $y = \sqrt[3]{3t \sin t + 3 \cos t + C}$

    $y(0) = 2 = \sqrt[3]{3 + C}$ so $C = 5$ and the solution is

    $y = \sqrt[3]{3t \sin t + 3 \cos t + 5}$.

23. $3y^2 y' = -\sin t$, $y\left(\dfrac{\pi}{2}\right) = 1$

    $3y^2 \dfrac{dy}{dt} = -\sin t \Rightarrow \displaystyle \int 3y^2 dy = \int (-\sin t)\, dt$

    $y^3 + C_1 = \cos t + C_2 \Rightarrow y = \sqrt[3]{\cos t + C}$

    $y\left(\dfrac{\pi}{2}\right) = 1 = \sqrt[3]{\cos \frac{\pi}{2} + C}$

    so $C = 1$ and the solution is $y = \sqrt[3]{\cos t + 1}$.

25. $\dfrac{dy}{dt} = \dfrac{t+1}{ty}$, $t > 0$, $y(1) = -3$.

    $y \dfrac{dy}{dt} = \dfrac{t+1}{t} \Rightarrow \displaystyle \int y\ dy = \int \left(1 + \dfrac{1}{t}\right) dt$

    $\dfrac{y^2}{2} = t + \ln|t| + C_1$

    $y = \pm\sqrt{2t + 2\ln|t| + C}$

Since $y(1) = -3$, the negative solution applies and $-3 = -\sqrt{2 + C}$, so $C = 7$ and the solution is $y = -\sqrt{2t + 2\ln t + 7}$.

27. $y' = 5ty - 2t$, $y(0) = 1$

    $y' = 5t\left(y - \dfrac{2}{5}\right)$

    $\left(y - \dfrac{2}{5}\right)^{-1} \dfrac{dy}{dt} = 5t$

    $\displaystyle \int \left(y - \dfrac{2}{5}\right)^{-1} dy = \int 5t\ dt$

    $\ln\left|y - \dfrac{2}{5}\right| = \dfrac{5}{2} t^2 + C$

    $y = \dfrac{2}{5} + Ae^{(5/2)t^2}$

    $y(0) = 1 = \dfrac{2}{5} + A$, so $A = \dfrac{3}{5}$ and the solution is

    $y = \dfrac{2}{5} + \dfrac{3}{5} e^{(5/2)t^2}$.

29. $\dfrac{dy}{dx} = \dfrac{\ln x}{\sqrt{xy}}$, $y(1) = 4$

    $\displaystyle \int \sqrt{y}\ dy = \int \dfrac{\ln x}{\sqrt{x}}\ dx = \int \dfrac{2 \ln \sqrt{x}}{\sqrt{x}}\ dx$

    $\dfrac{2}{3} y^{3/2} = 2\sqrt{x} \ln \sqrt{x} - 4\sqrt{x} + C_1$

    $y = \left(3\sqrt{x} \ln \sqrt{x} - 6\sqrt{x} + C\right)^{2/3}$

    $y(1) = 4 = (-6 + C)^{2/3} \Rightarrow C = 14$ and the

    solution is $y = \left(3\sqrt{x} \ln x - 6\sqrt{x} + 14\right)^{2/3}$.

31. $\dfrac{dy}{dp} = -\dfrac{1}{2}\left(\dfrac{y}{p+3}\right) \Rightarrow y^{-1} \dfrac{dy}{dp} = -\dfrac{1}{2(p+3)}$

    (Assuming $y \neq 0$)

    $\displaystyle \int y^{-1} dy = -\dfrac{1}{2} \int \dfrac{1}{p+3}\ dp \Rightarrow$

    $\ln|y| = -\dfrac{1}{2} \ln|p+3| + C \Rightarrow y = \dfrac{A}{\sqrt{p+3}}$

    $A > 0$ because $y > 0$.

33. $\dfrac{dp}{dt} = k(1 - p)$; $0 \le p \le 1$, $p(0) = 0$.

    $\displaystyle \int (1-p)^{-1} dp = \int k\ dt$ (Assuming $p \neq 1$)

    $-\ln|1 - p| = kt + C_1 \Rightarrow 1 - p = e^{-kt+C} \Rightarrow$

    $p = 1 - e^{-kt+C} = 1 - Ae^{-kt}$

    $p(0) = 0 = 1 - A$, so $A = 1$ and the solution is

    $y = 1 - e^{-kt}$.

35. $\dfrac{dV}{dt} = kV^{2/3}$, $k < 0$ (since V decreases),

$V(0) = 27$, $V(4) = 15.625$. Solving the differential equation gives

$\displaystyle\int V^{-2/3} dV = \int k\,dt \Rightarrow 3V^{1/3} = kt + C_1$, or

$V = \left(\dfrac{kt}{3} + C\right)^3$. Using the conditions

$V(0) = 27 \Rightarrow 27 = C^3$ or $C = 3$.

$V(4) = 15.625 \Rightarrow 15.625 = \left(\dfrac{4k}{3} + 3\right)^3$, or

$k = -\dfrac{3}{8}$. Hence $V = \left(3 - \dfrac{t}{8}\right)^3$, $V = 0$ when

$t = 24$ weeks.

37. $\dfrac{dy}{dt} = -ay \ln\left(\dfrac{y}{b}\right) \Rightarrow \displaystyle\int \dfrac{1}{y \ln\left(\frac{y}{b}\right)}\, dy = \int -a\, dt$

$(y \neq 0, b)$

$\ln\left|\ln\left(\dfrac{y}{b}\right)\right| = -at + C \Rightarrow \ln\left(\dfrac{y}{b}\right) = Ce^{-at} \Rightarrow$

$y = be^{Ce^{-at}}$

($y = 0$ is not a solution and the solution $y = b$ corresponds to $C = 0$.)

39. a. The slope lines are pointing downward at $(0, 6000)$, so an initial population of 6000 will decrease and approach 5000.

    b. The slope lines are pointing upward at $(0, 1000)$, so an initial population of 1000 will increase and approach 5000.

    c.

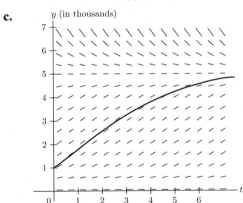

The solution shows that an initial population of 1000 fish will increase over time, asymptotically approaching a maximum of 5000 fish.

10.3 First-Order Linear Differential Equations

1. $y' - 2y = t \Rightarrow a(t) = -2$

$A(t) = \displaystyle\int (-2)\,dt = -2t; \, e^{A(t)} = e^{-2t}$

The integrating factor is e^{-2t}.

3. $t^3 y' + y = 0 \Rightarrow y' + \dfrac{1}{t^3} y = 0$ (since $t > 0$)

$a(t) = \dfrac{1}{t^3}; \, A(t) = \displaystyle\int \dfrac{1}{t^3}\, dt = -\dfrac{1}{2} t^{-2}$

$e^{A(t)} = e^{-t^{-2}/2}$

The integrating factor is $e^{-1/(2t^2)}$.

5. $y' - \dfrac{y}{10+t} = 2 \Rightarrow a(t) = -\dfrac{1}{10+t}$

$A(t) = \displaystyle\int \left(-\dfrac{1}{10+t}\right) dt = -\int \dfrac{1}{10+t}\, dt$

$\qquad = -\ln|10+t|$

$\qquad = -\ln(10+t)$

$\qquad\qquad \left[|10+t| = 10+t, \text{ since } t > 0\right]$

$e^{A(t)} = e^{-\ln(10+t)} = \dfrac{1}{10+t}$

The integrating factor is $\dfrac{1}{10+t}$.

7. $y' + y = 1 \Rightarrow a(t) = 1; b(t) = 1$

$A(t) = \displaystyle\int 1\,dt = t \Rightarrow e^{A(t)} = e^t; e^{-A(t)} = e^{-t}$

$y = e^{-A(t)} \left[\displaystyle\int e^{A(t)} b(t)\,dt + C\right]$

$\quad = e^{-t}\left[\displaystyle\int (1)e^t\, dt + C\right]$

$\quad = e^{-t}\left[\displaystyle\int e^t\, dt + C\right] = e^{-t}[e^t + C]$

$\quad = 1 + Ce^{-t}$

9. $y' - 2ty = -4t \Rightarrow a(t) = -2t; b(t) = -4t$

$A(t) = \displaystyle\int (-2t)\,dt = -t^2$

$y = e^{-A(t)}\left[\displaystyle\int e^{A(t)} b(t)\,dt + C\right]$

$\quad = e^{-(-t^2)}\left[\displaystyle\int e^{-t^2}(-4t)\,dt + C\right]$

$\quad = e^{t^2}\left[-\displaystyle\int 4te^{-t^2}\,dt + C\right] = e^{t^2}\left[2e^{-t^2} + C\right]$

$\quad = 2 + Ce^{t^2}$

11. $y' = 0.5(35 - y) \Rightarrow y' = 17.5 - 0.5y$
$y' + 0.5y = 17.5$

$a(t) = 0.5; \, b(t) = 17.5; \, A(t) = \int 0.5 \, dt = 0.5t$

$y = e^{-A(t)} \left[\int e^{A(t)} b(t) dt + C \right]$

$= e^{-0.5t} \left[\int 17.5 e^{0.5t} dt + C \right]$

$= e^{-0.5t} [35 e^{0.5t} + C] = 35 + Ce^{-0.5t}$

13. $y' + \dfrac{y}{10+t} = 0 \Rightarrow a(t) = \dfrac{1}{10+t}; \, b(t) = 0$

$A(t) = \int \dfrac{1}{10+t} dt = \ln|10+t| = \ln(10+t), \text{ since}$

$t > 0$

$y = e^{-A(t)} \left[\int e^{A(t)} b(t) dt + C \right]$

$= e^{-\ln(10+t)} \left[\int e^{\ln(10+t)} (0) dt + C \right]$

$= \dfrac{1}{10+t} [0 + C] = \dfrac{C}{10+t}$

15. $(1+t)y' + y = -1 \Rightarrow y' + \dfrac{1}{1+t} y = -\dfrac{1}{1+t},$

since $t > 0$

$a(t) = \dfrac{1}{1+t}; \, b(t) = -\dfrac{1}{1+t}$

$A(t) = \int \dfrac{1}{1+t} dt = \ln|1+t| = \ln(1+t) \text{ since } t > 0$

$e^{A(t)} = e^{\ln(1+t)} = 1+t; \, e^{-A(t)} = e^{-\ln(1+t)} = \dfrac{1}{1+t}$

$y = e^{-A(t)} \left[\int e^{A(t)} b(t) dt + C \right]$

$= \dfrac{1}{1+t} \left[\int (1+t) \left(\dfrac{-1}{1+t} \right) dt + C \right]$

$= \dfrac{1}{1+t} \left[\int (-1) dt + C \right] = \dfrac{1}{1+t} [-t + C]$

$= \dfrac{C-t}{1+t}$

17. $6y' + ty = t \Rightarrow y' + \dfrac{1}{6} ty = \dfrac{1}{6} t$

$a(t) = \dfrac{1}{6} t; \, b(t) = \dfrac{1}{6} t; \, A(t) = \int \dfrac{1}{6} t \, dt = \dfrac{1}{12} t^2$

$y = e^{-A(t)} \left[\int e^{A(t)} b(t) dt + C \right]$

$= e^{-t^2/12} \left[\int e^{t^2/12} \left(\dfrac{1}{6} t \right) dt + C \right]$

$= e^{-t^2/12} \left[e^{t^2/12} + C \right]$

$= 1 + Ce^{-t^2/12}$

19. $y' + y = 2 - e^t \Rightarrow a(t) = 1; \, b(t) = 2 - e^t$

$A(t) = \int 1 \, dt = t$

$y = e^{-A(t)} \left[\int e^{A(t)} b(t) dt + C \right]$

$= e^{-t} \left[\int e^t (2 - e^t) dt + C \right]$

$= e^{-t} \left[\int (2e^t - e^{2t}) dt + C \right]$

$= e^{-t} \left[2e^t - \dfrac{1}{2} e^{2t} + C \right]$

$= 2 - \dfrac{1}{2} e^t + Ce^{-t}$

21. $y' + 2y = 1, \, y(0) = 1; \, a(t) = 2; \, b(t) = 1$

$A(t) = \int 2 \, dt = 2t$

$y = e^{-A(t)} \left[\int e^{A(t)} b(t) dt + C \right]$

$= e^{-2t} \left[\int e^{2t} (1) dt + C \right]$

$= e^{-2t} \left[\dfrac{1}{2} e^{2t} + C \right] = \dfrac{1}{2} + Ce^{-2t}$

$y(0) = 1 = \dfrac{1}{2} + Ce^0; \text{ so } C = \dfrac{1}{2} \text{ and the solution}$

is $y = \dfrac{1}{2} \left(1 + e^{-2t} \right).$

23. $y' + \dfrac{y}{1+t} = 20, \, y(0) = 10 \Rightarrow$

$a(t) = \dfrac{1}{1+t}; \, b(t) = 20$

$A(t) = \int \dfrac{1}{1+t} dt = \ln|1+t| = \ln(1+t), \text{ assuming}$

$t \geq 0.$

$e^{A(t)} = e^{\ln(1+t)} = 1+t; \, e^{-A(t)} = e^{-\ln(1+t)} = \dfrac{1}{1+t}$

$y = e^{-A(t)} \left[\int e^{A(t)} b(t) dt + C \right]$

$= \dfrac{1}{1+t} \left[\int (1+t) 20 \, dt + C \right]$

$= \dfrac{1}{1+t} \left[\int (20t + 20) dt + C \right]$

$= \dfrac{1}{1+t} [10t^2 + 20t + C]$

$y(0) = 10 = \dfrac{1}{1+0} [10(0)^2 + 20(0) + C], \text{ so}$

$C = 10$ and the solution is

$y = \dfrac{1}{1+t} [10t^2 + 20t + 10]$

$= \dfrac{1}{1+t} \left[10(t+1)^2 \right] = 10t + 10.$

25. $y' + y = e^{2t}, y(0) = -1; a(t) = 1; b(t) = e^{2t}$

$A(t) = \int (1)dt = t$

$y = e^{-A(t)}\left[\int e^{A(t)}b(t)dt + C\right]$

$= e^{-t}\left[\int e^t e^{2t} dt + C\right]$

$= e^{-t}\left[\int e^{3t}dt + C\right] = e^{-t}\left[\frac{1}{3}e^{3t} + C\right]$

$= \frac{1}{3}e^{2t} + Ce^{-t}$

$y(0) = -1 = \frac{1}{3}e^0 + Ce^0$, so $C = -\frac{4}{3}$ and the

solution is $y = \frac{1}{3}e^{2t} - \frac{4}{3}e^{-t}$.

27. $y' + 2y\cos(2t) = 2\cos(2t), y\left(\frac{\pi}{2}\right) = 0$

$a(t) = 2\cos(2t); b(t) = 2\cos(2t)$

$A(t) = \int 2\cos(2t)dt = \sin(2t)$

$y = e^{-A(t)}\left[\int e^{A(t)}b(t)dt + C\right]$

$= e^{-\sin(2t)}\left[\int e^{\sin(2t)}(2\cos(2t))dt + C\right]$

$= e^{-\sin(2t)}\left[e^{\sin(2t)} + C\right] = 1 + Ce^{-\sin(2t)}$

$y\left(\frac{\pi}{2}\right) = 0 = 1 + Ce^{-\sin(0)} = 1 + Ce^0$, so $C = -1$

and the solution is $y = 1 - e^{-\sin(2t)}$.

29. a. $y'(0) = -\frac{y(0)}{1+0} + 10 = -\frac{50}{1} + 10 = -40$

Since $y'(0)$ is negative, the solution $y(t)$
is decreasing when $t = 0$.

b. $y' = -\frac{y}{1+t} + 10; y(0) = 50$

$y' + \frac{1}{1+t}y = 10; a(t) = \frac{1}{1+t}; b(t) = 10$

$A(t) = \int \frac{1}{1+t}dt = \ln|1+t| = \ln(1+t),$

assuming $t > 0$.

$y = e^{-A(t)}\left[\int e^{A(t)}b(t)dt + C\right]$

$= e^{-\ln(1+t)}\left[\int e^{\ln(1+t)}(10)dt + C\right]$

$= \frac{1}{1+t}\left[\int (10t+10)dt + C\right]$

$= \frac{1}{1+t}[5t^2 + 10t + C]$

$y(0) = 50 = \frac{1}{1+0}[5(0)^2 + 10(0) + C] = C,$

so $C = 50$ and the solution is

$y = \frac{5t^2 + 10t + 50}{1+t}.$

10.4 Applications of First-Order Linear Differential Equations

1. a. $y' = 0.06y + 2400,$ so

$y'\big|_{y=30,000} = .06(30,000) + 2400 = 4200$

The account was growing at \$4200 per
year.

b. Set $y' = 2(2400) = 4800,$ and solve for $y.$
$.06y + 2400 = 4800 \Rightarrow y = 40,000$
The account contained \$40,000.

c. Solve $P(t) = 40,000$ for $t.$

$P(t) = -40,000 + 41,000e^{0.06t} = 40,000$

$.06t = \ln\left(\frac{80}{41}\right); t = \frac{1}{.06}\ln\left(\frac{80}{41}\right) \approx 11.14$

About 11 years and 2 months.

3. a. $y' = .05y + 3600$

b. $y' - .05y = 3600, y(0) = 0; a(t) = -.05;$
$b(t) = 3600$

$A(t) = \int (-.05)dt = -.05t$

$y = e^{-(-0.05t)}\left[\int e^{-0.05t}(3600)dt + C\right]$

$= e^{0.05t}\left[-72,000e^{-0.05t} + C\right]$

$= -72,000 + Ce^{0.05t}$

$f(0) = 0 = -72,000 + Ce^0$, so $C = 72,000$
and the solution is

$f(t) = -72,000 + 72,000e^{0.05t}$

After 25 years, the account will contain

$f(25) = -72,000 + 72,000e^{0.05(25)}$

$\approx \$179,305$

5. Both accounts satisfy the differential equation
    $y' = 0.05y + A$, $y(0) = 0$, where A is the
    yearly contribution. The solution is

    $$y = \frac{A}{0.05}(e^{0.05t} - 1).$$

    At her retirement, Kelly will have

    $$y(20) = \frac{1200}{0.05}(e^{0.05(20)} - 1) \approx \$41,238.76$$

    John will have

    $$y(10) = \frac{2400}{0.05}(e^{0.05(10)} - 1) \approx \$31,138.62$$

    Kelly has more than John.

7. Let $f(t)$ be the amount owed at time t. $f(t)$
    satisfies the differential equation
    $y' = .075 - A$, $y(0) = 100,000$, where A is the
    annual rate of payment. The solution is

    $$f(t) = 100,000e^{0.075t} + \frac{A}{.075}\left(1 - e^{0.075t}\right)$$

    For the loan to be paid in 10 years, $f(10) = 0$.
    Set $f(10) = 0$ and solve for A.
    $f(10)$

    $$= 100,000e^{0.075(10)} + \frac{A}{.075}\left(1 - e^{0.075(10)}\right)$$

    $$= 0$$

    $$A = \frac{7500e^{0.75}}{e^{0.75} - 1} \approx \$14,214.41$$

    The person should pay approximately
    \$14,214.41 per year.

9. **a.** Since the person paid 10% down, the
    amount of the mortgage is
    $.9(278,900) = \$251,010.$

    $$y' = .031y - A, \ y(0) = 251,010$$

    b. The solution is

    $$f(t) = 251,010e^{0.031t} + \frac{A}{.031}\left(1 - e^{0.031t}\right)$$

    To find A, set $f(30) = 0$ and solve for A.
    $f(30)$

    $$= 251,010e^{0.031(30)} + \frac{A}{0.031}\left(1 - e^{0.031(30)}\right)$$

    $$= 0$$

    $$A = \frac{0.031(251,010)e^{0.93}}{e^{0.93} - 1} \approx 12,852.19$$

    The monthly payment will be \$12,852.19/12
    or about \$1071.

    c. Subtract the amount borrowed from the
    total amount paid.
    $30(12,852.19) - 251,010 \approx \$134,556$

11. **a.** Substitute $E(p) = p + 1$ and $q = f(p)$ in

    $$E(p) = \frac{-pf'(p)}{f(p)}, \text{ which yields the result.}$$

    $$-pq' = (p+1)q$$
    $$(\text{or } - py' = (p+1)y \text{ with } y = q)$$

    b. $-p\dfrac{dq}{dp} = (p+1)q; \dfrac{1}{q}\dfrac{dq}{dp} = -1 - \dfrac{1}{p}$

    $$\int \frac{1}{q}\,dq = -\int\left(1 + \frac{1}{p}\right)dp$$

    $$\ln|q| = -(p + \ln|p|) + C$$

    q and p are positive, so we have

    $$\ln q = -(p + \ln p) + C; q = Ae^{-(p+\ln p)}$$

    So $f(p) = Ae^{-(p+\ln p)} = \dfrac{Ae^{-p}}{p}$.

    $$f(1) = 100 = Ae = Ae^{-1}$$

    so $A = 100e$, and the solution is

    $$f(p) = 100\frac{e^{1-p}}{p}.$$

13. $y' = .1(10 - y)$, $y(0) = 350$
    $y' + .1y = 1 \Rightarrow a(t) = .1; b(t) = 1$
    $A(t) = \int (.1)dt = .1t$

    $$y = e^{-0.1t}\left[\int (1)e^{0.1t}dt + C\right] = 10 + Ce^{-0.1t}$$

    $y(0) = 350 = 10 + Ce^0$; so $C = 340$ and the
    solution is $f(t) = 10 + 340e^{-0.1t}$.

15. **a.** $T = 70$

    b. Let $y' = -5$, $T = 70$, and $y = 80$ in
    $y' = k(T - y)$, and solve for k.
    $-5 = k(70 - 80) \Rightarrow k = .5$

    c. The initial value problem is
    $y' = .5(70 - y)$, $y(0) = 98$.
    $y' + .5y = 35 \Rightarrow a(t) = .5; b(t) = 35$
    $A(t) = \int .5\,dt = .5t$

    $$y = e^{-0.5t}\left[\int 35e^{0.5t}dt + C\right] = 70 + Ce^{-0.5t}$$

    $y(0) = 98 = 70 + Ce^0$; so $C = 28$ and the
    solution is $f(t) = 70 + 28e^{-0.5t}$.

d. Set $f(t) = 85$ and solve for t.

$$f(t) = 85 = 70 + 28e^{-0.5t} \Rightarrow t = -2\ln\left(\frac{15}{28}\right) \approx 1.25$$

The person died about 1 hour and 15 minutes before being discovered.

17. $y' = .45y + e^{0.03t} + 2$

19. a. Substitute $y' = 10$ and $y = 75$ into $y' = k(110 - y)$ and solve for k.

$$10 = k(110 - 75) \Rightarrow k = \frac{10}{35} = \frac{2}{7}$$

b. $y'(0) = \frac{2}{7}(110 - y(0)) = \frac{2}{7}(110) = \frac{220}{7} \approx 31.43$ grams/liter/hour.

This is more than 3 times the rate at the end of a four-hour session. Draining and replacing the solution after 4 hours will greatly shorten the amount of time spent in dialysis.

21. a. $y' - .04y = -2000 - 500t,\ y(0) = 100,000$

b. $a(t) = -0.04,\ b(t) = -2000 - 500t;\ A(t) = \int(-.04)\,dt = -.04t$

$$y = e^{-(-0.04t)}\left[\int(-2000 - 500t)e^{-0.04t}\,dt + C\right]$$

$$= e^{0.04t}\left[\frac{1}{(-0.04)^2}e^{-0.04t}\left(0.04(500)t + 0.04(2000) + 500\right) + C\right] = 12,500t + 362,500 + Ce^{0.04t}$$

$y(0) = 100,000 = 12,500(0) + 362,500 + Ce^0$, so $C = -262,500$ and the solution is

$f(t) = -262,500e^{0.04t} + 12,500t + 362,500$.

c.

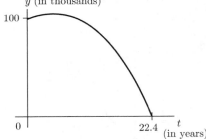

The account will be empty in about 22 years and 4 months.

23. a. $3000 - 500t = 0 \Rightarrow t = 6$
The person contributed for 6 years before starting to withdraw.

b. $y' - 0.04y = 3000 - 500t,\ y(0) = 10,000$

25. a. $y' + .35y = t$

b. Solve the initial value problem $y' + 0.35y = t$, $y(0) = 0$.

$$a(t) = .35; \ b(t) = t; \ A(t) = \int .35 \, dt = .35t$$

$$y = e^{-0.35t} \left[\int te^{0.35t} \, dt + C \right] = e^{-0.35t} \left[\frac{1}{(0.35)^2} e^{0.35t} (.35(1)t + .35(0) - 1) + C \right] = \frac{1}{.35} t - \frac{1}{.1225} + Ce^{-0.35t}$$

$$y(0) = 0 = \frac{1}{0.35}(0) - \frac{1}{0.1225} + Ce^0, \text{ so } C = \frac{1}{0.1225} \text{ and the solution is}$$

$$f(t) = \frac{1}{0.35} t + \frac{1}{0.1225} (e^{-0.35t} - 1), t \leq 8.$$

$f(8) \approx 15.2$ milligrams

10.5 Graphing Solutions of Differential Equations

1.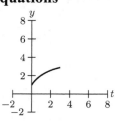

3.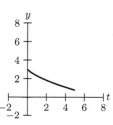

5.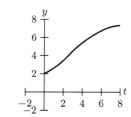

7.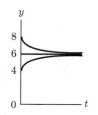

9.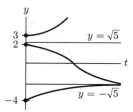

11.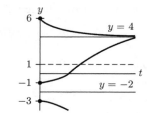

13.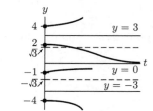

15.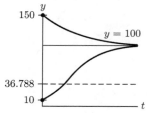

17.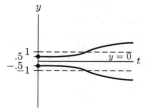

19.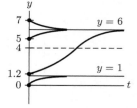

21.

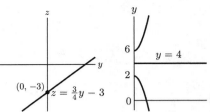

23.

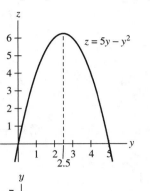

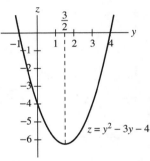

25.

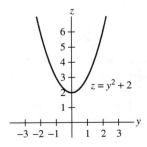

27.

$z = y^2 + 2$

29.

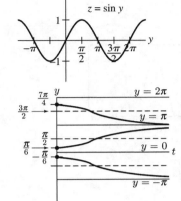

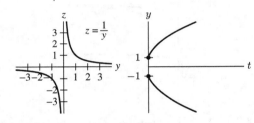

31.

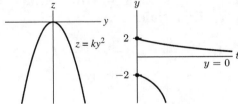

33.

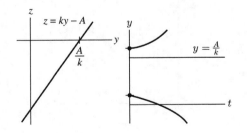

35.

37. $y' = ky(H - y), k > 0$

H = height at maturity
$y = f(t)$

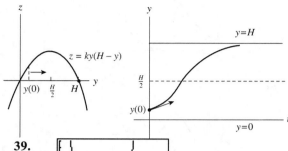

39.

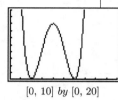

[0, 10] by [0, 20]

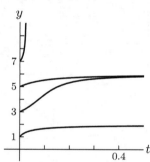

10.6 Applications of Differential Equations

1. a. $\dfrac{dN}{dt} = N(1 - N) = \dfrac{r}{K} N(K - N)$

Carrying capacity $K = 1$, intrinsic rate $r = 1$

b.

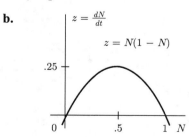

c., d.

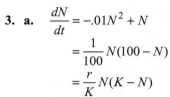

3. a. $\dfrac{dN}{dt} = -.01N^2 + N$

$\qquad = \dfrac{1}{100} N(100 - N)$

$\qquad = \dfrac{r}{K} N(K - N)$

Carrying capacity $K = 100$, intrinsic rate $r = 1$.

b.

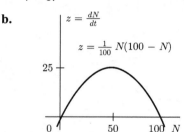

c., d.

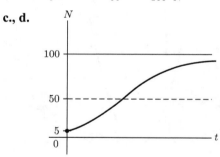

5. $\dfrac{dN}{dt} = \dfrac{r}{1000} N(1000 - N)$

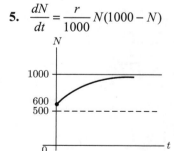

This graph starts out at 600, which is much higher than the graph in Example 2. Both graphs, however, are concave down and approach $N = 1000$ asymptotically from below.

7. $y' = k(100 - y), k > 0$

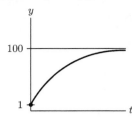

9. $y' = ky(M - y), k > 0$

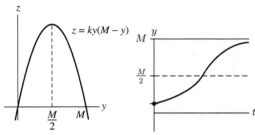

The reaction is fastest when $y = \dfrac{M}{2}$.

11. $y' = k(c - y), k > 0, c > 0$

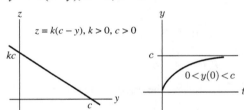

13. $y' = ky^2, k < 0$

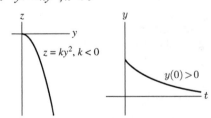

15. $y' = k(E - y), k > 0$

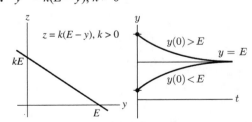

17. a. $\dfrac{dN}{dt} = \dfrac{.4}{1000} N(1000 - N) - 75$

b. $\dfrac{dN}{dt} = 0$ is a quadratic equation with solutions $N = 250$ and $N = 750$. So the new upper limit on the fish population is 750.

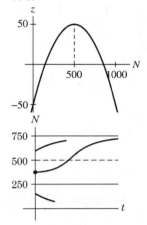

c. Catching 75 fish per year is sustainable. The fish population will continue to grow but will never come close to the pond's carrying capacity.

19. a. $y' = .05y + 10,000, y(0) = 0$

b. $\dfrac{dy}{dt} = .05y + 10,000 \Rightarrow 20\dfrac{dt}{dy} = y + 200,000$

$$\int \dfrac{20}{y + 200,000}\, dy = \int dt$$
$$20\ln(y + 200,000) = t + C$$

$y + 200,000 = Ae^{.05t}; \; y = Ae^{.05t} - 200,000$
Since $y(0) = 0$, $A = 200,000$; so the solution is $y = 200,000(e^{.05t} - 1)$.

$y(5) = 200,000(e^{.05(5)} - 1) \approx \$56,805$

21. a. $y' = .05 - .2y, y(0) = 6.25$

b. $y' = .13 - .2y, y(0) = 6.25$

23. $y' = -.14y$

25.

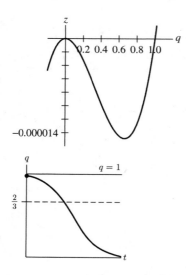

10.7 Numerical Solution of Differential Equations

1. The slope of the graph $= y' = ty - 5$
$= 2 \cdot 4 - 5 = 3$.

3. $y' = y^2 + ty - 7$, $y(0) = 3$
$y'(0) = 3^2 + 0(3) - 7 = 2$
Thus $f(t)$ is increasing at $t = 0$.

5. $g(t, y) = y' = t^2 y$; $y(0) = -2$; $h = .5$.
The iterates are

t_1	y_1
0	−2
.5	−2
1	−2.25

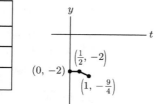

So $y(1) \approx -\dfrac{9}{4}$.

7. $g(t, y) = 2t - y + 1$; $y(0) = 5$; $h = .5$.
The iterates are

t_1	y_1
0	5
.5	3
1	2.5
1.5	2.75
2	3.375

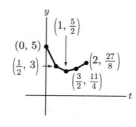

So $f(2) \approx \dfrac{27}{8}$.

9. *Euler's Method*: $g(t, y) = -(t+1)y^2$; $y(0) = 1$, $h = .2$. The iterates are

t_1	y_1
0	1
.2	.8
.4	.6464
.6	.52941
.8	.43972
1	.37011

So $f(1) \approx 3.7011$.
Exact Solution:
$$\int y^{-2} dy = \int -(t+1)\ dt$$
$$-\frac{1}{y} = \frac{-t^2}{2} - t + C_1$$
$$y = \frac{1}{\left(\frac{t^2}{2}\right) + t + C_2} = \frac{2}{t^2 + 2t + C}$$
$y(0) = 1 = \dfrac{2}{C}$, so $C = 2$ and the exact solution
is $f(t) = \dfrac{2}{t^2 + 2t + 2}$.
$$f(1) = \frac{2}{5} = .4\ .$$
Therefore, the error in the estimate is
$\left| .4 - .37011 \right| = .02989$.

11. a. $y' = .1(1 - y)$, $y(0) = 0$

b. Using Euler's method with $n = 3$, $h = 1$ gives the sequence of iterates

t_1	y_1
0	0
1	$0 + .1(1 - 0) = .1$
2	$0.1 + .1(1 - .1) = .19$
3	$0.19 + .1(1 - .19) = .271$

c.
$$\int (y - 1)^{-1} dy = -0.1 \int dt$$
$$\ln|y - 1| = -0.1t + C$$
$$1 - y = Ae^{-0.1t}$$
$$y = 1 - Ae^{-0.1t}$$
$y(0) = 0 = 1 - A$, so $A = 1$.
So $y = 1 - e^{-0.1t}$ and
$$y(3) = 1 - e^{-0.3} \approx .25918.$$

d. Euler's method gives $y(3) = .271$.
The exact value is

$$y(3) = 1 - e^{-0.3} \approx .25918.$$

The error in Euler's method is about

$$|.271 - .25918| = .01182.$$

13. $y' = 0.5(1 - y)(4 - y)$

To generate the following graphs, set the TI83/TI84 in sequence mode and then use the following in the sequence **Y=** editor.

Set v**Min** equal to the value of $y(0)$.

a. $y(0) = -1$; solution is type C: increasing, concave down, and asymptotic to the line $y = 1$.

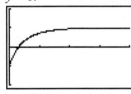

[0, 4] by [−2, 2]

b. $y(0) = 1$; solution is type A: constant solution.

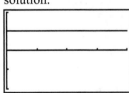

[0, 4] by [−2, 2]

c. $y(0) = 2$; solution is type E: decreasing, concave up, and asymptotic to the line $y = 1$.

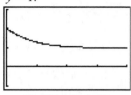

[0, 4] by [−1, 3]

d. $y(0) = 3.9$; solution is type B: decreasing, has an inflection point, and asymptotic to the line $y = 1$.

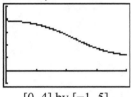

[0, 4] by [−1, 5]

e. $y(0) = 4.1$; solution is type D: concave up and increasing indefinitely.

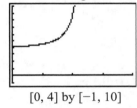

[0, 4] by [−1, 10]

15. $y' = e^t - 2y,\ y(0) = 1,\ y = \dfrac{1}{3}(2e^{-2t} + e^t)$

t_i	0	.25	.5	.75	1	1.25	1.5	1.75	2
y_i	1	.75	.6960	.7602	.9093	1.1342	1.4397	1.8403	2.3588
y	1	.8324	.7948	.8544	.9963	1.2182	1.5271	1.9383	2.4752

The greatest difference between y and y_i is $2.4752 - 2.3588 = .1164$.

Chapter 10 Review Exercises

1. $y^2 y' = 4t^3 - 3t^2 + 2$

$\int y^2 dy = \int (4t^3 - 3t^2 + 2)\ dt$

$\dfrac{y^3}{3} = t^4 - t^3 + 2t + C_1$

$y = (3t^4 - 3t^3 + 6t + C)^{1/3}$

2. $\dfrac{y'}{t+1} = y+1$

$\int (y+1)^{-1} dy = \int (t+1)\ dt \ \ (y \neq -1)$

$\ln|y+1| = \dfrac{t^2}{2} + t + C$

$y = -1 + Ae^{t^2/2+t}$

3. $y' = \dfrac{y}{t} - 3y, t > 0$

$\int y^{-1} dy = \int \left(\dfrac{1}{t} - 3\right) dt \ \ (y \neq 0)$

$\ln|y| = \ln|t| - 3t + C \Rightarrow y = Ate^{-3t}$

4. $(y')^2 = t \Rightarrow y' = \pm\sqrt{t}$

$\int dy = \pm\int t^{1/2} dt \Rightarrow y = \pm\dfrac{2}{3}t^{3/2} + C$

5. $y = 7y' + ty', y(0) = 3$

$\int y^{-1} dy = \int (7+t)^{-1} dt \ \ (y \neq 0)$

$\ln|y| = \ln|7+t| + C$

$y = A(7+t)$

$y(0) = 3 = 7A$ so $A = \dfrac{3}{7}$ and the solution is

$y = \dfrac{3}{7}(7+t) = 3 + \dfrac{3}{7}t$.

6. $y' = te^{t+y}, y(0) = 0$

$\int e^{-y} dy = \int te^t dt$

$-e^{-y} = te^t - e^t + C_1$

$e^{-y} = -te^t + e^t + C$

$y = -\ln(-te^t + e^t + C)$

$y(0) = 0 = -\ln(1 + C)$, so $C = 0$ and the

solution is $y = -\ln(-te^t + e^t)$.

7. $yy' + t = 6t^2, y(0) = 7$

$\int y\ dy = \int (6t^2 - t)\ dt$

$\dfrac{y^2}{2} = 2t^3 - \dfrac{t^2}{2} + C_1$

$y = \pm\sqrt{4t^3 - t^2 + C}$

$y(0) = 7 = \sqrt{C}$, so $C = 49$ and the solution is

$y = \sqrt{4t^3 - t^2 + 49}$.

8. $y' = 5 - 8y, y(0) = 1$

$y' = -8\left(-\dfrac{5}{8} + y\right)$

$\int \left(y - \dfrac{5}{8}\right)^{-1} dy = \int -8\ dt$

$\ln\left|y - \dfrac{5}{8}\right| = -8t + C \Rightarrow y = \dfrac{5}{8} + Ae^{-8t}$

$y(0) = 1 = \dfrac{5}{8} + A$, so $A = \dfrac{3}{8}$ and the solution is

$y = \dfrac{5}{8} + \dfrac{3}{8}e^{-8t}$.

9. $y' - \dfrac{2}{1-t}y = (1-t)^4 \Rightarrow$

$a(t) = -\dfrac{2}{1-t}, b(t) = (1-t)^4$

$A(t) = -\int \dfrac{2}{1-t} dt = 2\ln|1-t| = \ln(1-t)^2$

$e^{A(t)} = e^{\ln(1-t)^2} = (1-t)^2$

$e^{-A(t)} = e^{-\ln(1-t)^2} = \dfrac{1}{(1-t)^2}$

$y = e^{-A(t)}\left[\int e^{A(t)}b(t)dt + C\right]$

$= \dfrac{1}{(1-t)^2}\left[\int (1-t)^2(1-t)^4 dt + C\right]$

$= \dfrac{1}{(1-t)^2}\left[\int (1-t)^6 dt + C\right]$

$= \dfrac{1}{(1-t)^2}\left[-\dfrac{1}{7}(1-t)^7 + C\right]$

$= -\dfrac{1}{7}(1-t)^5 + \dfrac{C}{(1-t)^2}$

or $y = \dfrac{1}{7}(t-1)^5 + \dfrac{C}{(t-1)^2}$

10. $y' - \dfrac{1}{2(1+t)}y = 1+t \Rightarrow$

$a(t) = -\dfrac{1}{2(1+t)}, b(t) = 1+t$

$A(t) = -\dfrac{1}{2}\int \dfrac{1}{1+t}dt = -\dfrac{1}{2}\ln|1+t|$

$\qquad = -\dfrac{1}{2}\ln(1+t), \ (t \geq 0)$

$\qquad = \ln\left[(1+t)^{-1/2}\right]$

$e^{A(t)} = e^{\ln\left[(1+t)^{-1/2}\right]} = (1+t)^{-1/2}$

$e^{-A(t)} = e^{-\ln\left[(1+t)^{-1/2}\right]} = (1+t)^{1/2}$

$y = e^{-A(t)}\left[\int e^{A(t)}b(t)dt + C\right]$

$\quad = (1+t)^{1/2}\left[\int (1+t)^{-1/2}(1+t)dt + C\right]$

$\quad = (1+t)^{1/2}\left[\int (1+t)^{1/2}dt + C\right]$

$\quad = (1+t)^{1/2}\left[\dfrac{2}{3}(1+t)^{3/2} + C\right]$

$\quad = \dfrac{2}{3}(1+t)^2 + C\sqrt{1+t}$

11. slope $= \dfrac{dy}{dx} = x+y; \ y' = x+y; \ y' - y = x$

$a(x) = -1; \ b(x) = x$

$A(x) = \int(-1)dx = -x; \ e^{A(x)} = e^{-x}; \ e^{-A(x)} = e^x$

$y = e^{-A(x)}\left[\int e^{A(x)}b(x)dx + C\right]$

$\quad = e^x\left[\int e^{-x}(x)dx + C\right]$

$\quad = e^x\left[(-x-1)e^{-x} + C\right] = -x-1+Ce^x$

$y(0) = 0 = 0 - 1 + Ce^0; C = 1$

The solution is then $y = -x - 1 + e^x$.

12. a. $\dfrac{dP}{dt} = k(D-S) = k(12 - 3.3P)$

$\dfrac{dP}{dt}\bigg|_{D=10, S=20} = -1 = k(10-20); k = .1$

$\dfrac{dP}{dt} = .1(12 - 3.3P) = 1.2 - .33P$

b. $\dfrac{1}{1.2 - .33P}\dfrac{dP}{dt} = 1$

$\int \dfrac{1}{1.2 - .33P}dP = \int dt$

$-\dfrac{1}{.33}\ln|1.2 - .33P| = t + C$

$1.2 - .33P = A_1 e^{-0.33t} \ (A_1 \neq 0)$

$P = \dfrac{1.2}{.33} - Ae^{-0.33t}$

$P(0) = 1 = \dfrac{1.2}{.33} - Ae^0$

$A = \dfrac{.87}{.33}$

The solution is

$P = \dfrac{1}{.33}\left(1.2 - 0.87e^{-0.33t}\right)$, or

$P = \dfrac{1}{11}\left(40 - 29e^{-0.33t}\right)$.

13. If $f(t) = y(t) = 3$, then

$y'(t) = (2-3)e^{-3} = -e^{-3} < 0$, so $f(t)$ is decreasing at this point.

14. Note that the constant solution $y = 1$ is a solution to the given equation. Since this solution satisfies the initial condition $y(0) = 1$, it must be the desired particular solution.

15. $z = 2\cos y$

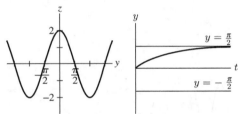

16. $z = 5 + 4y - y^2$

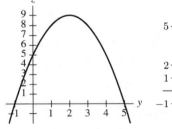

17. $z = y^2 + y$

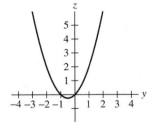

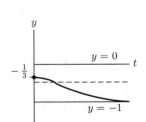

18. $z = y^2 - 2y + 1$

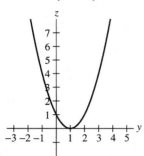

19. $z = \ln y$

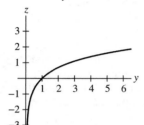

20. $z = \cos y + 1$

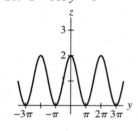

21. $z = \dfrac{1}{y^2 + 1}$

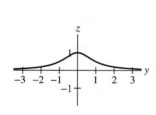

22. $z = \dfrac{3}{y + 3}$

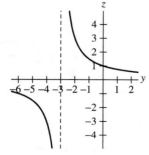

23. $z = .4y^2(1 - y)$

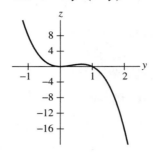

24. $z = y^3 - 6y^2 + 9y$

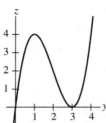

25. a. $N' = .015N - 3000$

b. There is a constant solution $N = 200{,}000$, but it is unstable. It is unlikely a city would have such a constant population.

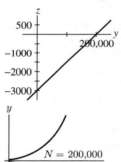

26. a. $\left(10 - \dfrac{1}{4}y\right)$ represents the amount of unreacted substance A present and $\left(15 - \dfrac{3}{4}y\right)$ that of B.

b. $k > 0$ since the amount of C is increasing.

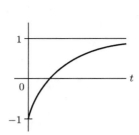
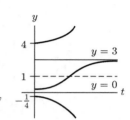
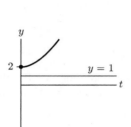
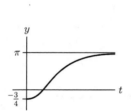
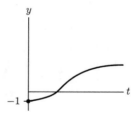
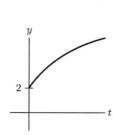

c.

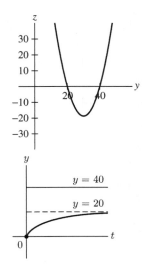

27. Let $f(t)$ be the balance in the account after t years. Then $f(t)$ satisfies the differential equation $y' = .05y - 2000$, $y(0) = 20,000$.

Thus $y' = .05(y - 40,000)$.

$$\int (y - 40,000)^{-1} \, dy = \int .05 \, dt$$
$$\ln|y - 40,000| = .05t + C$$
$$y = 40,000 + Ae^{0.05t}$$

$y(0) = 20,000 = 40,000 + A$, so $A = -20,000$ and $f(t) = 40,000 - 20,000e^{0.05t}$.

We want to find t so that $f(t) = 0$, i.e.

$40,000 = 20,000e^{0.05t}$,

$2 = e^{0.05t} \Rightarrow \ln 2 = 0.05t \Rightarrow$
$t = 20 \ln 2 \approx 13.86294$ years.

28. Let $f(t)$ be the balance in the savings account after t years and let $M = f(0)$ be the initial amount. Then $f(t)$ satisfies the differential equation

$y' = .06y - 12,000 = .06(y - 200,000)$,

$y(0) = M$. Thus,

$$\int (y - 200,000)^{-1} dy = \int .06 \, dt$$
$$\ln|y - 200,000| = .06t + C$$
$$y = 200,000 + Ae^{0.06t}$$

Now $y(0) = M = 200,000 + A$, so $A = M - 200,000$ and

$$f(t) = 200,000 + (M - 200,000)e^{0.06t}.$$

a. If the initial amount M is to fund the endowment forever, we must have $f(t) > 0$ for all t. This happens first in case $M - 200,000 \geq 0 \Rightarrow M \geq \$200,000$. (A \$200,000 endowment would give the constant solution $f(t) = 200,000$.)

b. Using the expression for $f(t)$ above, setting $f(20) = 0$ gives

$$-200,000 = (M - 200,000)e^{1.2} \Rightarrow$$
$$-\frac{200,000}{e^{1.2}} + 200,000 = M \Rightarrow$$
$$M \approx \$139,761.16.$$

29. *Euler's Method*: $y' = 2e^{2t-y}$, $y(0) = 0$, $n = 4$; $h = .5$. The iterates are

t_1	y_1
0	0
.5	1
1	2
1.5	3
2	4

So the estimate is $f(2) \approx 4$.
Exact Solution:

$$\int e^y dy = \int 2e^{2t} dt \Rightarrow e^y = e^{2t} + C \Rightarrow$$
$$y = \ln(e^{2t} + C)$$

$y(0) = 0 = \ln(1 + C)$, so $C = 0$ and the exact solution is $y = \ln(e^{2t}) = 2t$. Then $f(2) = 4$ and the estimate above is exact.

30. $y' = \dfrac{t+1}{y}$, $y(0) = 1$

Euler's Method: $n = 3$, $h = \frac{1}{3}$. The iterates are

t_1	y_1
0	1
$\frac{1}{3}$	$\frac{4}{3}$
$\frac{2}{3}$	$\frac{5}{3}$
1	2

So $y(1) \approx 2$.
Exact Solution:

$$\int y \, dy = \int (t+1) \, dt$$
$$\frac{y^2}{2} = \frac{t^2}{2} + t + C_1$$
$$y = \pm\sqrt{t^2 + 2t + C}$$

$y(0) = 1 = \sqrt{C}$, so $C = 1$ and the exact solution is $y = \sqrt{t^2 + 2t + 1} = \sqrt{(t+1)^2} = |t+1|$.
Thus $y(1) = 2$ and the above estimate is exact.

31. $y' = .1y(20 - y), y(0) = 2; n = 6, h = .5.$

t_1	y_1
0	2
.5	3.8
1	6.878
1.5	11.39066
2	16.29396
2.5	19.31326
3	19.97642

$y(3) \approx 19.97642$

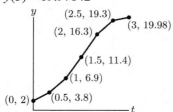

32. $y' = \dfrac{1}{2}y(y - 10); y(0) = 9; n = 5, h = 0.2$

t_1	y_1
0	9
.2	8.1
.4	6.561
.6	4.30467
.8	1.85302
1	.34337

$y(1) \approx .34337$

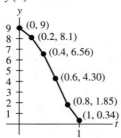

Chapter 11 Taylor Polynomials and Infinite Series

11.1 Taylor Polynomials

1. $p_3(x) = \sin 0 + \cos(0)x + \dfrac{-\sin 0}{2!}x^2 + \dfrac{-\cos 0}{3!}x^3 = x - \dfrac{1}{6}x^3$

3. $p_3(x) = 5e^{2(0)} + 10e^{2(0)}x + 20e^{2(0)}\dfrac{x^2}{2!} + 40e^{2(0)}\dfrac{x^3}{3!} = 5 + 10x + 10x^2 + \dfrac{20}{3}x^3$

5. $p_3(x) = 1 + 2x - \dfrac{4x^2}{2!} + 24\dfrac{x^3}{3!} = 1 + 2x - 2x^2 + 4x^3$

7. $f(x) = xe^{3x};\ f'(x) = e^{3x} + 3xe^{3x}$

$f''(x) = 3e^{3x} + 3e^{3x} + 9xe^{3x} = 6e^{3x} + 9xe^{3x}$

$f'''(x) = 18e^{3x} + 9e^{3x} + 27xe^{3x} = 27e^{3x} + 27xe^{3x}$

$p_3(x) = 0 + (1 + 0)x + (6 + 0)\dfrac{x^2}{2!} + (27 + 0)\dfrac{x^3}{3!} = x + 3x^2 + \dfrac{9}{2}x^3$

9. $p_4(x) = e^0 + e^0 x + e^0\dfrac{x^2}{2!} + e^0\dfrac{x^3}{3!} + e^0\dfrac{x^4}{4!} = 1 + x + \dfrac{x^2}{2} + \dfrac{x^3}{6} + \dfrac{x^4}{24}$

$e^{0.01} \approx 1 + .01 + \dfrac{(.01)^2}{2} + \dfrac{(.01)^3}{6} + \dfrac{(.01)^4}{24} \approx 1.01005$

11. See Example 3 on page 515 in the text for the derivation of the Taylor polynomials at $x = 0$.

$f(x)$ and $p_1(x) = 1 + x$ $\qquad\qquad$ $f(x)$ and $p_2(x) = 1 + x + x^2$ $\qquad\qquad$ $f(x)$ and $p_3(x) = 1 + x + x^2 + x^3$

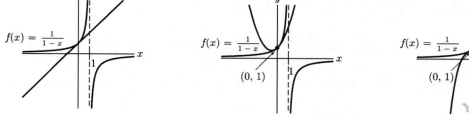

13. $p_n(x) = e^0 + e^0 x + e^0\dfrac{x^2}{2!} + \cdots + e^0\dfrac{x^n}{n!} = 1 + x + \dfrac{x^2}{2} + \dfrac{x^3}{6} + \cdots + \dfrac{x^n}{n!}$

15. $f(x) = \ln(1 + x^2);\ f'(x) = \dfrac{2x}{1 + x^2};\ f''(x) = \dfrac{2(1 + x^2) - 4x^2}{(1 + x^2)^2} = \dfrac{-2x^2 + 2}{(1 + x^2)^2}$

$p_2(x) = \ln(1) + (0)x + 2\dfrac{x^2}{2!} = x^2$

$\displaystyle\int_0^{1/2} \ln(1 + x^2)\,dx \approx \int_0^{1/2} x^2\,dx = \dfrac{x^3}{3}\Big|_0^{1/2} = \dfrac{1}{24} \approx .0417$

17. $f(x) = \dfrac{1}{5-x}$; $f'(x) = \dfrac{1}{(5-x)^2}$; $f''(x) = \dfrac{2}{(5-x)^3}$; $f'''(x) = \dfrac{6}{(5-x)^4}$

$p_3(x) = \dfrac{1}{5-4} + \dfrac{1}{(5-4)^2}(x-4) - \dfrac{2}{(5-4)^3}\dfrac{(x-4)^2}{2!} + \dfrac{6}{(5-4)^4}\dfrac{(x-4)^3}{3!}$

$\qquad = 1 + (x-4) + (x-4)^2 + (x-4)^3$

19. $f(x) = \cos x$; $f'(x) = -\sin x$; $f''(x) = -\cos x$; $f'''(x) = \sin x$; $f^{(4)}(x) = \cos x$

$p_3(x) = \cos\pi - \sin\pi(x-\pi) - \cos\pi\dfrac{(x-\pi)^2}{2!} + \sin\pi\dfrac{(x-\pi)^3}{3!} = -1 + \dfrac{1}{2}(x-\pi)^2$

$p_4(x) = p_3(x) + \cos\pi\dfrac{(x-\pi)^4}{4!} = -1 + \dfrac{1}{2}(x-\pi)^2 - \dfrac{1}{24}(x-\pi)^4$

21. $f(x) = \sqrt{x}$; $f'(x) = \dfrac{1}{2\sqrt{x}}$; $f''(x) = -\dfrac{1}{4}x^{-3/2}$

$p_2(x) = \sqrt{9} + \dfrac{1}{2\sqrt{9}}(x-9) - \dfrac{1}{4}9^{-3/2}\dfrac{(x-9)^2}{2!} = 3 + \dfrac{1}{6}(x-9) - \dfrac{1}{216}(x-9)^2$

$p_2(9.3) = 3 + \dfrac{.3}{6} - \dfrac{(.3)^2}{216} \approx 3.04958$

23. $f(x) = x^4 + x + 1$; $f'(x) = 4x^3 + 1$; $f''(x) = 12x^2$; $f'''(x) = 24x$; $f^{(4)}(x) = 24$;

$f^{(n)}(x) = 0$ all $n > 4$.

$p_0(x) = 2^4 + 2 + 1 = 19$

$p_1(x) = 19 + (4(2)^3 + 1)(x-2) = 19 + 33(x-2)$

$p_2(x) = 19 + 33(x-2) + 12(2)^2\dfrac{(x-2)^2}{2!} = 19 + 33(x-2) + 24(x-2)^2$

$p_3(x) = 19 + 33(x-2) + 24(x-2)^2 + 24(2)\dfrac{(x-2)^3}{3!} = 19 + 33(x-2) + 24(x-2)^2 + 8(x-2)^3$

$p_4(x) = p_3(x) + 24\dfrac{(x-2)^4}{4!} = 19 + 33(x-2) + 24(x-2)^2 + 8(x-2)^3 + (x-2)^4$; $p_n(x) = p_4(x)$ for all $n \geq 4$.

25. The expression on the right must be the Taylor expansion of $f(x)$ at $x = 0$. Therefore, $f''(0) = -5$ and $f'''(0) = 7$.

27. a. If $f(x) = \cos x$, then $f^{(4)}(x) = \cos x$ as well, so $|f^{(4)}(c)| \leq |\cos(c)| \leq 1$ for all c.

b. $R_3(0.12) = \dfrac{\cos(c)}{4!}(0.12)^4$ for some c between 0 and 0.12. From part (a), it follows that

$|R_3(0.12)| = |\text{error in the approximation}|$

$\leq \dfrac{1}{4!}(.12)^4 = 8.64 \times 10^{-6}$

29. a. $f'''(x) = \dfrac{3}{8}x^{-5/2}$

$R_2(x) = \dfrac{f'''(c)}{3!}(x-9)^3$

$= \dfrac{\frac{3}{8}c^{-5/2}}{3!}(x-9)^3 = \dfrac{c^{-5/2}}{16}(x-9)^3$

for some c between 9 and x.

b. The function $f^{(3)}(c) = \dfrac{3}{8}c^{-5/2}$ is positive and decreasing for $c > 0$. Thus for all

$c \geq 9$, $|f^{(3)}(c)| \leq f^{(3)}(9) = \dfrac{3}{8 \cdot 3^5} = \dfrac{1}{648}$.

c. $|\text{error}| = R_2(9.3) \le \dfrac{\frac{1}{648}}{3!}(.3)^3$ (from part b)

$$= \frac{1}{144} \times 10^{-3} < 7 \times 10^{-6}$$

31. $y_1 = \dfrac{1}{1-x}$, $y_2 = 1 + x + x^2 + x^3 + x^4$

When $b = .55$, the difference is approximately
$2.22 - 2.11 = .11$.
When $b = -.68$, the difference is
approximately $.682 - .595 = .087$.

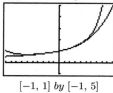

$[-1, 1]$ *by* $[-1, 5]$

33. $y_1 = e^x$, $y_2 = 1 + x + \dfrac{1}{2}x^2 + \dfrac{1}{6}x^3 + \dfrac{1}{24}x^4$

Choose $b = 1.85$. Then the difference is
approximately $6.3598 - 6.1046 = .2552$. When
$x = 3$, the difference is approximately
$20.0855 - 16.375 = 3.7105$.

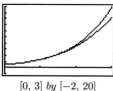

$[0, 3]$ *by* $[-2, 20]$

11.2 The Newton-Raphson Algorithm

1. Let $f(x) = x^2 - 5$, $f'(x) = 2x$.

$$x_0 = 2; \quad x_1 = 2 - \frac{2^2 - 5}{2(2)} = 2.25$$

$$x_2 = 2.25 - \frac{(2.25)^2 - 5}{2(2.25)} \approx 2.2361$$

$$x_3 \approx 2.2361 - \frac{(2.2361)^2 - 5}{2(2.2361)} \approx 2.23607$$

3. Let $f(x) = x^3 - 6$; $f'(x) = 3x^2$.

$$x_0 = 2; \quad x_1 = 2 - \frac{2^3 - 6}{3(2^2)} \approx 1.8333$$

$$x_2 \approx 1.8333 - \frac{1.8333^3 - 6}{3(1.8333)^2} \approx 1.81726$$

$$x_3 = 1.81726 - \frac{1.81726^3 - 6}{3(1.81726)^2} \approx 1.81712$$

5. $f(x) = x^2 - x - 5$; $f'(x) = 2x - 1$

$$x_0 = 2; \quad x_1 = \frac{2^2 - 2 - 5}{2(2) - 1} = 3$$

$$x_2 = 3 - \frac{3^2 - 3 - 5}{2(3) - 1} = 2.8$$

$$x_3 = 2.8 - \frac{2.8^2 - 2.8 - 5}{2(2.8) - 1} \approx 2.7913$$

7. $f(x) = \sin x + x^2 - 1$; $f'(x) = \cos x + 2x$

$x_0 = 0$, $x_1 = 1$, $x_2 \approx .66875$, $x_3 \approx .63707$

9.

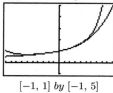

$(0, 2)$

$x_0 = -1$; $x_1 = -.8$; $x_2 \approx -.77143$;
$x_3 \approx -.77092$

11. $f(x) = e^{-x} - x^2$; $f'(x) = -e^{-x} - 2x$

$x_0 = 1$, $x_1 \approx .73304$, $x_2 \approx .70381$,
$x_3 \approx .70347$

13. The internal rate of return i satisfies the equation

$$500(1+i)^3 - 100(1+i)^2 - 200(1+i) - 300 = 0.$$

Putting $x = 1 + i$,

$$f(x) = 500x^3 - 100x^2 - 200x - 300;$$

$$f'(x) = 1500x^2 - 200x - 200$$

If $x_0 = 1.1$, we have

$x_1 \approx 1.08244$, $x_2 \approx 1.08208$, $x_3 \approx 1.08208$.

Thus $i \approx .0821$ or 8.21% per month.

15. The interest rate i satisfies the equation

$f(i) = 0$, where $f(i) = 563i + 116((1+i)^{-5} - 1)$;

$f'(i) = 563 - 580(1+i)^{-6}$. Starting with

$i_0 = 0.02$ gives $i_1 \approx 0.01323$, $i_2 \approx 0.01062$,

$i_3 \approx 0.01003$.

Thus the monthly interest rate is about 1%.

17. $x_1 \approx 3.5, x_2 \approx 3.0$

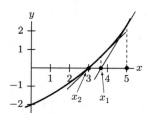

19. $m = 4$(slope) $= f'(x)$ at $x = 3, f(3) = 17$.

$$x_1 = x_0 - \frac{f(x_0)}{f'(x_0)} = 3 - \frac{17}{4} = -\frac{5}{4}$$

21. $x_0 > 0$

23. Given any x_0, x_1 will be the zero of $f(x)$.

$$x_1 = x_0 - \frac{mx_0 + b}{m},$$

$$f(x_1) = m\left(x_0 - \frac{mx_0 + b}{m}\right) + b = 0$$

25. $f(x) = x^{1/3}$; $f'(x) = \frac{1}{3}x^{-2/3}$

$$x_0 = 1; \ x_1 = 1 - \frac{1}{\left(\frac{1}{3}\right)} = -2 ;$$

$$x_2 = -2 - \frac{\sqrt[3]{-2}}{\left(\frac{1}{3}\right)(-2)^{-2/3}} = 4; \ x_3 = 4 - 12 = -8$$

The iterates diverge.

27. Assume that the calculator displays up to 10 digits. For $f(x) = x^2 - 4$, $x_n = 2$ when $n = 4$. For $g(x) = (x - 2)^2$, $x_n = 2$ when $n = 31$. This difference is due to the derivatives: $f'(x) = 2x$, whereas $g'(x) = 2(x - 2) = 2x - 4$.

$[-6, 6]$ *by* $[-5, 10]$

29.

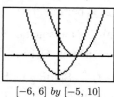

$[-2, 2]$ *by* $[-2, 2]$

a. $x_0 = 1.1$ leads to the zero at $x = \sqrt{2}$.

b. $x_0 = 0.95$ leads to the zero at $x = -\sqrt{2}$.

c. $x_0 = 0.9$ leads to the zero at $x = 0$.

11.3 Infinite Series

1. The series is geometric with $a = 1$, $r = \frac{1}{6}$, so the sum is $\dfrac{1}{1 - \frac{1}{6}} = \dfrac{6}{5}$.

3. $a = 1$, $r = -\dfrac{1}{9}$; sum $= \dfrac{1}{1 + \frac{1}{9}} = \dfrac{9}{10}$

5. $a = 2$, $r = \dfrac{1}{3}$; sum $= \dfrac{2}{1 - \frac{1}{3}} = 3$

7. $a = \dfrac{1}{5}$, $r = \dfrac{\frac{1}{5^4}}{\frac{1}{5}} = \dfrac{1}{5^3} = \dfrac{1}{125}$

sum $= \dfrac{\frac{1}{5}}{1 - \frac{1}{125}} = \dfrac{25}{124}$

9. $a = 3$; $r = \dfrac{-\frac{3^2}{7}}{3} = -\dfrac{3}{7}$

sum $= \dfrac{3}{1 + \frac{3}{7}} = \dfrac{21}{10}$

11. $a = \dfrac{2}{5^4} = \dfrac{2}{625}$; $r = \dfrac{\frac{-2^4}{5^5}}{\frac{2}{5^4}} = \dfrac{-2^3}{5} = -\dfrac{8}{5}$

Since $|r| > 1$, the series diverges.

13. $a = 5$; $r = \dfrac{4}{5}$; sum $= \dfrac{5}{1 - \frac{4}{5}} = 25$

15. $.27\overline{27} = \dfrac{27}{100} + \dfrac{27}{100^2} + \dfrac{27}{100^3} + \cdots$

This is a geometric series with $a = \dfrac{27}{100}$;

$r = \dfrac{1}{100}$. Therefore $.27\overline{27} = \dfrac{\frac{27}{100}}{1 - \frac{1}{100}} = \dfrac{3}{11}$.

17. $0.2\overline{2} = \dfrac{2}{10} + \dfrac{2}{10^2} + \dfrac{2}{10^3} + \cdots$

This is a geometric series with $a = \dfrac{1}{5}$; $r = \dfrac{1}{10}$.

Therefore, $.2\overline{2} = \dfrac{\frac{1}{5}}{1 - \frac{1}{10}} = \dfrac{2}{9}$.

19. $0.01\overline{1011} = \dfrac{11}{1000} + \dfrac{11}{1000^2} + \cdots$

This is a geometric series with $a = \dfrac{11}{1000}$ and $r = \dfrac{1}{1000}$. Its sum is $\dfrac{\frac{11}{1000}}{1 - \frac{1}{1000}} = \dfrac{11}{999}$. Therefore,

$4.01\overline{1011} = 4 + \dfrac{11}{999} = \dfrac{4007}{999}$.

21. $0.99\overline{9} = \dfrac{9}{10} + \dfrac{9}{10^2} + \dfrac{9}{10^3} + \cdots$

This is a geometric series with $a = \dfrac{9}{10}$, $r = \dfrac{1}{10}$. Therefore, $.99\overline{9} = \dfrac{\frac{9}{10}}{1 - \frac{1}{10}} = 1$.

23. The additional spending would be $10(0.95) + 10(0.95)^2 + 10(0.95)^3 + \cdots = \dfrac{10(0.95)}{1 - 0.95} = 190$ billion dollars

25. a. $\displaystyle\sum_{k=0}^{\infty} 100(1.01)^{-k}$

b. $a = 100;\ r = \dfrac{1}{1.01}$

$\text{sum} = \dfrac{100}{1 - \frac{100}{101}} = \$10{,}100$

27. Amount $= 1{,}000{,}000[1 + (0.396) + (0.396)^2 + (0.396)^3 + \ldots + (0.396)^n + \ldots]$

$= 1{,}000{,}000 \left(\dfrac{1}{1 - 0.396} \right) \approx \$1{,}655{,}629$

29. $6 + 6(0.7) + 6(0.7)^2 + 6(0.7)^3 + \cdots = \dfrac{6}{1 - 0.7} = 20$ mg

31. $M + \dfrac{3}{4}M + \left(\dfrac{3}{4}\right)^2 M + \cdots = M\left(\dfrac{1}{1 - 0.75}\right) = 4M$

$4M = 20$, so $M = 5$ mg.

33. a. $S_{10} = 3 - \dfrac{5}{10} = 2.5$

b. Yes, since $\displaystyle\lim_{n \to \infty}\left(3 - \dfrac{5}{n}\right) = 3$.

35. $\dfrac{1}{1 - \frac{5}{6}} = 6$

37. $\displaystyle\sum_{j=1}^{\infty} 5^{-2j} = \sum_{j=1}^{\infty}\left(\dfrac{1}{25}\right)^j = \dfrac{1}{25}\sum_{j=0}^{\infty}\left(\dfrac{1}{25}\right)^j = \dfrac{\frac{1}{25}}{1 - \frac{1}{25}} = \dfrac{1}{24}$

39. $\displaystyle\sum_{k=0}^{\infty}(-1)^k \dfrac{3^{k+1}}{5^k} = 3\sum_{k=0}^{\infty}\left(-\dfrac{3}{5}\right)^k = \dfrac{3}{1 + \frac{3}{5}} = \dfrac{15}{8}$

41. a. $(1-r)(a+ar+ar^2+\cdots+ar^n) = a+ar+ar^2+\cdots+ar^n-ar-ar^2-\cdots-ar^n-ar^{n+1} = a-ar^{n+1}$

Thus $a+ar+ar^2+\cdots+ar^n = \dfrac{a-ar^{n+1}}{1-r} = \dfrac{a}{1-r} - \dfrac{ar^{n+1}}{1-r}$.

b. As $n\to\infty$, $\dfrac{ar^{n+1}}{1-r}$ approaches 0 if $|r|<1$. Hence, in this case,

$$\sum_{k=0}^{\infty} ar^k = \lim_{n\to\infty}\sum_{k=0}^{n} ar^k = \lim_{n\to\infty}\left[\frac{a}{1-r} - \frac{ar^{n+1}}{1-r}\right] = \frac{a}{1-r}.$$

c. If $|r|>1$, $\left|\dfrac{ar^{n+1}}{1-r}\right|$ increases without bound as $n\to\infty$. Thus in this case, the series diverges.

d. If $r=1$, then the series is $a+a+a+\cdots$ which clearly diverges. If $r=-1$, then the expression in part (a) is $\dfrac{a}{2} - \dfrac{a(-1)^n}{2}$. Thus the partial sums alternate between 0 and a and $\lim_{n\to\infty}\sum_{k=0}^{\infty} ar^k$ does not exist.

43. $a=1,\ r=\dfrac{2}{3}$

$$\sum_{x=0}^{\infty}\left(\frac{2}{3}\right)^x = \frac{1}{1-\frac{2}{3}} = \frac{1}{\frac{1}{3}} = 3$$

45. $\displaystyle\sum_{x=1}^{5}\frac{(-1)^{2x}}{2^{(x+1)}} = \frac{(-1)^2}{2^2} + \frac{(-1)^4}{2^3} + \frac{(-1)^3}{2^4} + \frac{(-1)^8}{2^5} + \frac{(-1)^{10}}{2^6} = \frac{1}{4}+\frac{1}{8}+\frac{1}{16}+\frac{1}{32}+\frac{1}{64}$

$\displaystyle\sum_{x=1}^{\infty}\frac{(-1)^{2x}}{2^{(x+1)}} = 1 + \frac{1}{2} + \frac{1}{4} + \frac{1}{8} + \cdots - \left(1+\frac{1}{2}\right) = 2 - \left(\frac{3}{2}\right) = \frac{1}{2}$

47. $\displaystyle\sum_{x=1}^{n} x = \frac{n(n+1)}{2}$

$\displaystyle\sum_{x=1}^{10} x = 55 \Rightarrow \frac{10(10+1)}{2} = 55$

$\displaystyle\sum_{x=1}^{50} x = 1275 \Rightarrow \frac{50(50+1)}{2} = 1275$

$\displaystyle\sum_{x=1}^{100} x = 5050 \Rightarrow \frac{100(101)}{2} = 5050$

49.
```
sum(seq(1/(X²),X
,1,999,1)
         1.643933567
π²/6
         1.644934067
■
```

11.4 Series with Positive Terms

1. $\displaystyle\int_1^b \frac{3}{\sqrt{x}}\,dx = 6x^{1/2}\Big|_1^b = 6b^{1/2} - 6$

Thus $\displaystyle\int_1^\infty \frac{3}{\sqrt{x}}\,dx$ diverges so $\displaystyle\sum_{k=1}^{\infty}\frac{3}{\sqrt{k}}$ diverges.

3. $\displaystyle\int_2^b \frac{1}{(x-1)^3}\,dx = -\frac{1}{2}(x-1)^{-2}\Big|_2^b = \frac{1}{2} - \frac{1}{2}(b-1)^{-2}$

Thus $\displaystyle\int_2^\infty \frac{1}{(x-1)^3}\,dx = \frac{1}{2}$ and $\displaystyle\sum_{k=2}^{\infty}\frac{1}{(k-1)^3}$ converges.

5. $\displaystyle\int_1^b \frac{2}{5x-1}\,dx = \frac{2}{5}\ln(5x-1)\Big|_1^b$

$= \frac{2}{5}\ln(5b-1) - \frac{2}{5}\ln 4$

Thus $\displaystyle\int_1^\infty \frac{2}{5x-1}\,dx$ diverges, so $\displaystyle\sum_{k=1}^{\infty}\frac{2}{5k-1}$ diverges.

7. $\displaystyle\int_2^\infty \frac{x}{(x^2+1)^{3/2}}\,dx = -(x^2+1)^{-1/2}\Big|_2^b$

$= \frac{1}{\sqrt{5}} - (b^2+1)^{-1/2}$

$\displaystyle\sum_{k=2}^{\infty}\frac{k}{(k^2+1)^{3/2}}$ converges.

9. $\int_2^b \dfrac{1}{x(\ln x)^2}\,dx = -(\ln x)^{-1}\Big|_2^b = \dfrac{1}{\ln 2} - \dfrac{1}{\ln b}$

so $\displaystyle\sum_{k=2}^{\infty} \dfrac{1}{k(\ln k)^2}\,dx$ converges.

11. $\int_1^b e^{3-x}\,dx = -e^{3-x}\Big|_1^b = e^2 - e^{3-b}$

so $\displaystyle\sum_{k=1}^{\infty} e^{3-k}$ converges.

13. $\int_1^b xe^{-x^2}\,dx = -\dfrac{1}{2}e^{-x^2}\Big|_1^b = \dfrac{1}{2}e^{-1} - \dfrac{1}{2}e^{-b^2}$

so $\displaystyle\sum_{k=1}^{\infty} ke^{-k^2}$ converges.

15. $\int_1^b \dfrac{2x+1}{x^2+x+2}\,dx = \ln(x^2+x+2)\Big|_1^b$

$\qquad\qquad = \ln(b^2+b+2) - \ln 4$

so $\displaystyle\sum_{k=1}^{\infty} \dfrac{2k+1}{k^2+k+2}$ diverges.

17. The series is $\displaystyle\sum_{k=0}^{\infty} \dfrac{3}{9+k^2}$. Let $f(x) = \dfrac{3}{9+x^2}$.

Then $f(x) > 0$ for all $x \ge 0$, $f(x)$ is continuous, $f'(x) = -6x(9+x^2)^{-2} < 0$ for all $x > 0$, so $f(x)$ is decreasing. Therefore, the series converges.

19. Let $f(x) = \dfrac{x}{e^x} = xe^{-x}$. Then $f(x) > 0$ for all $x \ge 1$, $f(x)$ is continuous and

$f'(x) = \dfrac{e^x - xe^x}{e^{2x}} < 0$ for all $x > 1$, so $f(x)$ is decreasing for $x \ge 1$.

Integrating by parts,

$\int_1^b xe^{-x}\,dx = -xe^{-x} - e^{-x}\Big|_1^b$

$\qquad\qquad = 2e^{-1} - be^{-b} - e^{-b}.$

Thus $\displaystyle\sum_{k=1}^{\infty} ke^{-k}$ converges.

21. For all $k \ge 2$, $\dfrac{1}{k^2+5} < \dfrac{1}{k^2}$. The series $\displaystyle\sum_{k=2}^{\infty} \dfrac{1}{k^2}$ is shown in the text to be convergent. Thus

$\displaystyle\sum_{k=2}^{\infty} \dfrac{1}{k^2+5}$ converges by the comparison test.

23. For $k \ge 1$, $\dfrac{1}{2^k+k} < \dfrac{1}{2^k}$. $\displaystyle\sum_{k=1}^{\infty} \dfrac{1}{2^k}$ converges. It is geometric with $r = \dfrac{1}{2}$, so $\displaystyle\sum_{k=1}^{\infty} \dfrac{1}{2^k+k}$ converges.

25. For $k \ge 1$, $\dfrac{1}{5^k}\cos^2\left(\dfrac{k\pi}{4}\right) \le \dfrac{1}{5^k}(1)$. $\displaystyle\sum_{k=1}^{\infty} \dfrac{1}{5^k}$ converges (geometric with $r = 1$), so

$\displaystyle\sum_{k=1}^{\infty} \dfrac{1}{5^k}\cos^2\left(\dfrac{k\pi}{4}\right)$ converges.

27. No; in order for the comparison test to yield any information we would need $\dfrac{1}{k \ln k} > \dfrac{1}{k}$ for $k \ge 2$, which is false.

29. Area of top set of rectangles: T
Area of bottom set of rectangles: S
Area of combined set of rectangles: $S + T$

31. $\displaystyle\sum_{k=0}^{\infty} \dfrac{8^k + 9^k}{10^k} = \sum_{k=0}^{\infty}\left(\dfrac{8}{10}\right)^k + \sum_{k=0}^{\infty}\left(\dfrac{9}{10}\right)^k$

$\qquad = \dfrac{1}{1 - \frac{8}{10}} + \dfrac{1}{1 - \frac{9}{10}} = 5 + 10 = 15$

11.5 Taylor Series

1. $f(x) = \dfrac{1}{2x+3}$; $f'(x) = -\dfrac{2}{(2x+3)^2}$;

$f''(x) = \dfrac{2^2 \cdot 2}{(2x+3)^3}$; $f'''(x) = \dfrac{-2^3 \cdot 2 \cdot 3}{(2x+3)^4}$

The Taylor series at $x = 0$ is

$f(x) = \dfrac{1}{3} - \dfrac{2}{9}x + \dfrac{2^2 \cdot 2}{3^3 \cdot 2!}x^2 - \dfrac{2^3 \cdot 3!}{3^4 \cdot 3!}x^3 + \cdots$

$\quad = \dfrac{1}{3} - \dfrac{2}{9}x + \dfrac{4}{27}x^2 - \dfrac{2^3}{3^4}x^3 + \dfrac{2^4}{3^5}x^4 - \cdots$

3. $f(x) = (1+x)^{1/2}$; $f'(x) = \dfrac{1}{2}(1+x)^{-1/2}$; $f''(x) = -\dfrac{1}{2^2}(1+x)^{-3/2}$; $f'''(x) = \dfrac{3}{2^3}(1+x)^{-5/2}$;

$f^{(4)}(x) = \dfrac{-3\cdot 5}{2^4}(1+x)^{-7/2}$.

The Taylor series at $x = 0$ is

$$f(x) = 1 + \frac{1}{2}x - \frac{1}{2^2\cdot 2!}x^2 + \frac{1\cdot 3}{2^3\cdot 3!}x^3 - \frac{1\cdot 3\cdot 5}{2^4\cdot 4!}x^4 + \cdots$$

5. $\dfrac{1}{1-3x} = 1 + 3x + (3x)^2 + (3x)^3 + \cdots$

7. $\dfrac{1}{1+x^2} = 1 - x^2 + x^4 - x^6 + x^8 - \cdots$

(using Exercise 6)

9. $\dfrac{1}{(1+x)^2} = -\dfrac{d}{dx}\left[\dfrac{1}{1+x}\right] = 1 - 2x + 3x^2 - 4x^3 + \cdots$

(using Exercise 6)

11. $5e^{x/3} = 5 + 5\left(\dfrac{x}{3}\right) + \dfrac{5}{2!}\left(\dfrac{x}{3}\right)^2 + \dfrac{5}{3!}\left(\dfrac{x}{3}\right)^3 + \dfrac{5}{4!}\left(\dfrac{x}{3}\right)^4 + \cdots$

13. $1 - e^{-x} = 1 - \left[1 - x + \dfrac{(-x)^2}{2!} + \dfrac{(-x)^3}{3!} + \dfrac{(-x)^4}{4!} + \cdots\right] = x - \dfrac{x^2}{2!} + \dfrac{x^3}{3!} - \dfrac{x^4}{4!} + \cdots$

15. $\ln(1+x) = \displaystyle\int \dfrac{1}{(1+x)}\,dx = \int (1 - x + x^2 - x^3 + x^4 - \cdots)\,dx = C + x - \dfrac{x^2}{2} + \dfrac{x^3}{3} - \dfrac{x^4}{4} + \dfrac{x^5}{5} - \cdots$

For $x = 0$, $\ln(1+x) = 0 = C + 0 + 0 + \cdots$; so $C = 0$ and $\ln(1+x) = x - \dfrac{x^2}{2} + \dfrac{x^3}{3} - \dfrac{x^4}{4} + \dfrac{x^5}{5} - \cdots$

17. $\cos 3x = 1 - \dfrac{1}{2!}(3x)^2 + \dfrac{1}{4!}(3x)^4 - \cdots$

19. $\sin 3x = 3x - \dfrac{1}{3!}(3x)^3 + \dfrac{1}{5!}(3x)^5 - \cdots$

21. $xe^{x^2} = x\left[1 + x^2 + \dfrac{x^4}{2!} + \dfrac{x^6}{3!} + \dfrac{x^8}{4!} + \cdots\right] = x + x^3 + \dfrac{x^5}{2!} + \dfrac{x^7}{3!} + \dfrac{x^9}{4!} + \cdots$

23. a. $f(x) = \dfrac{1}{2}(e^x + e^{-x})$; $f'(x) = \dfrac{1}{2}(e^x - e^{-x})$; $f''(x) = \dfrac{1}{2}(e^x + e^{-x})$

The Taylor expansion at $x = 0$ is

$$\cosh x = \frac{1}{2}(2) + \frac{1}{2}(2)\frac{x^2}{2!} + \frac{1}{2}(2)\frac{x^4}{4!} + \frac{1}{2}(2)\frac{x^6}{6!} + \cdots = 1 + \frac{x^2}{2!} + \frac{x^4}{4!} + \frac{x^6}{6!} + \cdots$$

b. $\cosh x = \dfrac{1}{2}\left(e^x + e^{-x}\right) = \dfrac{1}{2}\left(\left[1 + x + \dfrac{x^2}{2!} + \dfrac{x^3}{3!} + \dfrac{x^4}{4!} + \dfrac{x^5}{5!} + \cdots\right] + \left[1 - x + \dfrac{x^2}{2!} - \dfrac{x^3}{3!} + \dfrac{x^4}{4!} - \dfrac{x^5}{5!} + \cdots\right]\right)$

$$= 1 + \frac{x^2}{2!} + \frac{x^4}{4!} + \frac{x^6}{6!} + \cdots$$

25. Substituting $-x$ for x in the given series yields

$$\frac{1}{\sqrt{1-x}} = 1 + \frac{1}{2}x + \frac{1\cdot 3}{2\cdot 4}x^2 + \frac{1\cdot 3\cdot 5}{2\cdot 4\cdot 6}x^3 + \frac{1\cdot 3\cdot 5\cdot 7}{2\cdot 4\cdot 6\cdot 8}x^4 + \cdots \text{ at } x = 0.$$

27. Substituting x^2 for x in the given series in the statement of Exercise 25 gives

$$\frac{1}{\sqrt{1+x^2}} = 1 - \frac{1}{2}x^2 + \frac{1\cdot 3}{2\cdot 4}x^4 - \frac{1\cdot 3\cdot 5}{2\cdot 4\cdot 6}x^6 + \frac{1\cdot 3\cdot 5\cdot 7}{2\cdot 4\cdot 6\cdot 8}x^8 - \cdots.$$

Since $\ln\left(x + \sqrt{1+x^2}\right) + C = \int \frac{1}{\sqrt{1+x^2}}\,dx,$ it follows that

$$\ln\left(x + \sqrt{1+x^2}\right) + C = x - \frac{1}{2\cdot 3}x^3 + \frac{1\cdot 3}{2\cdot 4\cdot 5}x^5 - \frac{1\cdot 3\cdot 5}{2\cdot 4\cdot 6\cdot 7}x^7 + \frac{1\cdot 3\cdot 5\cdot 7}{2\cdot 4\cdot 6\cdot 8\cdot 9}x^9 - \cdots.$$

Since $\ln(0+1) + C = 0 + C = 0 + 0 + \cdots,$ $C = 0.$

29. $e^x = 1 + x + \dfrac{x^2}{2!} + \dfrac{x^3}{3!} + \cdots$

$$\frac{d}{dx}[e^x] = 0 + 1 + \frac{2x}{2!} + \frac{3x^2}{3!} + \cdots = 1 + x + \frac{x^2}{2!} + \frac{x^3}{3!} + \cdots = e^x$$

31. The coefficient of x^5 in the series must equal $\dfrac{f^{(5)}(0)}{5!}.$ Thus $f^{(5)}(0) = 5!\left(\dfrac{2}{5}\right) = 48.$

33. The coefficient x^4 in the series must equal $\dfrac{f^{(4)}(0)}{4!}.$ Thus $f^{(4)}(0) = 4!(0) = 0.$

35. $\displaystyle \int e^{-x^2}\,dx = \int\left[1 - x^2 + \frac{x^4}{2!} - \frac{x^6}{3!} + \frac{x^8}{4!} - \cdots\right]dx = \left[x - \frac{x^3}{3} + \frac{x^5}{5\cdot 2!} - \frac{x^7}{7\cdot 3!} + \frac{x^9}{9\cdot 4!} - \cdots\right] + C$

37. $\displaystyle \int \frac{1}{1+x^3} = \int[1 - x^3 + x^6 - x^9 + \cdots]dx = \left[x - \frac{x^4}{4} + \frac{x^7}{7} - \frac{x^{10}}{10} + \cdots\right] + C$

39. $\displaystyle \int_0^1 e^{-x^2}\,dx = \int_0^1\left[1 - x^2 + \frac{x^4}{2!} - \frac{x^6}{3!} + \frac{x^8}{4!} - \cdots\right]dx = \left[x - \frac{x^3}{3} + \frac{x^5}{5\cdot 2!} - \frac{x^7}{7\cdot 3!} + \frac{x^9}{9\cdot 4!} - \cdots\right]_0^1$

$$= 1 - \frac{1}{3} + \frac{1}{5\cdot 2!} - \frac{1}{7\cdot 3!} + \frac{1}{9\cdot 4!} - \cdots$$

41. **a.** $e^x = 1 + x + \dfrac{x^2}{2!} + \dfrac{x^3}{3!} + \cdots$

Since the above expansion is valid for all x and all of the terms are positive for $x > 0$, it follows that

$$e^x > 1 + x + \frac{x^2}{2!} > \frac{x^2}{2}.$$

b. For $x > 0$, e^x and $\dfrac{x^2}{2}$ are both positive. Thus from $\dfrac{x^2}{2} < e^x$, it follows that $\dfrac{1}{\left(\frac{x^2}{2}\right)} > \dfrac{1}{e^x}$; or $\dfrac{2}{x^2} > e^{-x}.$

c. For $x > 0$, from (b) we have $xe^{-x} < \dfrac{2x}{x^2} = \dfrac{2}{x}.$ Now $xe^{-x} > 0$ for all $x > 0$. Thus $0 < xe^{-x} < \dfrac{2}{x}$ for all $x > 0$. Since $\dfrac{2}{x} \to 0$ as $x \to \infty$, it follows that $\displaystyle\lim_{x\to\infty} xe^{-x} = 0.$

43. The expression

$$e^x = 1 + x + \frac{x^2}{2!} + \frac{x^3}{3!} + \frac{x^4}{4!} + \cdots \text{ is valid for all}$$

x. For $x > 0$, all terms in the series are positive. Therefore, for $x > 0$,

$$e^x > 1 + x + \frac{x^2}{2} + \frac{x^3}{6} > \frac{x^3}{6}. \text{ Thus for } x > 0,$$

$$\frac{1}{e^x} < \frac{1}{\left(\frac{x^3}{6}\right)}; \text{ or } e^{-x} < \frac{6}{x^3}, \text{ which implies}$$

$$x^2 e^{-x} < \frac{6x^2}{x^3} = \frac{6}{x}. \text{ Therefore, for } x > 0,$$

$$0 < x^2 e^{-x} < \frac{6}{x}. \text{ Since } \lim_{x \to \infty} \frac{6}{x} = 0, \text{ it follows}$$

that $\lim_{x \to \infty} x^2 e^{-x} = 0$.

45. For any fixed value of x there exists a value c such that $|R_n(x)| = \frac{|\sin(c)|}{(n+1)!}|x|^{n+1}$ when n is

even and $|R_n(x)| = \frac{|\cos(c)|}{(n+1)!}|x|^{n+1}$ when n is

odd. For any value c, $|\cos(c)| \le 1$ and $|\sin(c)| \le 1$ so in either case we have

$|R_n(x)| \le \frac{|x|^{n+1}}{(n+1)!}$. Now as

$n \to \infty$, $\frac{|x|^{n+1}}{(n+1)!} \to 0$ and therefore

$|R_n(x)| \to 0$.

Chapter 11 Review Exercises

1. $f(x) = x(x+1)^{3/2}$

$$f'(x) = (x+1)^{3/2} + \frac{3}{2}x(x+1)^{1/2}$$

$$f''(x) = \frac{3}{2}(x+1)^{1/2} + \frac{3}{2}(x+1)^{1/2} + \frac{3}{4}x(x+1)^{-1/2}$$

$$= 3(x+1)^{1/2} + \frac{3}{4}x(x+1)^{-1/2}$$

$$p_2(x) = 0 + x + \frac{3}{2!}x^2 = x + \frac{3}{2}x^2$$

2. $f(x) = (2x+1)^{3/2}; \ f'(x) = 3(2x+1)^{1/2}$

$$f''(x) = 3(2x+1)^{-1/2};$$

$$f'''(x) = -3(2x+1)^{-3/2};$$

$$f^{(4)}(x) = 9(2x+1)^{-5/2}$$

$$p_4(x) = 1 + 3x + \frac{3x^2}{2!} - \frac{3x^3}{3!} + \frac{9x^4}{4!}$$

$$= 1 + 3x + \frac{3}{2}x^2 - \frac{1}{2}x^3 + \frac{3}{8}x^4$$

3. For all $n \ge 3$, $p_n(x) = x^3 - 7x^2 + 8$.

4. $f(x) = \frac{2}{2-x} = \frac{1}{1-\frac{x}{2}}$

$$= 1 + \frac{x}{2} + \left(\frac{x}{2}\right)^2 + \left(\frac{x}{2}\right)^3 + \cdots + \left(\frac{x}{2}\right)^n + \cdots$$

So $p_n(x) = 1 + \frac{x}{2} + \frac{1}{2^2}x^2 + \frac{1}{2^3}x^3 + \cdots \frac{1}{2^n}x^n$.

5. $f(x) = x^2, \ f'(x) = 2x, \ f''(x) = 2,$

$$f^{(n)}(x) = 0 \text{ for } n \ge 3.$$

$$p_3(x) = 3^2 + 2(3)(x-3) + \frac{2}{2!}(x-3)^2 + 0$$

$$= 9 + 6(x-3) + (x-3)^2$$

6. $f(x) = e^x = f^{(n)}(x)$ for all $n \ge 1$.

$$p_3(x) = e^2 + e^2(x-2) + \frac{e^2}{2!}(x-2)^2 + \frac{e^2}{3!}(x-2)^3$$

7. $f(t) = -\ln(\cos 2t), \ f'(t) = \frac{2\sin 2t}{\cos 2t} = 2\tan 2t,$

$$f''(t) = 4\sec^2 2t$$

$$p_2(t) = 0 + \frac{0}{1!}t + \frac{4}{2!}t^2 = 2t^2$$

$$\int_0^{1/2} f(t)dt \approx \int_0^{1/2} 2t^2 dt = \left(\frac{2}{3}t^3\right)\Big|_0^{\frac{1}{2}}$$

$$= \frac{2}{3}\left(\frac{1}{8}\right) = \frac{1}{12}$$

8. $f(x) = \tan x, \ f'(x) = \sec^2 x,$

$$f''(t) = (2\sec x)\sec x \tan x = 2\sec^2 x \tan x$$

$$p_2(x) = 0 + x + 0 = x$$

$$\tan(.1) \approx p_2(.1) = .1$$

9. a. $f(x) = x^{1/2}$; $f'(x) = \frac{1}{2}x^{-1/2}$;

$f''(x) = -\frac{1}{4}x^{-3/2}$

$p_2(x) = 3 + \frac{1}{6}(x-9) - \frac{1}{216}(x-9)^2$

b. $p_2(8.7) = 3 + \frac{1}{6}(-.3) - \frac{1}{216}(-.3)^2$

≈ 2.949583

c. $f(x) = x^2 - 8.7$; $f'(x) = 2x$

$x_0 = 3$, $x_1 = 3 - \frac{.3}{6} = 2.95$,

$x_2 = 2.95 - \frac{(2.95)^2 - 8.7}{2(2.95)} \approx 2.949576$

10. a. $f(x) = \ln(1-x)$; $f'(x) = -\frac{1}{1-x}$;

$f''(x) = -\frac{1}{(1-x)^2}$; $f'''(x) = -\frac{2}{(1-x)^3}$

$p_3(x) = 0 - (1)x - \frac{1}{2!}x^2 - \frac{2}{3!}x^3$

$= -x - \frac{1}{2}x^2 - \frac{1}{3}x^3$

$\ln(1.3) = f(-.3) \approx p_3(-.3)$

$= -(-.3) - \frac{1}{2}(-.3)^2 - \frac{1}{3}(-.3)^3$

$= .264$

b. $f(x) = e^x - 1.3$; $f'(x) = e^x$; $x_0 = 0$

$x_1 = 0 - \frac{e^0 - 1.3}{e^0} = .3$

$x_2 = 0.3 - \frac{e^{0.3} - 1.3}{e^{0.3}} \approx .2631$

11. $f(x) = x^2 - 3x - 2$, $f'(x) = 2x - 3$

$x_0 = 4$

$x_1 = 4 - \frac{4^2 - 3(4) - 2}{2(4) - 3} = \frac{18}{5} = 3.6$

$x_2 = 3.6 - \frac{3.6^2 - 3(3.6) - 2}{2(3.6) - 3} = \frac{374}{105} \approx 3.5619$

12. $f(x) = e^{2x} - e^{-x} - 1$, $f'(x) = 2e^{2x} + e^{-x}$

$x_0 = 0$

$x_1 = 0 - \frac{e^{2(0)} - e^{-0} - 1}{2e^{2(0)} + e^{-0}} = \frac{1}{3}$

$x_2 \approx .2832$

13. The series is geometric with $a = 1$, $r = -\frac{3}{4}$.

The sum is $\frac{1}{1 + \frac{3}{4}} = \frac{4}{7}$.

14. The series is geometric with

$a = \frac{5^2}{6} = \frac{25}{6}$, $r = \frac{5}{6}$. The sum is $\frac{\frac{25}{6}}{1 - \frac{5}{6}} = 25$.

15. The series is geometric with $a = \frac{1}{8}$, $r = \frac{1}{8}$.

The sum is $\frac{\frac{1}{8}}{1 - \frac{1}{8}} = \frac{1}{7}$.

16. The series is geometric with $a = \frac{4}{7}$, $r = -\frac{8}{7}$,

so it diverges.

17. The series is geometric with

$a = \frac{1}{m+1}$, $r = \frac{m}{m+1}$, so (since $m > 0$) it

converges to $\frac{\frac{1}{m+1}}{1 - \frac{m}{m+1}} = \frac{1}{m+1}\left(\frac{m+1}{1}\right) = 1$.

18. The series is geometric with $a = \frac{1}{m}$, $r = -\frac{1}{m}$,

so it converges if $m > 1$. In this case, the sum

is $\frac{\frac{1}{m}}{1 + \frac{1}{m}} = \frac{1}{m+1}$. It diverges if $m \leq 1$.

19. This is the Taylor series for e^x with $x = 2$.

Thus the sum is e^2.

20. This is the Taylor series for e^x with $x = \frac{1}{3}$.

Thus the sum is $e^{1/3}$.

21. Since $\sum_{k=0}^{\infty} \frac{1}{3^k}$ converges to $\frac{1}{1 - \frac{1}{3}} = \frac{3}{2}$ and

$\sum_{k=0}^{\infty}\left(\frac{2}{3}\right)^k$ converges to $\frac{1}{1 - \frac{2}{3}} = 3$,

$\sum_{k=0}^{\infty}\left[\frac{1}{3^k} + \left(\frac{2}{3}\right)^k\right] = \sum_{k=0}^{\infty}\frac{1 + 2^k}{3^k} = \frac{3}{2} + 3 = \frac{9}{2}$.

22. $\displaystyle\sum_{k=0}^{\infty}\frac{3^k+5^k}{7^k}=\sum_{k=0}^{\infty}\left(\frac{3}{7}\right)^k+\sum_{k=0}^{\infty}\left(\frac{5}{7}\right)^k=\frac{1}{1-\frac{3}{7}}+\frac{1}{1-\frac{5}{7}}=\frac{7}{4}+\frac{7}{2}=\frac{21}{4}$

23. $\displaystyle\int_1^{\infty}\frac{1}{x^3}\,dx=\lim_{b\to\infty}\left[-\frac{1}{2}x^{-2}\Big|_1^b\right]=\lim_{b\to\infty}\left[\frac{1}{2}-\frac{1}{2b^2}\right]=\frac{1}{2}$

The given series converges by the integral test.

24. The series is geometric with $a=\dfrac{1}{3}$ and $r=\dfrac{1}{3}$, so it converges.

25. $\displaystyle\int_1^{\infty}\frac{\ln x}{x}\,dx=\lim_{b\to\infty}\left[\frac{(\ln x)^2}{2}\Big|_1^b\right]=\lim_{b\to\infty}\left[\frac{(\ln b)^2}{2}\right]=\infty$

Thus the series diverges by the integral test.

26. $\dfrac{k^3}{(k^4+1)^2}=\dfrac{k^3}{k^8+2k^4+1}\le\dfrac{k^3}{k^8}=\dfrac{1}{k^5}$ for $k\ge 1$. Thus since $\displaystyle\sum_{k=0}^{\infty}\frac{1}{k^5}$ converges by the integral test,

$\displaystyle\sum_{k=1}^{\infty}\frac{k^3}{(k^4+1)^2}$ converges by the comparison test.

27. The series converges if $\displaystyle\int_1^{\infty}\frac{1}{x^p}\,dx$ converges (by the integral test).

$\displaystyle\int_1^{\infty}\frac{1}{x^p}\,dx=\lim_{b\to\infty}\left[\frac{1}{-p+1}x^{-p+1}\Big|_1^b\right]=\lim_{b\to\infty}\left[\frac{1}{1-p}(b^{1-p}-1)\right]$

This limit is finite if $p>1$.

28. This is a geometric series with $r=\dfrac{1}{p}$. Thus it converges when $\left|\dfrac{1}{p}\right|<1$ or $|p|>1$.

29. Replacing x by $-x^3$ in the series for $\dfrac{1}{1-x}$ gives $\dfrac{1}{1+x^3}=1-x^3+x^6-x^9+x^{12}-\cdots$

30. $\dfrac{d}{dx}[\ln(1+x^3)]=\dfrac{3x^2}{1+x^3}$, so

$\ln\left(1+x^3\right)=\int\left[3x^2-3x^5+3x^8-3x^{11}+3x^{14}-\cdots\right]dx=\left[x^3-\dfrac{3}{6}x^6+\dfrac{3}{9}x^9-\dfrac{3}{12}x^{12}+\cdots\right]+C$

$\left(\text{using the expansion of }\dfrac{1}{1+x^3}\text{ in Exercise 29.}\right)$

$\ln(1)=0=0+0+\cdots+C;$ so $C=0.$

$\ln(1+x^3)=x^3-\dfrac{1}{2}x^6+\dfrac{1}{3}x^9-\dfrac{1}{4}x^{12}+\dots,\ |x|<1$

31. $\dfrac{1}{(1-3x)^2}=\dfrac{1}{3}\dfrac{d}{dx}\left[\dfrac{1}{1-3x}\right]=\dfrac{1}{3}\dfrac{d}{dx}[1+3x+3^2x^2+3^3x^3+\cdots]=1+6x+27x^2+108x^3+\cdots$

32. $\dfrac{e^x-1}{x}=\dfrac{1}{x}\left[x+\dfrac{x^2}{2!}+\dfrac{x^3}{3!}+\cdots\right]=1+\dfrac{1}{2!}x+\dfrac{1}{3!}x^2+\dfrac{1}{4!}x^3+\cdots$

33. **a.** $\cos 2x = 1 - \dfrac{1}{2!}(2x)^2 + \dfrac{1}{4!}(2x)^4 - \dfrac{1}{6!}(2x)^6 + \cdots = 1 - \dfrac{2^2}{2!}x^2 + \dfrac{2^4}{4!}x^4 - \dfrac{2^6}{6!}x^6 + \cdots$

b. $\sin^2 x = \dfrac{1}{2}(1 - \cos 2x) = \dfrac{1}{2} - \dfrac{1}{2}\cos 2x = \dfrac{1}{2} - \dfrac{1}{2}\left[1 - \dfrac{2^2}{2!}x^2 + \dfrac{2^4}{4!}x^4 - \dfrac{2^6}{6!}x^6 + \cdots \right]$

$\qquad = \dfrac{2}{2!}x^2 - \dfrac{2^3}{4!}x^4 + \dfrac{2^5}{6!}x^6 - \cdots = x^2 - \dfrac{2^3}{4!}x^4 + \dfrac{2^5}{6!}x^6 - \cdots$

34. **a.** $\cos 3x = 1 - \dfrac{1}{2!}(3x)^2 + \dfrac{1}{4!}(3x)^4 - \dfrac{1}{6!}(3x)^6 + \cdots = 1 - \dfrac{3^2}{2!}x^2 + \dfrac{3^4}{4!}x^4 - \dfrac{3^6}{6!}x^6 + \cdots$

b. Adding the first three terms above to the corresponding terms in the expansion of $3\cos x$ and multiplying

by $\dfrac{1}{4}$ gives $p_4(x) = \dfrac{1}{4}\left[(1+3) - \left(\dfrac{3^2}{2!} + \dfrac{3}{2!} \right)x^2 + \left(\dfrac{3^4}{4!} + \dfrac{3}{4!} \right)x^4 \right] = 1 - \dfrac{3}{2}x^2 + \dfrac{7}{8}x^4.$

35. $\dfrac{1+x}{1-x} = \dfrac{1}{1-x} + \dfrac{x}{1-x} = [1 + x + x^2 + x^3 + \cdots] + [x + x^2 + x^3 + x^4 + \cdots] = 1 + 2x + 2x^2 + 2x^3 + \cdots$

36. Using Exercise 32,

$\displaystyle \int_0^{1/2} \dfrac{e^x - 1}{x}\, dx = \int_0^{1/2}\left[1 + \dfrac{1}{2}x + \dfrac{1}{3!}x^2 + \dfrac{1}{4!}x^3 + \cdots \right] dx = \left[x + \dfrac{1}{4}x^2 + \dfrac{1}{3!\cdot 3}x^3 + \dfrac{1}{4!\cdot 4}x^4 + \cdots \right]_0^{1/2}$

$\qquad = \dfrac{1}{2} + \dfrac{1}{4\cdot 2^2} + \dfrac{1}{3!\cdot 3\cdot 2^3} + \dfrac{1}{4!\cdot 4\cdot 2^4} + \cdots$

37. **a.** x^2

b. 0

c. $\displaystyle \int_0^1 \sin x^2\, dx \approx \int_0^1\left(x^2 - \dfrac{1}{6}x^6 \right) dx = \dfrac{x^3}{3} - \dfrac{1}{42}x^7 \bigg|_0^1 = \dfrac{1}{3} - \dfrac{1}{42} = \dfrac{13}{42} \approx .3095$

(exact value to four decimal places: .3103)

38. $p_4(x) = x + \dfrac{1}{3!}x^3$

39. **a.** $f'(x) = 2x + 4x^3 + 6x^5 + \cdots$

b. The series given for $f(x)$ is the Taylor series of $\dfrac{1}{1-x^2}$. Thus $f(x) = \dfrac{1}{1-x^2}$ and $f'(x) = \dfrac{2x}{(1-x^2)^2}$.

40. **a.** $\displaystyle \int f(x)\, dx = \int \left[x - 2x^3 + 4x^5 - 8x^7 + 16x^9 - \cdots \right] dx = \left[\dfrac{1}{2}x^2 - \dfrac{1}{2}x^4 + \dfrac{2}{3}x^6 - x^8 + \dfrac{8}{5}x^{10} - \cdots \right] + C$

b. The series given for $f(x) = x[1 - 2x^2 + 2^2 x^4 - 2^3 x^6 + 2^4 x^8 - \cdots]$ is the Taylor expansion of $\dfrac{x}{1+2x^2}$.

Thus $f(x) = \dfrac{x}{1+2x^2}$ and $\displaystyle \int f(x)\, dx = \dfrac{1}{4}\ln(1+2x^2) + C.$

41. $100 + 100(0.85) + 100(0.85)^2 + 100(0.85)^3 + \cdots = \dfrac{100}{1-0.85} \approx 666.666667$

The amount beyond the original 100 million dollars is $566,666,667.

42. $100 + (0.85)(100) + (0.80)(0.85)^2(100) + (0.80)^2(0.85)^3(100) + \cdots$

$$= 100 + 85 + (0.80)(0.85)(85) + (0.80)^2(0.85)^2(85) + \cdots$$

$$= 100 + \frac{85}{1-(0.80)(0.85)} = 365.625 \text{ million dollars}$$

43. $\displaystyle\sum_{k=1}^{\infty} 10,000 e^{-0.08k} = \sum_{k=1}^{\infty} 10,000 \left(e^{-0.08}\right)^k = \frac{10,000(e^{-0.08})}{1-e^{-0.08}} \approx \$120,066.66$

44. $\displaystyle\sum_{k=1}^{\infty} 10,000(0.9)^k e^{-0.08k} = \sum_{k=1}^{\infty} 10,000 \left(0.9 e^{-0.08}\right)^k = \frac{10,000(0.9)e^{-0.08}}{1-0.9e^{-0.08}} \approx \$49,103.30$

45. $\displaystyle\sum_{k=1}^{\infty} 10,000(1.08)^k e^{-0.08k} = \sum_{k=1}^{\infty} 10,000 \left(1.08 e^{-0.08}\right)^k = \frac{10,000(1.08)e^{-0.08}}{1-1.08e^{-0.08}} \approx \$3,285,603.18$

Chapter 12 Probability and Calculus

12.1 Discrete Random Variables

1. $E(X) = 0\left(\frac{1}{5}\right) + 1\left(\frac{4}{5}\right) = \frac{4}{5}$

$V(X) = \left(0 - \frac{4}{5}\right)^2\left(\frac{1}{5}\right) + \left(1 - \frac{4}{5}\right)^2\left(\frac{4}{5}\right)$

$= \frac{20}{125} = 0.16$

Standard deviation $= \sqrt{0.16} = 0.4$

3. a. $E(X) = 4(0.5) + 6(0.5) = 5$

$V(X) = (4-5)^2(0.5) + (6-5)^2(0.5) = 1$

b. $E(X) = 3(0.5) + 7(0.5) = 5$

$V(X) = (3-5)^2(0.5) + (7-5)^2(0.5) = 4$

c. $E(X) = 1(0.5) + 9(0.5) = 5$

$V(X) = (1-5)^2(0.5) + (9-5)^2(0.5) = 16$

As the difference between maximum and minimum values increases, so does the variance.

5. a.

Outcome	0	1	2	3
Probability	$\frac{11}{52}$	$\frac{26}{52}$	$\frac{13}{52}$	$\frac{2}{52}$

b. $E(X) = 0\left(\frac{11}{52}\right) + 1\left(\frac{26}{52}\right) + 2\left(\frac{13}{52}\right) + 3\left(\frac{2}{52}\right)$

$= \frac{58}{52} = \frac{29}{26} \approx 1.12$

c. $E(X)$ is the average number of accidents per week in the given year.

7. a. $\dfrac{\text{area within } \left(\frac{1}{2}\right) \text{ unit of center}}{\text{Total area}} = \dfrac{\pi\left(\frac{1}{2}\right)^2}{\pi(1)^2} = \dfrac{1}{4}$

$= 0.25$

Thus 25% of the points in the circle are within $\frac{1}{2}$ unit of the center.

b. $100 \times \dfrac{\pi c^2}{\pi(1)^2} = 100c^2\%$

9. Let X be the profit that the grower makes if he does not protect the fruit. Then
$E(X) = 100{,}000(0.75) + 60{,}000(0.25)$
$= 90{,}000 < 95{,}000.$
Therefore he *should* spend the $5000 to protect the fruit.

12.2 Continuous Random Variables

1. I. $\frac{1}{18}x \geq 0$ for all $0 \leq x \leq 6$

II. $\int_0^6 \frac{1}{18}x\,dx = \frac{1}{18}\left(\frac{x^2}{2}\right)\Big|_0^6 = 1 - 0 = 1$

3. I. $\frac{1}{4} \geq 0$

II. $\int_1^5 \frac{1}{4}\,dx = \frac{1}{4}x\Big|_1^5 = \frac{5}{4} - \frac{1}{4} = 1$

5. I. $5x^4 \geq 0$ for all $0 \leq x \leq 1$.

II. $\int_0^1 5x^4\,dx = x^5\Big|_0^1 = 1 - 0 = 1$

7. $\int_1^3 kx\,dx = \frac{k}{2}x^2\Big|_1^3 = \frac{9k}{2} - \frac{k}{2} = 4k = 1;$ so $k = \frac{1}{4}$.

9. $\int_5^{20} k\,dx = kx\Big|_5^{20} = 20k - 5k = 15k = 1;$ so $k = \frac{1}{15}.$

11. $\int_0^1 kx^2(1-x)\,dx = \int_0^1 (kx^2 - kx^3)\,dx$

$= \frac{k}{3}x^3 - \frac{k}{4}x^4\Big|_0^1 = \frac{k}{12} = 1;$

so $k = 12.$

13.

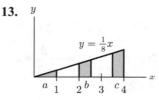

15. $\int_1^2 \frac{1}{18}x\,dx = \frac{x^2}{36}\Big|_1^2 = \frac{1}{12}$

17. $\int_1^3 \frac{1}{4}\,dx = \frac{1}{2}$

19. $\int_{35}^{50} \frac{1}{20}\,dx = \frac{x}{20}\Big|_{35}^{50} = \frac{15}{20} = \frac{3}{4}$

21. $f(x) = F'(x) = \frac{1}{4}(x-1)^{-1/2}$

23. $F(x) = \frac{1}{5}x + C$; $F(2) = \frac{1}{5}(2) + C = 0$; so

$C = -\frac{2}{5}$ and $F(x) = \frac{1}{5}x - \frac{2}{5}$.

25. a. $\int_2^3 \frac{1}{21}x^2 dx = \frac{x^3}{63}\Big|_2^3 = \frac{19}{63}$

b. $F(1) = \frac{x^3}{63} + C$; $F(1) = \frac{1}{63} + C = 0$;

$F(x) = \frac{x^3}{63} - \frac{1}{63}$

c. $F(3) - F(2) = \frac{19}{63}$

27. Points whose largest coordinate has value $\leq x$ lie within the square with vertices (0, 0), (0, x),

(x, x), and (x, 0). The area of this square is x^2. Thus $F(x)$ = the probability that a randomly selected point has maximum coordinate $\leq x$

$= \frac{x^2}{4}$.

29. Points whose coordinates sum to a value $\leq x$ lie within the triangle with vertices (0, 0), (0,

x), and (x, 0). The area of this triangle is $\frac{x^2}{2}$.

Thus $F(x)$ = the probability that a randomly selected point has coordinates summing to a

value $\leq x = \frac{\left(\frac{x^2}{2}\right)}{2} = \frac{x^2}{4}$.

31. $\int_0^5 2ke^{-kx} dx = -2e^{-kx}\Big|_0^5 = -2e^{-5k} + 2$

$= -2(e^{\ln 2})^{-1/2} + 2 = -\frac{2}{\sqrt{2}} + 2$

$= 2 - \sqrt{2} \approx 0.59$

33. $\int_0^b \frac{1}{3} dx = \frac{1}{3}b = 0.6$, so $b = 1.8$.

35. $F(b) = \frac{b^2}{4} = .09$, so $b = .6$.

37. a. I. $4x^{-5} \geq 0$ for all $x \geq 1$.

   II. $\int_1^\infty 4x^{-5} dx = \lim_{b \to \infty} \int_1^b 4x^{-5} dx$

   $= \lim_{b \to \infty}\left[-x^{-4}\Big|_1^b\right]$

   $= \lim_{b \to \infty}\left[1 - \frac{1}{b^4}\right] = 1$

b. $F(x) = -x^{-4} + C$; $F(1) = -1 + C = 0$; so

$C = 1$ and $F(x) = 1 - x^{-4}$.

c. $\Pr(1 \leq X \leq 2) = F(2) - F(1) = 1 - \frac{1}{16} = \frac{15}{16}$

$\Pr(2 \leq X) = 1 - \Pr(1 \leq X \leq 2) = \frac{1}{16}$

12.3 Expected Value and Variance

1. $E(X) = \int_0^6 \frac{1}{18}x^2 dx = \frac{x^3}{54}\Big|_0^6 = 4$;

$V(X) = \int_0^6 \frac{1}{18}x^3 dx - 4^2 = \frac{x^4}{72}\Big|_0^6 - 16 = 2$

3. $E(X) = \int_1^5 \frac{1}{4}x\, dx = \frac{x^2}{8}\Big|_1^5 = 3$

$V(X) = \int_1^5 \frac{1}{4}x^2 dx - 3^2 = \frac{x^3}{12}\Big|_1^5 - 9$

$= \frac{31}{3} - 9 = \frac{4}{3}$

5. $E(X) = \int_0^1 5x^5 dx = \frac{5}{6}x^6\Big|_0^1 = \frac{5}{6}$

$V(X) = \int_0^1 5x^6 dx - \left(\frac{5}{6}\right)^2$

$= \frac{5}{7}x^7\Big|_0^1 - \left(\frac{5}{6}\right)^2 = \frac{5}{252}$

7. $E(X) = \int_0^1 12x^2(1-x)^2 dx$

$= \int_0^1 (12x^4 - 24x^3 + 12x^2) dx$

$= \frac{12}{5}x^5 - 6x^4 + 4x^3\Big|_0^1 = \frac{2}{5}$

$V(X) = \int_0^1 (12x^5 - 24x^4 + 12x^3) dx - \frac{4}{25}$

$= 2x^6 - \frac{24}{5}x^5 + 3x^4\Big|_0^1 - \frac{4}{25} = \frac{1}{25}$

9. a. $f(x) = 30x^2(1-x)^2 = 30x^4 - 60x^3 + 30x^2,$

so $F(x) = 6x^5 - 15x^4 + 10x^3 + C;$

$F(0) = C = 0,$ so $F(x) = 6x^5 - 15x^4 + 10x^3.$

b. $F(0.25) = \dfrac{53}{512}$

c. $E(X) = \int_0^1 (30x^5 - 60x^4 + 30x^3)dx$

$= 5x^6 - 12x^5 + \dfrac{15}{2}x^4 \Big|_0^1 = \dfrac{1}{2}$

On average, the newspaper devotes $\dfrac{1}{2}$ of its space to advertising.

d. $V(X) = \int_0^1 (30x^6 - 60x^5 + 30x^4)dx - \dfrac{1}{4}$

$= \dfrac{30}{7}x^7 - 10x^6 + 6x^5 \Big|_0^1 - \dfrac{1}{4} = \dfrac{1}{28}$

11. a. Since $F(x) = \dfrac{1}{9}x^2,\ 0 \le x \le 3,\ f(x) = \dfrac{2}{9}x$

and $E(X) = \int_0^3 \dfrac{2}{9}x^2 dx = \dfrac{2}{27}x^3 \Big|_0^3 = 2.$

This means that the average useful life of the component is 200 hrs.

b. $V(X) = \int_0^3 \dfrac{2}{9}x^3 dx - 4 = \dfrac{1}{18}x^4 \Big|_0^3 - 4 = \dfrac{1}{2}$

13. $E(X) = \int_0^{12} \dfrac{1}{72}x^2 dx = \dfrac{x^3}{72 \cdot 3}\Big|_0^{12} = 8$ (minutes)

15. a. $f(x) = \dfrac{6x - x^2}{18},\ 3 \le x \le 6,$ so

$F(x) = \dfrac{x^2}{6} - \dfrac{x^3}{54} + C.$

$F(3) = \dfrac{3}{2} - \dfrac{1}{2} + C = 0,$ so $C = -1$ and

$F(x) = \dfrac{x^2}{6} - \dfrac{x^3}{54} - 1.$

b. $F(5) = \dfrac{25}{6} - \dfrac{125}{54} - 1 = \dfrac{23}{27}$

c. $E(X) = \dfrac{1}{18}\int_3^6 (6x^2 - x^3)dx$

$= \dfrac{1}{18}\left(2x^3 - \dfrac{x^4}{4}\right)\Big|_3^6 = \dfrac{1}{18}\left(108 - \dfrac{135}{4}\right)$

$= 4.125$

Thus the mean completion time is 412.5 worker-hrs.

d. $V(X) = \dfrac{1}{18}\int_3^6 (6x^3 - x^4)dx - (4.125)^2$

$= \dfrac{1}{18}\left(\dfrac{3}{2}x^4 - \dfrac{x^5}{5}\right)\Big|_3^6 - (4.125)^2$

$\approx .5344$

17. $E(X) = \int_1^\infty 4x^{-4}dx = \lim_{b \to \infty}\left[-\dfrac{4}{3}x^{-3}\Big|_1^b\right]$

$= \lim_{b \to \infty}\left[\dfrac{4}{3} - \dfrac{4}{3}b^{-3}\right] = \dfrac{4}{3}$

$V(X) = \int_1^\infty 4x^{-3}dx - \left(\dfrac{4}{3}\right)^2$

$= \lim_{b \to \infty}\left[-2x^{-2}\Big|_1^b\right] - \dfrac{16}{9}$

$= \lim_{b \to \infty}[2 - 2b^{-2}] - \dfrac{16}{9} = \dfrac{2}{9}$

19. $\int_0^M \dfrac{1}{18}x\,dx \Rightarrow \dfrac{x^2}{36}\Big|_0^M = \dfrac{1}{2} \Rightarrow$

$\dfrac{M^2}{36} = \dfrac{1}{2} \Rightarrow M = \sqrt{18} = 3\sqrt{2}$

21. $F(M) = \dfrac{1}{9}x^2 = \dfrac{1}{2} \Rightarrow$

$x = \dfrac{3\sqrt{2}}{2}$ (hundred hours)

23. $\int_0^T \dfrac{11}{10(x+1)^2}dx = \dfrac{1}{2} \Rightarrow -\dfrac{11}{10}(x+1)^{-1}\Big|_0^T = \dfrac{1}{2} \Rightarrow$

$\dfrac{11}{10} - \dfrac{11}{10(T+1)} = \dfrac{1}{2} \Rightarrow T = \dfrac{5}{6}$ minutes

25. By definition, $E(X) = \int_A^B x f(x)\,dx$ and $F'(x) = f(x)$. Using integration by parts,

$$\int_A^B x f(x)\,dx = x F(x)\Big|_A^B - \int_A^B F(x)\,dx$$

$$= B F(B) - A F(A) - \int_A^B F(x)\,dx$$

$$= B(1) - A(0) - \int_A^B F(x)\,dx$$

$$= B - \int_A^B F(x)\,dx$$

12.4 Exponential and Normal Random Variables

1. $E(X) = \dfrac{1}{3};\ V(X) = \dfrac{1}{9}$

3. $E(X) = 5;\ V(X) = 25$

5. The probability density function is $f(x) = 2e^{-2x}$. Thus

$$\Pr\left(\frac{1}{2} < X < 1\right) = \int_{1/2}^1 2e^{-2x}\,dx = -e^{-2x}\Big|_{1/2}^1$$

$$= e^{-1} - e^{-2}$$

7. The probability density function is $f(x) = \dfrac{1}{3}e^{-(1/3)x}$. Thus

$$\Pr(X < 2) = \int_0^2 \frac{1}{3}e^{-(1/3)x}\,dx = -e^{-(1/3)x}\Big|_0^2$$

$$= 1 - e^{-2/3}$$

9. The probability density function is $f(x) = \dfrac{1}{20}e^{-(1/20)x}$. Thus

$$\Pr(X > 60) = 1 - \Pr(X \le 60)$$

$$= 1 - \int_0^{60} \frac{1}{20}e^{-(1/20)x}\,dx$$

$$= 1 + \left[e^{-(1/20)x}\Big|_0^{60}\right] = e^{-3}$$

11. The probability density function is $f(x) = \dfrac{1}{2}e^{-(1/2)x}$. Thus

$$\Pr(X < 4) = \int_0^4 \frac{1}{2}e^{-(1/2)x}\,dx = -e^{-(1/2)x}\Big|_0^4$$

$$= 1 - e^{-2}$$

13. The probability density function is $f(x) = \dfrac{1}{72}e^{-(1/72)x}$.

    a. $\Pr(X > 24) = 1 - \Pr(X \le 24)$

$$= 1 - \int_0^{24} \frac{1}{72}e^{-(1/72)x}\,dx$$

$$= 1 + \left[e^{-(1/72)x}\Big|_0^{24}\right] = e^{-(1/3)}$$

    b. $r(t) = \Pr(X > t) = 1 - \Pr(X \le t)$

$$= 1 - \int_0^t \frac{1}{72}e^{-(1/72)x}\,dx$$

$$= 1 + \left[e^{-(1/72)x}\Big|_0^t\right] = e^{-t/72}$$

15. $\mu = 4,\ \sigma = 1$

17. $\mu = 0,\ \sigma = 3$

19. $f(x) = e^{-x^2/2};\ f'(x) = -xe^{-x^2/2}$

Thus $f'(0) = 0$.

$$f''(x) = -\left[e^{-x^2/2} - x^2 e^{-x^2/2}\right]\ \text{so}$$

$f''(0) = -1 < 0$.

Therefore $f(x)$ has a relative maximum at $x = 0$.

21. $f(x) = e^{-x^2/2}$

$f''(x) = x^2 e^{-x^2/2} - e^{-x^2/2}$ (see Exercise 19)

$f''(\pm 1) = 0$, but $f'(\pm 1) = \pm 1 e^{-1/2} > 0$, so $f(x)$ has inflection points at $x = \pm 1$.

23. a. $\Pr(-1.3 \le Z \le 0) = A(1.3) = 0.4032$

    b. $\Pr(0.25 \le Z) = 1 - \Pr(Z \le 0.25)$
$= 1 - (0.5 + A(0.25)) = 0.4013$

    c. $\Pr(-1 \le Z \le 2.5) = A(1) + A(2.5)$
$= 0.3413 + 0.4938 = 0.8351$

    d. $\Pr(Z \le 2) = 0.5 + A(2) = 0.9772$

25. $\mu = 6,\ \sigma = \dfrac{1}{2}$

    a. $\Pr(6 \le x \le 7) = A\left(\dfrac{7-6}{\left(\frac{1}{2}\right)}\right) = A(2) = 0.4772$

So 47.72% of births occur between 6 and 7 months.

b. $\Pr(5 \le x \le 6) = A\left(\dfrac{5-6}{\left(\frac{1}{2}\right)}\right)$

$= A(-2) = A(2) = 0.4772$

So 47.72% of births occur between 5 and 6 months.

27. $M = 128.2$, $\sigma = .2$

$\Pr(X < 128) = 1 - \left[0.5 + A\left(\dfrac{128.2 - 128}{0.2}\right)\right]$

$= 1 - [0.5 + A(1)] = 0.1587$

29. Let B be the amount of time the Beltway route takes and let L be the time it takes on the local route. Then $\mu_B = 25$, $\sigma_B = 5$, $\mu_L = 28$, $\sigma_L = 3$ and

$\Pr(B < 30) = 0.5 + A\left(\dfrac{30 - 25}{5}\right) = 0.5 + A(1)$

$= 0.8413$

$\Pr(L < 30) = 0.5 + A\left(\dfrac{30 - 28}{3}\right) = 0.5 + A\left(\dfrac{2}{3}\right)$

$= 0.7475$

Therefore the student should take the Beltway route.

31. Let X be the diameter of a randomly selected bolt.

$\Pr(X > 20) = 1 - \Pr(X \le 20)$

$= 1 - \left[0.5 + A\left(\dfrac{20 - 18.2}{.8}\right)\right]$

$= 1 - [0.5 + A(2.25)] = 0.0122$

Therefore about 1.22% of the bolts will be discarded.

33. If X has density $f(x) = ke^{-kx}$, then

$\dfrac{\Pr(a \le X \le a+b)}{\Pr(a \le X)} = \dfrac{\int_a^{a+b} ke^{-kx}\,dx}{1 - \int_0^a ke^{-kx}\,dx}$

$= \dfrac{-e^{-kx}\big|_a^{a+b}}{1 + e^{-kx}\big|_0^a}$

$= \dfrac{e^{-ak} - e^{-ak-bk}}{e^{-ak}} = 1 - e^{-bk}$

$= -e^{-kx}\big|_0^b = \int_0^b ke^{-kx}\,dx$

$= \Pr(0 \le X \le b)$

35. Let X be the lifetime of a light bulb. $\Pr(0 \le X \le 100) = 0.8$ so solve the following for k:

$0.8 = \int_0^{100} ke^{-kx}\,dx = -e^{-kx}\big|_0^{100} = -e^{-100k} + 1$

So $e^{-100k} = 0.2 \Rightarrow \ln(e^{-100k}) = \ln(0.2) \Rightarrow$

$-100k \approx -1.609 \Rightarrow k \approx 0.0160944$

Then the average lifetime is

$E(X) = \dfrac{1}{k} \approx 62.13$ weeks.

37. The screen below shows that

$\int_{-8}^{8} x^2 f(x)\,dx = 1$, where $f(x)$ is the standard normal density function. By equation (2) on page 571, we have

$Var(X) = \int_{-\infty}^{\infty} [x - 0]^2 f(x)\,dx$

$= \int_{-\infty}^{\infty} x^2 f(x)\,dx = 1$

standard deviation $= \sqrt{Var(X)} = \sqrt{1} = 1$

```
fnInt(X²(1/√(2π)
)e^(-X²/2),X,-8,
8)
              1
```

12.5 Poisson and Geometric Random Variables

1. $p_6 = \dfrac{(3)^6}{6!} e^{-3} \approx 0.0504$

$p_7 = \dfrac{(3)^7}{7!} e^{-3} \approx 0.0216$

$p_8 = \dfrac{(3)^8}{8!} e^{-3} \approx 0.0081$

3. $p_0 = e^{-0.75} \approx 0.4724$,

$p_1 = \dfrac{0.75}{1!} e^{-0.75} \approx 0.3543$

$p_2 = \dfrac{(0.75)^2}{2!} e^{-0.75} \approx 0.1329$

$p_3 = \dfrac{(0.75)^3}{3!} e^{-0.75} \approx 0.0332$

$p_4 = \dfrac{(0.75)^4}{4!} e^{-0.75} \approx 0.0062$

$p_5 = \dfrac{(0.75)^5}{5!} e^{-0.75} \approx 0.0009$

(continued on next page)

(continued)

$$p_6 = \frac{(0.75)^6}{6!}e^{-0.75} \approx 0.0001$$

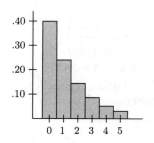

5. a. $p_0 = e^{-10} \approx 0.0000454$

b. $p_0 + p_1 + p_2 = e^{-10} + 10e^{-10} + \frac{(10)^2}{2}e^{-10}$
$$\approx 0.0027694$$

c. $1 - [p_0 + p_1 + p_2] \approx 0.9972306$

7. a. $p_0 = e^{-1.5} \approx 0.2231302$

b. $p_2 + p_3 = \frac{(1.5)^2}{2}e^{-1.5} + \frac{(1.5)^3}{6}e^{-1.5}$
$$\approx .3765321$$

c. $1 - [p_0 + p_1 + p_2 + p_3]$
$$= 1 - \left(e^{-1.5} + 1.5e^{-1.5} + \frac{(1.5)^2}{2}e^{-1.5}\right.$$
$$\left. + \frac{(1.5)^3}{6}e^{-1.5}\right)$$
$$\approx 0.0656425$$

9. If r is the number of raisins used in the batter and X is the number of raisins in a particular cookie, then $E(X) = \dfrac{r}{4800}$ and
$p_0 = e^{-r/4800} = 0.01$. So
$$-\frac{r}{4800} = \ln(0.01) \Rightarrow r \approx 22,105.$$

11. $p_0 = (0.6)^0(1 - 0.6) = 0.4$
$p_1 = (0.6)^1(1 - 0.6) = 0.24$
$p_2 = (0.6)^2(1 - 0.6) = 0.144$
$p_3 = (0.6)^3(1 - 0.6) = 0.0864$
$p_4 = (0.6)^4(1 - 0.6) = 0.05184$
$p_5 = (0.6)^5(1 - 0.6) = 0.031104$

13. a. $\Pr(X = n) = \left(\dfrac{3}{4}\right)^n\left(\dfrac{1}{4}\right)$

b. $\Pr(x \geq 3)$
$$= 1 - \Pr(x = 0) - \Pr(x = 1) - \Pr(x = 2)$$
$$= 1 - \left(\frac{3}{4}\right)^0\left(\frac{1}{4}\right) - \left(\frac{3}{4}\right)^1\left(\frac{1}{4}\right) - \left(\frac{3}{4}\right)^2\left(\frac{1}{4}\right)$$
$$\approx 0.4219$$

c. $E(X) = \dfrac{\frac{3}{4}}{\frac{1}{4}} = 3$

15. $\Pr(x < n) = (1 - p)\left(1 + p + p^2 + \cdots + p^{n-1}\right)$
$$= (1 - p)\left(\frac{1 - p^n}{1 - p}\right) = 1 - p^n$$

17. $\Pr(X = 4) = (0.95)^4(1 - 0.95) \approx 0.04073$

19. First derivative:

$$\lambda e^{-\lambda} - \frac{\lambda^2}{2}e^{-\lambda} = 0 \text{ when } \lambda e^{-\lambda} = \frac{\lambda^2}{2}e^{-\lambda} \text{ so}$$

$\lambda = 0$ is a possibility, then $2\lambda = \lambda^2 \Rightarrow 2 = \lambda$.

Second derivative: $e^{-\lambda} - 2\lambda e^{-\lambda} + \frac{1}{2}\lambda^2 e^{-\lambda}$

When $\lambda = 2$,

$$e^{-2} - 2(2)e^{-2} + \frac{1}{2}(2)^2 e^{-2} = -e^{-2}, \text{ which is}$$

negative, and when $\lambda = 0$, $e^0 - 0 + 0 = 1$. Therefore, the probability has a maximum at $\lambda = 2$.

21. $E(X) = 0 \cdot p_0 + 1 \cdot p_1 + 2 \cdot p_2 + 3 \cdot p_3 + \cdots = p(1-p) + 2p^2(1-p) + 3p^3(1-p) + \cdots$

$$= p(1-p)[1 + 2p + 3p^2 + \cdots] = p(1-p)\left(\frac{1}{(1-p)^2}\right) = \frac{p}{1-p}$$

23. $[p_0 + p_1 + p_2 + p_3 + p_4 + p_5 + p_6 + p_7 + p_8] - [p_0 + p_1] \approx 0.887058$

25. a. $p_7 - p_6$ is negative, so $p_6 > p_7$ and the answer is no.

b. $\Pr(X \le 15) = 0.9979061$

Chapter 12 Review Exercises

1. $f(x) = \frac{3}{8}x^2, \ 0 \le x \le 2$

a. $\Pr(X \le 1) = \int_0^1 \frac{3}{8}x^2 dx = \frac{1}{8}x^3\Big|_0^1 = \frac{1}{8}$

$\Pr(1 \le X \le 1.5) = \frac{1}{8}x^3\Big|_1^{1.5} = \frac{19}{64}$

b. $E(X) = \int_0^2 \frac{3}{8}x^3 dx = \frac{3}{32}x^4\Big|_0^2 = \frac{3}{2}$

$V(X) = \int_0^2 \frac{3}{8}x^4 dx - \frac{9}{4} = \frac{3}{40}x^5\Big|_0^2 - \frac{9}{4}$

$= \frac{12}{5} - \frac{9}{4} = \frac{3}{20}$

2. $f(x) = 2x - 6, \ 3 \le x \le 4$

a. $\Pr(3.2 \le X) = \int_{3.2}^4 (2x-6)dx$

$= x^2 - 6x\Big|_{3.2}^4 = 0.96$

$\Pr(3 \le x) = 1$ since the random variable is defined for $3 \le x \le 4$.

b. $E(X) = \int_3^4 (2x^2 - 6x)dx = \left(\frac{2}{3}x^3 - 3x^2\right)\Big|_3^4$

$= \frac{11}{3}$

$\int_3^4 (2x^3 - 6x^2)dx = \left(\frac{1}{2}x^4 - 2x^3\right)\Big|_3^4 = \frac{27}{2}$

$V(X) = \frac{27}{2} - \left(\frac{11}{3}\right)^2 = \frac{1}{18}$

3. I. $e^{A-x} \ge 0$ for all x.

II. $\int_A^\infty e^{A-x} dx = \lim_{b \to \infty}\left[-e^{A-x}\Big|_A^b\right]$

$= \lim_{b \to \infty}[1 - e^{A-b}] = 1$

Thus $f(x) = e^{A-x}, \ x \ge A$ is a density function.

$F(x) = \int_A^x e^{A-t} dt = -e^{A-t}\Big|_A^x = 1 - e^{A-x}$

4. $f(x) = \frac{kA^k}{x^{k+1}}, \ k > 0, A > 0, x \ge A$

I. Since k and A are > 0, $f(x) \ge 0$ for all $x \ge A$.

II. $\int_A^\infty \left(\frac{kA^k}{x^{k+1}}\right) dx = \lim_{b \to \infty}\left[\frac{-A^k}{x^k}\right]_A^b$

$= \lim_{b \to \infty}\left[1 - \frac{A^k}{b^k}\right] = 1$

Thus $f(x)$ is a density function.

$F(x) = \int_A^x \left(\frac{kA^k}{t^{k+1}}\right) dt = -\frac{A^k}{t^k}\Big|_A^x = 1 - \frac{A^k}{x^k}$

5. For $n \ge 2$, any choice of $c_n > 0$ will ensure $f_n(x) \ge 0$ for all $x \ge 0$. Thus we need only

$\int_0^\infty c_n x^{(n-2)/2} e^{-x/2} dx = 1$. If $n = 2$ this

becomes $c_2 \int_0^\infty e^{-x/2} dx = 1$

$c_2 \lim_{b \to \infty}\left[-2e^{-x/2}\Big|_0^b\right] = 1 \Rightarrow 2c_2 = 1 \Rightarrow c_2 = \frac{1}{2}$.

For $n = 4$, we have $c_4 \int_0^\infty x e^{-x/2} dx = 1$.

Integrating by parts twice gives

$\int_0^b x e^{-x/2} dx = e^{-x/2}(-4 - 2x)\Big|_0^b$.

Therefore, $c_4 \lim_{b \to \infty}\left[e^{-x/2}(-4 - 2x)\Big|_0^b\right] = 1 \Rightarrow$

$4c_4 = 1 \Rightarrow c_4 = \frac{1}{4}$.

6. **I.**   If $k > 0$, $\dfrac{1}{2k^3} x^2 e^{-x/k} \ge 0$ for all x.

II.   Integrating by parts twice,

$$\int_0^b x^2 e^{-x/k} dx = -e^{x/k}[kx^2 + 2k^2 x + 2k^3]\Big|_0^b.$$

Thus $\int_0^\infty \dfrac{1}{2k^3} x^2 e^{-x/k} dx = \dfrac{1}{2k^3}(2k^3) = 1.$

7. **a.** $E(X) = 1(.599) + 11(.401) = 5.01$

b.   200 samples is 20 batches of 10. Thus they can expect to run $20(5.01) \approx 100$ tests.

8. **a.** $E(X) = 1(.774) + 6(.226) = 2.13$

b.   200 samples is 40 batches of 5. Thus they can expect to run $40(2.13) \approx 85$ tests.

9. $F(x) = 1 - \dfrac{1}{4}(2 - x)^2$, $0 \le x \le 2$

a. $\Pr(X \le 1.6) = F(1.6) = .96$

b. $\Pr(X \le t) = 1 - \dfrac{1}{4}(2 - t)^2 = .99$

$t = 1.8$ (thousand gal.)

c. $f(x) = F'(x) = \dfrac{1}{2}(2 - x)$, $0 \le x \le 2$

10. $E(X) = \dfrac{1}{625} \int_0^5 x(x - 5)^4 dx$

$$= \dfrac{1}{625}\left[\dfrac{x}{5}(x - 5)^5 - \dfrac{1}{30}(x - 5)^6\right]\Big|_0^5$$

(Integration by parts)

$$= \dfrac{(-5)^6}{30 \cdot 5^4} = \dfrac{5}{6} = 0.8333 \text{ (hundred dollars)}$$

Thus on average the manufacturer can expect to make $100 - 83.33 = \$16.67$ on each service contract sold.

11. **a.** $E(X) = \int_{20}^{25} \dfrac{1}{5} x\, dx = \dfrac{1}{10} x^2 \Big|_{20}^{25} = 22.5$

$$V(X) = \int_{20}^{25} \dfrac{1}{5} x^2 dx - 22.5^2$$

$$= \dfrac{x^3}{15}\Big|_{20}^{25} - 22.5^2 \approx 2.0833$$

b. $\Pr(X \le b) = \int_{20}^b \dfrac{1}{5} dx = 0.3 \Rightarrow$

$\dfrac{1}{5} b - 4 = 0.3 \Rightarrow b = 21.5$

12. $F(x) = \dfrac{(x^2 - 9)}{16}$, $3 \le x \le 5$

a. $f(x) = F'(x) = \dfrac{x}{8}$, $3 \le x \le 5$

b. $\Pr(a \le X) = \dfrac{1}{4}$, so

$$F(a) = \dfrac{(a^2 - 9)}{16} = \dfrac{3}{4} \Rightarrow a = \sqrt{21}$$

13. $f(x) = kx$, $5 \le x \le 25$

a.   We need $k > 0$ and

$$\int_5^{25} kx\, dx = 1 \Rightarrow \dfrac{k}{2} x^2 \Big|_5^{25} = 1 \Rightarrow$$

$$\dfrac{k}{2}(600) = 1 \Rightarrow k = \dfrac{1}{300}$$

b. $\Pr(X \ge 20) = \int_{20}^{25} \dfrac{1}{300} x\, dx = \dfrac{1}{600} x^2 \Big|_{20}^{25} = \dfrac{3}{8}$

c. $E(X) = \int_5^{25} \dfrac{1}{300} x^2 dx = \dfrac{1}{900} x^3 \Big|_5^{25}$

≈ 17.222 thousand dollars

14. **a.** $F(x) = \Pr(3 \le X \le x)$

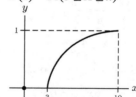

b. $F(7) - F(5) = \Pr(5 \le X \le 7)$

c. $\Pr(5 \le X \le 7) = \int_5^7 f(x)dx$

15.   Points (θ, y) satisfying the given condition are precisely those points under the curve $y = \sin \theta$, $0 \le \theta \le \pi$. This region has area

$\int_0^\pi \sin \theta\, d\theta = -\cos \theta \Big|_0^\pi = 2$. The area of the rectangle is π, so the probability that a randomly selected point falls in the region $y \le \sin \theta$ is $\dfrac{2}{\pi}$.

16. Since the length of the needle is 1 unit, $\sin\theta$ is the difference in the y-coordinates of the base and the end of the needle.

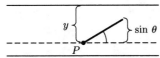

The needle will touch a ruled line if and only if this difference exceeds y, the vertical distance from the base to the next ruled line. To compute $\Pr(y \le \sin\theta)$, view dropping the needle as a random choice of a point (θ, y) from the square $0 \le \theta \le \pi$, $0 \le y \le 1$. Then Exercise 15 applies.

17. Let X be the lifetime of the computer monitor. Then

$$\Pr(Y = 0) = \Pr(X \le 3) = \int_0^3 \frac{1}{5} e^{-(1/5)x} dx$$

$$= -e^{-(1/5)x}\Big|_0^3 = 1 - e^{-3/5} \approx 0.45119$$

Thus $\Pr(Y = 100) \approx 0.54881$ and $E(Y) \approx \$54.88$.

18. Let Y be as in the hint and let X be the life span of the motor. Then

$$\Pr(Y = 300) = \Pr(X \le 1) = \int_0^1 \frac{1}{10} e^{-(1/10)x} dx$$

$$= -e^{-(1/10)x}\Big|_0^1 = 1 - e^{-1/10} \approx 0.09516$$

Thus $E(Y) \approx 300(0.09516) \approx \$28.55$. Since the insurance costs $25 to buy, you should buy it for the first year.

19. $\Pr(X \le 4) = \int_0^4 ke^{-kx} dx = -e^{-kx}\Big|_0^4$

$$= 1 - e^{-4k} = 0.75$$

so $e^{-4k} = 0.25 \Rightarrow -4k = \ln(0.25) \Rightarrow$

$$k = \frac{\ln(0.25)}{-4} \approx 0.35$$

20. $E(X) = .01\int_0^\infty x^2 e^{-x/10} dx$

Integrating by parts twice,

$$\int_0^b x^2 e^{-x/10} dx = -e^{-x/10}(10x^2 + 200x + 2000)\Big|_0^b.$$

So $E(X) = 2000(0.01) = 20$ (thousand hours) and the expected additional earnings from the machine are $20(5000) = \$100,000$. Since this amount exceeds the price, the machine should be purchased.

21. $f(x)$ is the density of a normal random variable X with $\mu = 50$, $\sigma = 8$. Thus

$$\Pr(30 \le X \le 50) = A\left(\frac{50 - 30}{8}\right)$$

$$= A(2.5) = 0.4938$$

22. Let X be the length of a randomly selected part. Then $\Pr(79.95 \le X \le 80.05)$

$$= A\left(\frac{79.99 - 79.95}{0.02}\right) + A\left(\frac{80.05 - 79.99}{0.02}\right)$$

$$= A(2) + A(3) = 0.4772 + 0.4987 = 0.9759$$

Hence out of a lot of 1000 parts, $1000(.9759) = 975.9$ should be within the tolerance limits, leaving about 24 defective parts.

23. Let X be the height of a randomly selected man in the city.

$$\Pr(X \ge 69) = 0.5 + A\left(\frac{70 - 69}{2}\right)$$

$$= 0.5 + A(0.5) = 0.6915$$

Thus about 69.15% of the men in the city are eligible.

24. Let Y be the height of a randomly selected woman from the city.

$$\Pr(Y \ge 69) = 0.5 - A\left(\frac{69 - 65}{1.6}\right)$$

$$= 0.5 - A(2.5) = 0.0062$$

So only about 0.62% of the women are eligible.

25. $\Pr(a \le Z) = 0.4$, then we must have $a > 0$ and $\Pr(0 \le a \le A) = A(a) = 0.5 - 0.4 = 0.1$. From the table, $A(0.25) = 0.0987 \approx 0.1$, so $a \approx 0.25$.

26. Using the result of exercise 25, the cutoff grade t must satisfy

$$\frac{t - 500}{100} = 0.25 \Rightarrow t = 525.$$

27. a. $\Pr(-1 \le Z \le 1) = 2A(1) = 0.6826$

   b. $\Pr(\mu - \sigma < X < \mu + \sigma) = \Pr(-1 < Z < 1) = 0.6826$

28. a. $\Pr(-2 \le Z \le 2) = 2A(2) = 0.9544$

   b. $\Pr(\mu - 2\sigma < X < \mu + 2\sigma) = \Pr(-2 < Z < 2) = 0.9544$

29. a. Let X be an exponential random variable with density $f(x) = ke^{-kx}$. Then $E(X) = \mu = \dfrac{1}{k}$ and

$V(X) = \sigma^2 = \dfrac{1}{k^2}$. Applying the inequality with $n = 2$ gives

$$\Pr\left(\frac{1}{k} - \frac{2}{k} \le X \le \frac{1}{k} + \frac{2}{k}\right) = \Pr\left(-\frac{1}{k} \le X \le \frac{3}{k}\right) = \Pr\left(0 \le X \le \frac{3}{k}\right) \ge 1 - \frac{1}{2^2} = \frac{3}{4}$$

b. $\Pr\left(0 \le X \le \dfrac{3}{k}\right) = \displaystyle\int_0^{3/k} ke^{-kx}\,dx = -e^{-kx}\Big|_0^{3/k} = 1 - e^{-3} \approx 0.9502$

30. Let X be a normal random variable with $E(X) = \mu$ and $V(X) = \sigma^2$. Applying the inequality with

$n = 2$ gives $\Pr(\mu - 2\sigma \le X \le \mu + 2\sigma) \ge 1 - \dfrac{1}{2^2} = \dfrac{3}{4}$.

The exact value is $\Pr(\mu - 2\sigma \le X \le \mu + 2\sigma) = 2A(2) = 0.9544$

31. $p_4 = \dfrac{(4)^4}{4!}e^{-4} \approx 0.1953668$

32. $1 - [p_0 + p_1 + p_2 + p_3 + p_4 + p_5 + p_6 + p_7] \approx 0.0511336$

33. $E(X) = 4$

34. $\left(\dfrac{2}{9}\right)\left(\dfrac{7}{9}\right)^n$

35. $E(X) = \dfrac{\frac{7}{9}}{1 - \frac{7}{9}} = \dfrac{7}{2}$

36. $1 - (p_0 + p_1 + p_2) \approx 0.4705075$